An Introduction to Algebra

2ND Edition

A Workbook for Reading, Writing, and Thinking about Mathematics

by
Brita Immergut
Jorge Perez
Dehlly Porras
Assad Thompson

LaGuardia Community College
of the City University
of New York

ARDSLEY HOUSE PUBLISHERS, INC., NEW YORK

Address orders and editorial
correspondence to:
Ardsley House, Publishers, Inc.
320 Central Park West
New York, NY 10025

ISBN: 0-912675-90-X

Printed in the United States of America

10 9 8 7 6 5 4 3

To Our Families

Preface

This is *not* a traditional elementary-algebra text either in content or in character. As the title implies, it is intended to teach students HOW TO READ, WRITE, AND THINK ABOUT MATHEMATICAL IDEAS and HOW TO EXPRESS THEM IN THEIR OWN WORDS.

The book builds on two basic strategies central to many mathematics classes, but nonexistent in most mathematics books. The first strategy is to have students use language to think about mathematics. Teacher after teacher knows it makes good sense to have mathematics students talk through their observations about numbers, operations, and real-world problems, talk through what they understand, and, equally important perhaps, talk through what they misunderstand. In short, the effective teacher knows when to keep quiet and let students take over. The groundbreaking peer study groups in calculus courses at the University of California, Berkeley (and their spread to high schools, popularized in the film *Stand and Deliver*) owe part of their success to Professor Yuri Treisman's understanding of student dialogue as a means to learning. Our book, we think, provides opportunities for dialogue—between text and student, between student and teacher, and between student and student.

The second tenet central to this book, and one again central to so many successful mathematics classes, is that deeper and longer-lasting learning of mathematics is made possible when students obtain their basic understanding not from some edict—teacher or text pronounced—but rather, from some discovery they make on their own. The book, therefore, by means of carefully sequenced activities and exercises, guides students to induce basic principles. Through their own writing, students transform their observations about mathematics into an abiding understanding. And the text urges students to bring into the classroom their questions and uncertainties so that the conversation begun in the book may continue in the class.

That writing, talking, and thinking are critically related is, of course, not a novel notion. What is new here, however, is the application of this very basic and commonplace truth to a mathematics text. The book's emergence at LaGuardia Community College of the City University of New York is not surprising. For many years LaGuardia has had a special faculty-development program, Integrated Skills Reinforcement (ISR), that brings together subject-area teachers from various disciplines to develop instructional materials that build literacy into the curriculum—language across the curriculum, if you will.

A group of us in the LaGuardia mathematics department began to write a textbook to help students with the transition from arithmetic to algebra. We felt that consistent emphasis on literacy in the classroom would be an enormous boon for students in basic mathematics courses. The ISR approach seemed critical to this venture and we were able to have our own ISR trainer and English professor meet with us every week for several years. Together, we devised a format that guides the students to discover the fundamental ideas of mathematics.

Over the years, pilot assessments of the materials that finally became the text have offered us encouraging results. Students using these materials scored significantly higher than students using an exercise book with no opportunities for extended reading, writing, or induction. More studies are needed, but we are convinced that we are on the right track.

The text begins with an introduction that covers the basic notions of symbols and operations, reviewing old concepts in fresh ways. It then proceeds with operations with signed numbers and variables, equations, word problems, and graphing. The Appendix provides a brief review of arithmetic. The book is decidedly not written for mathematicians in the traditional sense. The language is plain, but not condescending to the adult student who often has to take fundamental mathematics in college. All reading material comes in small portions in the so-called learning activities. We have tried to make the students actively involved in the learning process by asking them questions about mathematical concepts which they already know and then leading them through the discovery of a new rule. After each learning activity we give examples and then a few (mini-practice) problems, with answers provided immediately to reduce anxiety. Students then explain in their own words the rule they have discovered before we give the rule in more formal language to reinforce proper vocabulary. After each short section there are practice problems. A chapter generally contains a number of short sections, sometimes divided into several subsections, followed by a summary and by review problems from the whole chapter. A special feature of our book is that common mistakes are flagged by a subsection called **WATCH OUT!**

The 2nd Edition retains the basic format and writing style of the earlier edition. Various changes and additions have been made throughout the exposition.

- In Section 6.1 additional material has been added to clarify translations of algebraic expressions back into English. Section 10.4 has been rewritten to emphasize multiplication by powers of 10 when solving equations with decimals. More material on parabolas has been included in Section 15.3.

- Many new worked-out examples are included in the 2nd Edition.

- There are about twice as many student exercises in the 2nd Edition as in the 1st Edition.
- Each chapter begins by reminding the student of a number of things to be aware of while working through the chapter.
- On the first left-hand page after the chapter opening there is a list of Suggestions to the Student concerning either study habits or some key feature of the text. A listing of these suggestions is given on page xv, following the Table of Contents.

We hope that you and your students enjoy using this text, which we wrote with our students very much in mind.

In writing this book we received a great deal of help from Dr. Martin Moed, Dean of Faculty at the time, and Dr. JoAnn Anderson, Director of the ISR program. They gave support to our group effort in many ways, but primarily by lending us Dr. Nora Eisenberg, Associate Project Director of ISR. Most of all, we thank her for training us, encouraging us, and editing the manuscript several times. Dean Steven Brown made it possible for us to print and try out our materials. Dr. Roy McLeod, then Chairperson of the Department of Mathematics, supported us in critical ways, trusting in our belief in the students' capacities to learn actively. Diane Loweth typed a large part of the manuscript.

Professor Mary Fjeldstad, of the Department of Communication Skills, helped us with editing, Dr. Howard Everson, City University Graduate Center, with evaluations, and Dr. Harvey Wiener, City University Graduate Center, with planning.

Many members of the mathematics department at LaGuardia have used the book in different preliminary stages and have given us valuable feedback. Our students have been encouraging in their acceptance of the book.

A special thanks is due to Professor Peter E. Hurlemann of Citrus College for pointing out a number of typos in the first printing.

Brita Immergut
Jorge Perez
Dehlly Porras
Assad Thompson

Contents

Suggestions to the Student

What Is Algebra?

What Is Algebra?

The name "algebra" had its origin at the court of Caliph al-Mamun, who was the ruler of Baghdad around 800 B.C. Among the many learned men at his court was the greatest astronomer and mathematician of the land, Mahammad ibn Musa al-Kwarazmi. This man was best known for having written a mathematical paper titled "Hisab al-jabr wa'l mugabalsh," which means "the science of reduction and comparison." Over the centuries "al-jabr" was translated into other languages, most often Spanish and Latin, and was eventually transformed into the word "algebra."

Algebra has come to mean "extended arithmetic." In algebra, we work with negative as well as positive numbers. We also work with letters that "symbolically" represent numbers. **Symbols**, as you may know, are letters, or markings, or words that represent something. In this introductory chapter we will review four kinds of mathematical symbols: **number** symbols, **operation** symbols, **grouping** symbols, and **comparison** symbols.

NUMBER SYMBOLS

Give two examples of symbols used in daily life:

_____ and _____

You might have said

¢ is used for *cents* and °F is used for *degrees Fahrenheit*

for example.

Give two symbols that represent this many apples: ꙮꙮꙮ

_____ and _____

You might have used the mathematical symbols 3 or III or the words "three" or "tres" to represent or symbolize the "threeness" of the apples.

Strategies for Success in Mathematics

1. Give it your all!

2. Accept reality.
- Know your limitations. Don't schedule more courses than you can handle. Be careful about making too many outside commitments.
- Set aside definite periods of time for homework both during the week and on weekends.
- Review old material. Don't wait for the day before an exam.
- Memorize formulas when necessary.

3. Things you can do in class:
- Don't miss classes, except in an emergency.
- Don't come late. Usually the first part of class is set aside for review.
- Have the phone number of someone in the class in case you have questions or miss some work.
- Be well-prepared for class.
- Ask questions and participate in class.
- Ask your instructor questions after class, if possible.

4. Things you can do out of class:
- Expect to do at least an hour or two hours of homework for every hour of class.
- Know when your instructor holds office hours. You often get valuable help at that time.
- If your school offers a math lab, set aside at least two hours a week to work there.

This concept of threeness is what we call the **number** three. The symbol 3 and the other symbols that we use to represent numbers were introduced into Western Civilization during the 15th century. Different civilizations have used different symbols to represent "three." The ancient Egyptians would have written ||| . The Mayans of Mexico would have written ooo and the ancient Romans, III.

A *symbol* for a number is called a **numeral**. Mathematicians distinguish between numbers and numerals. In everyday language, however, we do not make such a distinction. From now on we will use the word "number" instead of the word "numeral." Thus, even though symbols such as 3, 49, $^8/_{13}$ are *numerals*, for convenience, we will call them *numbers*.

- Consult other textbooks and workbooks in the math lab or in the library.
- Find out if there are other classes that you can attend when you have missed a class or when you don't understand a topic.

5. **How to study math:**
  - Plan to spend a lot of time reading math problems. It is not the same as reading a novel.
  - Go slowly over each symbol and word; then go back and reread.
  - Read with pencil and paper nearby. Draw pictures that help you to "see" the problem.
  - Pause often while reading until you understand a concept fully; then move on. Allow enough time for each idea to sink in.

6. **Tests:**
  - A test is like a job interview . . . give it your all.
  - Start preparing for the test as soon as you know the date.
  - Ask the teacher for copies of old exams.
  - Exams reflect what has been done in class. Don't leave any questions unanswered. If you cannot solve the problem, write down what you think. Writing is a thinking process.
  - Do the problems that you know first.
  - Learn from mistakes. Next time, you should get them right.

7. **Don't get discouraged!**
  - If at first you don't succeed, TRY, TRY AGAIN!

Constants and Variables

Explain the meaning of the word "constant." _____

_____

Explain the meaning of the word "variable." _____

_____

In mathematics, a **constant** is a symbol that represents a number whose value is *fixed*. For example, (at sea level) the boiling point of water is always 100° Celsius; thus, on the Celsius scale, the constant 100 represents this boiling point. A constant can be a decimal, such as 0.5, a fraction, such as $\frac{2}{3}$, or a square root, such as $\sqrt{2}$.

In contrast, a **variable** is a symbol that represents *various numbers* in the course of a discussion. If the letter t represents the temperature during the course of a day, t takes on different values as the weather changes. Perhaps t is 60 degrees Fahrenheit at 8 AM, 78° at noon, and 72° at 4 PM. Thus t is a variable.

Part I of this book (Numbers and Operations) deals with constants and Part II (Variables and Operations) deals with letters as symbols for variable quantities. The Appendix contains a review of arithmetic operations on decimals and fractions.

OPERATION SYMBOLS

Basic Operations

We call the symbol " + " an **operation** symbol because it tells us to perform a specific mathematical operation.

List the four basic mathematical operations: _____

_____

In mathematics we say that the basic operations are addition, subtraction, multiplication, and division. For addition and subtraction we have the symbols " + " and " – ".

Which symbols have you used for multiplication? _____

Which symbols have you used for division? _____

Multiplication is symbolized by

$\times$, $\cdot$, $*$ (in computer language), and by parentheses

For division we use

$\div$, a fraction bar, or / (slash)

Use parentheses to write "three times four": _____

"3 times 4" can be written as (3)4, 3(4), or (3)(4). Of these three, the most common way of expressing this is 3(4).

Exponents

When a number is multiplied by itself, as in the case of 4×4, the resulting **product** is called a **squared number**.

Why do you think it got such a name? _____

_____

If the four-sided geometric figure called a "square" has a side of length 4 inches, its area is 4×4 square inches.

Instead of writing a square number as a repeated multiplication, 4×4 can be written as 4^2, which is read "four squared." This is called **exponential notation**. The "4" is called the **base** and the "2" is called the **exponent**.

Similarly, $2 \times 2 \times 2$ can be written in exponential form as 2^3. The base is 2 and the exponent is 3. We read 2^3 as "two to the third power" or simply "the cube of two."

Why is it called a "cube?" _____

A geometric "cube" with a side of length 2 inches has a volume of $2 \times 2 \times 2$ cubic inches.

Write the following repeated multiplication in exponential notation:

$2 \times 2 \times 2 \times 2 \times 2 =$ _____

The product of $2 \times 2 \times 2 \times 2 \times 2$, or 2^5, is 32, and is called "the fifth power of two."

COMMUTATIVE OPERATIONS

Addition

If you had two apples and got three more, how many apples would you have? _____

If you had started out with three apples and then got two more, how many apples would you have? _____

In both cases the resulting **sum** is five. We say that addition is a **commutative** operation because the order in which the numbers are written doesn't affect the result. The word "commutable" means "interchangeable."

Are the other basic operations also commutative? _____

_____

In the following sections we will find the answer to this question.

Subtraction

Look at these examples:

At 4 AM the temperature was 15 degrees, and over the next five hours the temperature dropped 10 degrees. What was the new temperature?

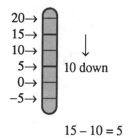

15 − 10 = 5

At 4 AM the temperature was 10 degrees, and over the next five hours the temperature dropped 15 degrees. What was the new temperature?

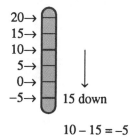

10 − 15 = −5

You can see from the example on the left that the new temperature was 5 above zero, whereas on the right, the new temperature was 5 below zero.

Is subtraction commutative? _____

Since 15 − 10 does not give the same result as 10 − 15, we can conclude that

subtraction is not commutative

Since subtraction is not commutative, the order in which we write the numbers is important.

Multiplication

What is the product of 6 and 3? _____

$6 \times 3 = 18$

Can we write 3×6 and get the same answer? _____

Yes, $3 \times 6 = 6 \times 3$. Therefore, we can say that

multiplication is commutative

Division   Do the following exercise:

$30 \div 6 =$ _____ $6 \div 30 =$ _____

30 divided by 6 equals 5. 6 divided by 30 results in 0.2 (or $\frac{1}{5}$). Therefore, we can conclude that

division is not commutative

Commutative Properties of Basic Operations

Addition and multiplication are commutative operations.
Subtraction and division are NOT commutative operations.

EXPONENTIAL NOTATION

Find the value of (or evaluate) the following:

$2^3 =$ _____ $3^2 =$ _____

Does $2^3 = 3^2$? _____

Explain why exponential notation is not commutative: _____

_____

Note that 2^3 equals 8, but 3^2 is 9.

$2^3 \neq 3^2$ (The symbol $\neq$ means is not equal to.)

Exponential notation is NOT commutative.

OPERATIONS WITH ZERO AND ONE

When performing the four basic mathematical operations we use two numbers.

What happens if one of the numbers is zero or one? _____

_____

Perform the following operations.

(a) Add zero to any number. For example: $5 + 0 =$ _____

(b) Subtract zero from any number. For example: $5 - 0 =$ _____

(c) Multiply zero by any number. For example: $5 \times 0 =$ _____

(d) Divide zero by any number. For example: $\frac{0}{5} =$ _____

(e) Divide any number by zero. (Do this problem using your calculator. Make sure to read the display carefully.)

For example: $\frac{5}{0} =$ _____

(f) Multiply one by any number. For example: $5 \times 1 =$ _____

(g) Divide any number by one. For example: $\frac{5}{1} =$ _____

(h) Divide any number by itself. For example: $\frac{5}{5} =$ _____

Go back to activity (e). Your calculator gave you an error message. This is because we are **NOT** allowed to divide by zero. In mathematics we say that division by zero is *undefined*.

Instead of selecting a particular number, we can use a letter to represent the number. If, for example, *a* stands for any number, the rules can be written as follows:

Properties of Zero and One

Let *a* be any number. Then

1. $a + 0 = a$ 2. $a - 0 = a$ 3. $a \times 0 = 0$ 4. $\frac{0}{a} = 0$

5. $\frac{a}{0}$ (undefined)   6. $a \times 1 = a$ 7. $\frac{a}{a} = 1$ 8. $\frac{a}{1} = a$

GROUPING SYMBOLS

Add punctuation marks to the sentence below to produce a grammatically correct and meaningful sentence:

Bill shouted Joe is missing

You might have written:

Bill shouted, "Joe is missing."

or

"Bill," shouted Joe, "is missing."

According to the first sentence Joe is missing; the second sentence indicates that Bill is missing. We can see that the difference in punctuation produces a complete change in meaning.

In mathematics we also need punctuation. However, we use **grouping symbols** instead of commas and quotation marks. The grouping symbols are the parentheses ( ), brackets [ ], or braces { }.

Mathematicians have agreed to a certain order among the different grouping symbols. For example, in expressions such as

$$6 - \{2 - [5 - (3 + 1)]\}$$

we always start to do the operations inside the parentheses:

$$(3 + 1)$$

which are the *innermost* grouping symbols.

$$3 + 1 = 4$$

We now have

$$6 - \{2 - [5 - 4]\}$$

Now we look inside the brackets:

$$[5 - 4]$$

Thus,

$$5 - 4 = 1$$

We now have

$$6 - \{2 - 1\}$$

Next we perform the operation inside the braces, which are the *outermost* grouping symbol:

$$\{2 - 1\}$$

Because

$$2 - 1 = 1$$

we now have

$$6 - 1 = 5$$

Chapter 3 (Order of Operations) contains many examples of working with grouping symbols.

COMPARISON SYMBOLS

In mathematics we sometimes need to compare the value of numbers. Two numbers are either of equal or unequal value. When two numbers are unequal, one number is either smaller or greater than the other number.

To indicate that two values are equal, we use the symbol

$$=$$

To indicate that two values are not equal we can use the symbol

$$\neq$$

We can indicate that one value is smaller than another with the symbol

$$< \text{ meaning "is less than"}$$

For example, $2 < 3$.

The symbol

$$> \text{ means "is greater than"}$$

Thus, $3 > 2$.

SUMMARY

Constants and Variables: A constant is a symbol that represents a number whose value is *fixed*. A variable is a symbol that represents *different* numbers during the course of a discussion.

Operation Symbols: Operations symbols tell us which basic operation to perform. The basic operations are:

addition with +

subtraction with –

multiplication with $\times$, $\cdot$, ( ), or *

and

division with $\div$, /, or —

When a number is multiplied by itself, possibly several times, **exponential notation** is used.

Grouping Symbols: Grouping symbols instruct us to keep certain numbers together. The most commonly used group symbols are

( ), [ ], { }

Comparison Symbols: Comparison symbols are used to compare two values. The most commonly used comparison symbols are

=, ≠, <, >

If you have questions about this introductory material, write them here:

_____

_____

_____

_____

_____

_____

PART I

Numbers and Operations

1 Signed Numbers

Remember to

     ✻ *DO ALL THE LEARNING ACTIVITIES*

     ✻ *WRITE ANSWERS TO ALL QUESTIONS*

     ✻ *WRITE OUT THE DEFINITIONS*

     ✻ *WRITE OUT EXPLANATIONS, WHEN CALLED FOR*

In this first chapter you will learn about the meaning of signed numbers, opposite numbers, absolute value, and the number line. You will also learn how to perform the four basic operations using signed numbers.

1.1 OPPOSITE NUMBERS AND ABSOLUTE VALUE

What meaning does the + sign convey to you? _____

What meaning does the – sign convey to you? _____

Maybe you said that + means addition and – mean subtraction. That is correct, but what would you say if the + or – sign is combined with a number — for example, +2 or –2?

What is the difference between these two numbers? _____

_____

You may have written that the plus sign indicates an increase or, perhaps, something coming to you (money, for example). You may have written that the minus sign means a decrease or a debt, or something going away from you.

Let us further consider the meaning of these signs. Look at the following:

Margaret gained weight
Margaret lost weight

Strategies for Using This Text

1. **Read the new material before each class.**
   - You will then be prepared to ask questions about your difficulties.
   - Do the learning activities. They are designed to help you to learn the lesson.
   - Answer all questions and write out your answers in the spaces provided.
   - You learn by writing as well as by reading.

2. **Write down your questions.**
   - This will help you recognize what you don't yet understand.
   - Bring your questions to class.
   - Ask your instructor to explain.
   - Discuss your questions with your classmates.
   - Talk about your math problems and listen to others. By talking and listening — as well as by reading and writing — you will learn.

3. **Go over each worked-out example.**
   - Make sure you understand every step.
   - If you don't follow a step, go back over the learning activities and the stated rules.

4. **Do all of the MINI-PRACTICES.**
   - Check the answers immediately.

How can you indicate that there is a difference between these two cases? _____

As you probably know, we show the difference by using positive and negative signs.

How do we write a gain of 3 pounds? _____

How do we write a loss of 6 pounds? _____

+3 is a gain and –6 is a loss. We call numbers with signs before them **signed numbers**. The + sign is usually omitted.

A number without a sign or with a plus sign is called a **positive number**, and a number with a minus sign is called a **negative number**.

Write in symbols: A gain of 10 pounds: _____

Write in symbols: A loss of 10 pounds: _____

- Find out what you have understood and where you need further help.

5. Write out the math rules you have discovered.
- Write these in the space provided next to IN YOUR OWN WORDS.
- This is difficult to do, but it will help you to understand.

6. Try the Practice Exercises.
- Do these when you think you have understood the text material.
- Go back to the learning activities and rules when you are stumped.
- Check your answers in the back of the book.
- List the numbers of the exercises you need help on.

7. Have you forgotten arithmetic concepts?
- Use the Appendix for review.

8. Are there math words you don't understand?
- Use the Index to locate these words.
- Always look up words in a dictionary.

9. Read the Chapter Summaries before a test.
- This triggers your memory process.
- It alerts you to topics you need to review in depth.
- Look back at your own writings. They will help you to remember key concepts.

+10 and −10 are called **opposite numbers**. For example, the opposite of 7 is −7 and the opposite of −7 is 7.

Give examples of two numbers with opposite signs: _____

Are these numbers also opposite numbers? _____

Numbers such as −3 and +2 have opposite signs, but are *not* opposite numbers.

−3 and +3 are opposite numbers
+2 and −2 are opposite numbers

Consider the opposite numbers 2 and −2.

What is the *same* in these two numbers? _____

Of course, the numeral 2 (here it is appropriate to use the word numeral) is the same. Another way of saying this is that the **magnitude** or the **absolute value** is the same.

In the earlier example of weight loss, the 10 pounds gained or lost involved the same *magnitude*. The *absolute value* of −10 is the same as the *absolute value* of +10, namely 10. The absolute value of a number is *always positive*.

Instead of writing the words "absolute value," we often use the symbol

$$|\quad|$$

For example,

| −6 | is read "the absolute value of negative six"

What is the value of | −6 |? _____

The answer is 6 since an absolute value is always positive (or 0). The signed numbers +6 and −6 are opposite numbers because they both have an absolute value of 6.

$$|+6| = |-6| = 6$$

What is the value of | 0 |? _____

What is the opposite of 0? _____

0 is its own opposite, and has an absolute value of 0.

Once again, the symbol

$$|\quad|$$

indicates absolute value.

Example 1   Find | −5 |.

Solution.

$$|-5| = 5$$

The − sign is ignored; we write only the magnitude of the number.   ■

Example 2   Find the value of − | −5 |.

Solution.

$$-|-5| = -(5) = -5$$
$$|-5| = 5$$

The absolute-value symbol acts like parentheses. We cannot do anything about the problem until we have taken care of the parentheses. In this case we translate | −5 | as (5).

$$-(5) = -5$$

The first − in − | −5 | is outside the absolute value symbol. Thus

$$-|-5| = -(5) = -5$$ ■

Example 3   Find the value of | −2 | · | −5 |.

Solution. $|-2| = 2,$ $|-5| = 5$

The operation is multiplication:

$$2 \cdot 5 = 10$$ ∎

MINI-PRACTICE

(Use this space for your work.)

1. Find the opposites of the following numbers:

(a) 17   (b) –8   (c) –4   (d) 15   (e) $-\dfrac{3}{5}$

Answers: (a) _____   (b) _____   (c) _____   (d) _____   (e) _____

2. Find the following absolute values:

(a) $|-15|$ (b) $|-201|$ (c) $|6|$ (d) $|0|$ (e) $|-10|$

Answers: (a) _____   (b) _____   (c) _____   (d) _____   (e) _____

3. Do the following operations using absolute values:

(a) $|-3| + |-5|$ (b) $3|-6|$ (c) $-|-8|$

(d) $|-7| - 3$ (e) $|-2| \cdot |2|$

Answers: (a) _____   (b) _____   (c) _____   (d) _____   (e) _____

IN YOUR OWN WORDS

state the definition of

1. *Signed Numbers:* _____

2. *Absolute Value:* _____

3. *Opposite Numbers:* _____

Answers:

1. *Signed Numbers* are numbers preceded by a + or a − sign.

2. *Absolute Value* is the magnitude of a number.

3. *Opposite Numbers* are two numbers with the same absolute value, but with different signs. In particular, the opposite of zero is zero itself. We say that zero has no sign.

Compare your definitions with the ones that appear here, and make any necessary changes in your own definitions.

Answers to Mini-Practice:

1. (a) −17   (b) 8   (c) 4   (d) −15   (e) $\dfrac{3}{5}$

2. (a) 15   (b) 201   (c) 6   (d) 0   (e) 10

3. (a) 8   (b) 18   (c) −8   (d) 4   (e) 4

Practice Exercises

In Exercises 1–17, find the opposite of each number.

1. −15   *Answer:* _____
2. +3   *Answer:* _____
3. −85   *Answer:* _____
4. −1000   *Answer:* _____
5. −0.02   *Answer:* _____
6. 1.2   *Answer:* _____
7. $-2\dfrac{1}{5}$ *Answer:* _____
8. 0.25   *Answer:* _____
9. −0.75   *Answer:* _____
10. 1.5   *Answer:* _____
11. −3.25   *Answer:* _____
12. 2.75   *Answer:* _____
13. $-\dfrac{1}{2}$ *Answer:* _____
14. $\dfrac{3}{4}$ *Answer:* _____
15. $-\dfrac{1}{8}$ *Answer:* _____
16. $3\dfrac{1}{4}$ *Answer:* _____
17. $-2\dfrac{1}{2}$ *Answer:* _____

In Exercises 18–32, find each absolute value.

18. $|-5|$ *Answer:* _____
19. $|2.3|$ *Answer:* _____
20. $|11|$ *Answer:* _____
21. $|-11|$ *Answer:* _____
22. $|5-3|$ *Answer:* _____
23. $|-0.5|$ *Answer:* _____
24. $|-0.75|$ *Answer:* _____
25. $|5.3-3.2|$ *Answer:* _____
26. $|4.07|$ *Answer:* _____
27. $|5.33|$ *Answer:* _____
28. $\left|\dfrac{1}{2}\right|$ *Answer:* _____
29. $\left|-\dfrac{1}{4}\right|$ *Answer:* _____
30. $\left|\dfrac{3}{4}\right|$ *Answer:* _____
31. $\left|\dfrac{7}{8}\right|$ *Answer:* _____
32. $\left|-\dfrac{1}{8}\right|$ *Answer:* _____

In Exercises 33–56, find each value.

33. $|-2|3$ *Answer:* _____
34. $|-3|2$ *Answer:* _____
35. $|-3|-2$ *Answer:* _____
36. $|-4|-|-4|$ *Answer:* _____
37. $|-4|\cdot|-4|$ *Answer:* _____
38. $|-8|+3$ *Answer:* _____
39. $|-27|-|3|$ *Answer:* _____
40. $|2.5|+|-0.5|$ *Answer:* _____
41. $|-0.7|+|0.5|$ *Answer:* _____
42. $|3.2|+|7.8|$ *Answer:* _____
43. $|0.8|\cdot|1.5|$ *Answer:* _____
44. $|5.5|\div|5|$ *Answer:* _____
45. $|-0.4|+|-0.9|$ *Answer:* _____
46. $|-3.9|-|-2.5|$ *Answer:* _____
47. $|0.4|\cdot|-2.3|$ *Answer:* _____
48. $\left|\dfrac{3}{4}\right|-\left|\dfrac{1}{4}\right|$ *Answer:* _____
49. $\left|\dfrac{3}{8}\right|+\left|-\dfrac{1}{4}\right|$ *Answer:* _____
50. $\left|\dfrac{1}{5}+\dfrac{7}{10}\right|$ *Answer:* _____
51. $\left|\dfrac{3}{4}\right|\cdot\left|\dfrac{-2}{9}\right|$ *Answer:* _____
52. $\left|\dfrac{-1}{4}\right|\div\left|\dfrac{5}{8}\right|$ *Answer:* _____
53. $\left|\dfrac{-8}{5}\right|-\left|\dfrac{-4}{5}\right|$ *Answer:* _____
54. $\left|\dfrac{3}{4}-\dfrac{1}{4}\right|$ *Answer:* _____
55. $\left|\dfrac{-5}{6}\right|\cdot\left|\dfrac{-12}{25}\right|$ *Answer:* _____
56. $\left|\dfrac{-3}{7}\right|-\left|\dfrac{-9}{28}\right|$ *Answer:* _____

In Exercises 57 and 58, fill in the missing word.

57. The *magnitude* of a number is its _____ _____ .

58. 9 and –9 are _____ numbers.

1.2 THE NUMBER LINE

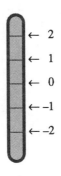

Look at the figure of a thermometer to the left.

Measure the distance between zero and +1.
Measure the distance between zero and –1.

Did you note that these distances are the same?

A thermometer is a scale just like a ruler is. The distance on a thermometer representing one degree is the same throughout the scale. On the thermometer, check that the distances between zero and 1.5 and between zero and –1.5 are the same.

Rotate your thermometer *clockwise* until it is horizontal. The thermometer should now look like this:

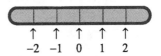

The following horizontal line, representing the thermometer, is an example of a number line. The number line is a device used by mathematicians to represent the relative position of numbers.

This is how we make a number line: First we select a starting point to represent the number 0. Then we decide how large one unit should be. Let's say that one unit should be 1/2 inch.

We mark one unit to the *right* of zero. Because this point is one distance unit from zero, we represent the point by the number 1.

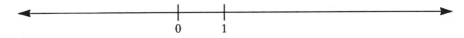

Continue to mark off 1/2-inch units on the number line to the right and to the left of zero. The numbers to the right of zero are positive and the numbers to the left of zero are negative.

The number line goes on forever in both directions, as the arrows suggest.

What is the distance between zero and 5? _____

What is the distance between zero and –5? _____

Since the opposite numbers –5 and 5 are each located five distance units from zero, both distances are represented by the same number, that is, by 5.

In the preceding section you learned that 5 is the *magnitude* or *absolute value* of both *+5 and –5*. The **absolute value** of a number can also be *defined* as the distance between zero and the number on the number line. Thus, +5 and –5 are each 5 distance units from the origin. Also, +3 and –3 are each 3 distance units from the origin.

Example 1   Suppose the unit distance is 1/4 inch, as on the following number line:

Complete the number line.

Solution. We can position 0 at any place we wish. 1/4 inch is marked to the right and to the left repeatedly. The numbers represented are listed below the markers. The number line now looks like this:

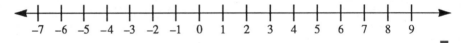

Example 2   Replace the question marks with numbers.

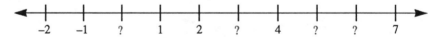

Solution.

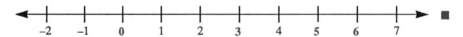

Example 3   Replace the question marks with numbers.

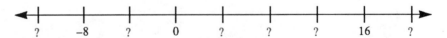

Solution.

Example 4   Replace the question marks with numbers.

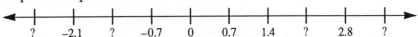

Solution.

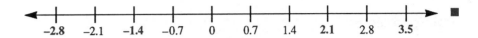 ■

MINI-PRACTICE

(Use this space for your work.)

1. Construct a number line with one unit equal to 1 inch.

2. Construct a number line with one unit equal to 1 cm.

3. Construct a number line with two units equal to 1 inch.

4. Complete the following number line:

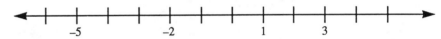

5. Complete the following number line:

IN YOUR OWN WORDS   explain how to construct a number line:

_____

_____

_____

_____

Answers to Mini-Practice:

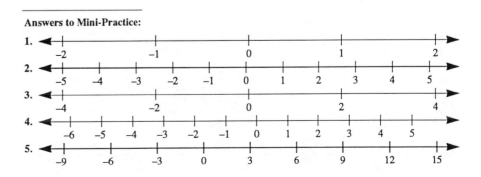

Answer:

When we construct a number line, we draw a straight line, select a point we call zero, and decide on the length of the unit distance (i.e., | 1 | = | −1 | = unit distance). Positive numbers are marked to the right and negative numbers to the left of 0.

Practice Exercises

1. Construct a number line with one unit equal to 1/2 inch.

$$\longleftrightarrow$$

2. Construct a number line with one unit equal to 2 cm.

$$\longleftrightarrow$$

3. Construct a number line with two units equal to 1/2 inch.

$$\longleftrightarrow$$

4. Construct a number line with five units equal to 1 inch.

$$\longleftrightarrow$$

5. Construct a number line with one unit equal to 5 cm.

$$\longleftrightarrow$$

6. Construct a number line with four units equal to 1 inch.

$$\longleftrightarrow$$

7. Construct a number line with one unit equal to 2 inches.

$$\longleftrightarrow$$

8. Construct a number line with two units equal to 1 cm.

$$\longleftrightarrow$$

9. Construct a number line with ten units equal to 1 inch.

$$\longleftrightarrow$$

In Exercises 10–25, complete each number line.

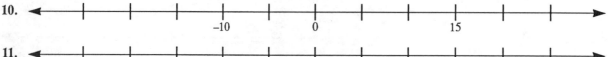

10.
  −10   0   15

11.
  −6   0   9

12.
  −1   0

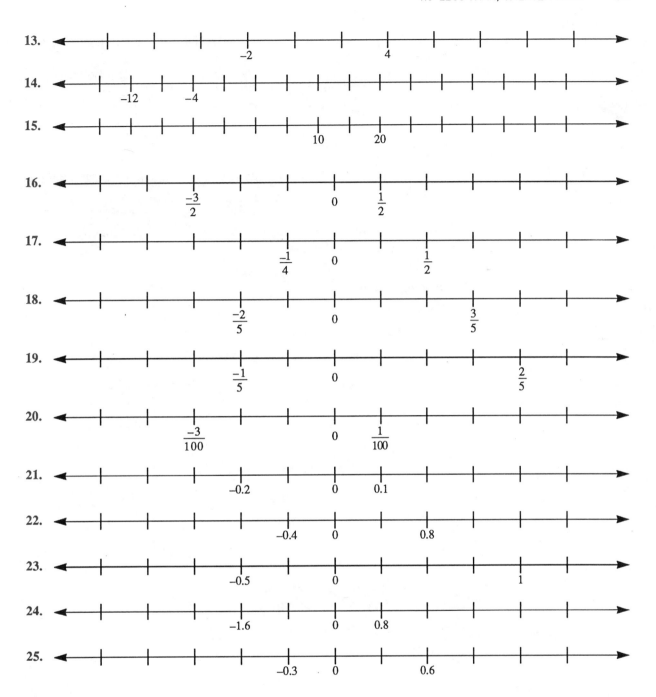

1.3 Less than, Greater than

In mathematics, what is the symbol for **is less than**? _____

What is the symbol for **is greater than**? _____

2 is less than 4 is written as 2 < 4

4 is greater than 2 is written as 4 > 2

In both cases the symbol "points to" the smaller number.

If you had problems with the symbols, go back to the introductory chapter for a review.

Which number is smaller, 2 or 4? _____

Which number is smaller, –2 or –4? _____

To understand the relation of "is less than" among negative numbers, think of money.

Are you better off financially if you have a *debt* of $4 or a *debt* of $2? _____

Because you would have *more* of a *debt*, you could say that you would have *less* money when you owe $4 than when you owe $2. In other words:

–4 is less than –2

In symbols:

$$-4 < -2$$

On the number line, where is 2 in relation to 4? _____

On the number line, where is –2 in relation to –4? _____

2 is to the left of 4, whereas –2 is to the right of –4 on the number line.

On the preceding number line, select two points.

Read the left point: _____   Read the right point: _____

Which number is the smaller: _____

If the number *a* is to the left of the number *b* on the number line, state the relationship between *a* and *b*: _____

As we have seen in the examples, when the number *a* is to the left of the number *b*, we know that the number *a* is smaller than the number *b* or

$$a < b$$

If *a* is to the right of *b*, state the relationship between the numbers *a* and *b*: _____

In this case the number *a* is larger than the number *b* or

$$a > b$$

Example 1   Insert the correct inequality symbol, < or >.

$$-5 \boxed{} 6$$

Solution.

$$-5 \text{ is smaller (less) than } 6,$$

$$-5 < 6$$

because −5 is to the *left* of 6 on the number line.   ∎

Example 2   Insert the correct inequality symbol, < or >.

$$-3 \boxed{\phantom{<}} -4$$

Solution.

$$-3 \text{ is larger (greater) than } -4,$$

$$-3 > -4$$

because −3 is to the *right* of −4 on the number line.   ∎

MINI-PRACTICE   Insert the correct inequality symbol, < or >, between the number pairs:
(Use this space for your work.)

(a)   15 $\boxed{\phantom{<}}$ 23   (b)   −7 $\boxed{\phantom{<}}$ −13   (c)   −5 $\boxed{\phantom{<}}$ 18

(d)   −30 $\boxed{\phantom{<}}$ −4   (e)   143 $\boxed{\phantom{<}}$ 79

IN YOUR OWN WORDS   state the relationship in size between numbers and their position on the number line: _____

_____

_____

Answer:

A number, *a*, to the left of another number, *b*, on the number line is the smaller of the two ($a < b$).

A number, *a*, to the right of another number, *b*, on the number line is the larger of the two ($a > b$).

Practice Exercises

In Exercises 1–38, insert the appropriate symbol, < or >, between the following pairs of numbers:

1. 2 $\boxed{\phantom{<}}$ 3   **2.** 3 $\boxed{\phantom{<}}$ 4   **3.** 4 $\boxed{\phantom{<}}$ 2   **4.** −2 $\boxed{\phantom{<}}$ 3

Answers to Mini-Practice:

(a) <   (b) >   (c) <   (d) <   (e) >

5. $3\boxed{}-4$ **6.** $3\boxed{}-2$ **7.** $1\boxed{}5$ **8.** $-3\boxed{}-1$

9. $0\boxed{}-2$ **10.** $-5\boxed{}-3$ **11.** $-4\boxed{}-5$ **12.** $-2\boxed{}2$

13. $4\boxed{}13$ **14.** $-9\boxed{}-12$ **15.** $-22\boxed{}19$ **16.** $7\boxed{}-14$

17. $3.5\boxed{}4.2$ **18.** $-2.3\boxed{}2.4$ **19.** $5.8\boxed{}5.75$ **20.** $-0.1\boxed{}-0.2$

21. $6.6\boxed{}5.9$ **22.** $-15.1\boxed{}15.4$ **23.** $-3.9\boxed{}-3.98$ **24.** $4.6\boxed{}-13.2$

25. $-52.1\boxed{}49.01$ **26.** $-0.12\boxed{}-0.13$ **27.** $\dfrac{1}{2}\boxed{}\dfrac{1}{4}$ **28.** $\dfrac{-1}{3}\boxed{}\dfrac{-2}{3}$

29. $-\dfrac{1}{10}\boxed{}-\dfrac{3}{10}$ **30.** $\dfrac{4}{7}\boxed{}\dfrac{3}{7}$ **31.** $-\dfrac{2}{5}\boxed{}-\dfrac{4}{5}$ **32.** $\dfrac{1}{6}\boxed{}\dfrac{5}{6}$

33. $-\dfrac{3}{4}\boxed{}-\dfrac{1}{4}$ **34.** $-\dfrac{5}{13}\boxed{}-\dfrac{3}{13}$ **35.** $\dfrac{4}{7}\boxed{}-\dfrac{2}{7}$ **36.** $-\dfrac{2}{5}\boxed{}\dfrac{1}{5}$

37. If a is to the left of 5, then $a\boxed{}5$.

38. If b is to the right of 8, then $b\boxed{}8$.

1.4 ADDING NUMBERS WITH THE SAME SIGN

In the following sections we will discuss the four basic operations on signed numbers. To simplify the discussion we will at first add only numbers with the same sign; then we will add numbers with opposite signs.

Rewrite the following phrase using operation symbols and signed numbers. (Every number should have a sign before it.)

$9 deposited in your account: _____

plus   _____

$5 deposited in your account: _____

Total amount deposited in your account: $14

Which symbol do you use to represent addition? _____

What is the sign of each of the numbers you added? _____

What is the sign of the sum?   _____

If you followed the instructions, you wrote

$$\begin{array}{r} +9 \\ +\ +5 \\ \hline +14 \end{array} \quad \text{or} \quad (+9) + (+5) = 14$$

However, we usually write this simply as

$$9 + 5 = 14$$

Now, rewrite this phrase using signs and numbers only. Again, each number should have a sign before it:

You owe your cousin $9:   _____

You owe your friend $5:   _____

Total amount you owe: −$14

Which operation did you use to solve this problem? _____

What is the sign of each of the numbers you **added**?_____

What is the sign of the **sum**?   _____

Probably you wrote:

$$\begin{array}{r} -9 \\ +\,-5 \\ \hline -14 \end{array}$$

If you were writing this horizontally, you would write it this way:

$$-9 + -5 = -14$$

or more commonly,

$$(-9) + (-5) = (-14)$$

We often drop the parentheses around the first number as well as in the answer:

$$-9 + (-5) = -14$$

Go over the following examples carefully so that you learn the procedure for adding negative numbers.

Example 1   Add: $-3 + (-4)$

Solution.

Determine the operation: +
Determine the common sign of the numbers: −
Find the absolute values of the numbers −3 and −4: 3 and 4
Add these absolute values: $3 + 4 = 7$
Put the common sign before the answer: −7

$$-3 + (-4) = -7 \qquad \blacksquare$$

Example 2   Add: $+6 + (+3)$

Solution.

Determine the operation: +
Determine the common sign of the numbers: +
Find the absolute values of the numbers: 6 and 3
Add these absolute values: $6 + 3 = 9$
Put the common sign before the answer: +9

$$+6 + (+3) = +9 \qquad \blacksquare$$

Example 3   Add: $\dfrac{-1}{9} + \left(\dfrac{-2}{9}\right)$

Solution.

Determine the operation: +
Determine the common sign of the numbers:

$$\dfrac{-1}{9} = -\dfrac{1}{9} \quad , \quad \dfrac{-2}{9} = -\dfrac{2}{9}$$

The common sign is −.

Find the absolute values of the numbers: $\dfrac{1}{9}$ and $\dfrac{2}{9}$

Add these absolute values: To add fractions with the same denominator, write down this denominator and add the numerators. Thus,

$$\dfrac{1}{9} + \dfrac{2}{9} = \dfrac{1+2}{9} = \dfrac{3}{9} = \dfrac{1}{3}$$

(The result is expressed in lowest terms.)

Put the common sign before the answer: $-\dfrac{1}{3}$. It follows that

$$\dfrac{-1}{9} + \left(\dfrac{-2}{9}\right) = -\dfrac{1}{3}$$ ■

If you need more practice in adding fractions see page 357 of the Appendix.

Example 4   Add: $\dfrac{3}{5} + \dfrac{1}{4}$

Solution. First find the least common denominator (l.c.d.)

The l.c.d. $= 5 \times 4 = 20$

$$\dfrac{3}{5} = \dfrac{3 \times 4}{5 \times 4} = \dfrac{12}{20}$$

$$\dfrac{1}{4} = \dfrac{1 \times 5}{4 \times 5} = \dfrac{5}{20}$$

Rewrite the problem: $\dfrac{12}{20} + \dfrac{5}{20}$

Determine the operation: +
Determine the common sign of the numbers: +

Add: $\dfrac{12}{20} + \dfrac{5}{20} = \dfrac{17}{20}$ ■

MINI-PRACTICE   Add:
(Use this space for your work.)

(a) $(-8) + (-9)$ (b) $(+8) + (+9)$ (c) $8 + 9$

(d) $7 + 12$ (e) $(+7) + (+12)$ (f) $(-7) + (-12)$

(g) $(-2) + (-3) + (-4)$ (h) $-0.5 + (-0.9)$ (i) $\dfrac{-1}{3} + \left(\dfrac{-1}{5}\right)$

Answers: (a) _____   (b) _____   (c) _____

(d) _____   (e) _____   (f) _____

(g) _____   (h) _____   (i) _____

IN YOUR OWN WORDS   state a rule for adding signed numbers with the same sign: _____

_____

_____

Answer:

To add numbers with a common sign, add their absolute values, and give this sum the sign appearing before each number (that is, the common sign).

Practice Exercises

In Exercises 1–44, add.

1. $(-9) + (-13)$ *Answer:* _____

2. $-3 + (-1)$ *Answer:* _____

3. $-5 + (-3)$ *Answer:* _____

4. $-4 + (-23)$ *Answer:* _____

5. $-4 + (-5)$ *Answer:* _____

6. $-12 + (-12)$ *Answer:* _____

7. $-12 + (-1)$ *Answer:* _____

8. $-9 + (-1)$ *Answer:* _____

9. $-13 + (-26)$ *Answer:* _____

10. $(-2) + (-17)$ *Answer:* _____

11. $-3 + (-12) + (-15)$ *Answer:* _____

12. $-27 + (-42) + (-18)$ *Answer:* _____

13. $-14 + (-3) + (-6)$ *Answer:* _____

14. $-22 + (-18) + (-5)$ *Answer:* _____

15. $-25 + (-31) + (-4)$ *Answer:* _____

16. $-2 + (-13) + (-9)$ *Answer:* _____

17. $-18 + (-9) + (-3)$ *Answer:* _____

18. $-1 + (-2) + (-1)$ *Answer:* _____

19. $-29 + (-19) + (-14)$ *Answer:* _____

20. $-5 + (-12) + (-23)$ *Answer:* _____

21. $-12 + (-31) + (-21)$ *Answer:* _____

22. $-2 + (-3) + (-4) + (-8)$ *Answer:* _____

23. $-9 + (-11) + (-13) + (-15)$ *Answer:* _____

24. $-29 + (-37) + (-53) + (-62)$ *Answer:* _____

25. $-13 + (-8) + (-5) + (-2)$ *Answer:* _____

26. $-0.7 + (-1.3)$ *Answer:* _____

27. $-0.12 + (-0.21)$ *Answer:* _____

28. $-15.6 + (-16.06)$ *Answer:* _____

29. $-8 + (-5.34)$ *Answer:* _____

30. $-2.8 + (-3)$ *Answer:* _____

31. $-10.5 + (-8.4)$ *Answer:* _____

32. $-19.2 + (-23.9)$ *Answer:* _____

33. $-2 + (-3.16)$ *Answer:* _____

34. $\frac{-1}{2} + \left(\frac{-1}{2}\right)$ *Answer:* _____

35. $\frac{1}{5} + \frac{2}{5}$ *Answer:* _____

36. $\frac{1}{4} + \frac{3}{4}$ *Answer:* _____

37. $\left(\dfrac{-5}{6}\right)+\left(\dfrac{-1}{6}\right)$ *Answer:* _____   **38.** $\left(\dfrac{-3}{8}\right)+\left(\dfrac{-1}{8}\right)$ *Answer:* _____

39. $\dfrac{-2}{3}+\left(\dfrac{-6}{11}\right)$ *Answer:* _____   **40.** $\dfrac{-7}{9}+\left(\dfrac{-5}{18}\right)$ *Answer:* _____

41. $\dfrac{-5}{7}+\left(\dfrac{-5}{14}\right)$ *Answer:* _____   **42.** $\dfrac{-2}{3}+\left(\dfrac{-1}{2}\right)$ *Answer:* _____

43. $\dfrac{-17}{10}+\left(\dfrac{-3}{5}\right)$ *Answer:* _____   **44.** $\dfrac{-4}{13}+\left(\dfrac{-1}{2}\right)$ *Answer:* _____

1.5 ADDING NUMBERS WITH DIFFERENT SIGNS

On the number line, perform the operation $2 + 3$.

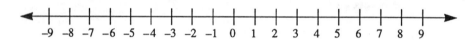

In which direction do you move when you add a *positive* number?

_____

From the figure,

we see that we move 3 steps to the right to obtain $2 + 3 = 5$.

Add: $-5 + 9$, using the preceding number line.

What starting point did you use? _____

Which direction did you move in? _____

At what point did you end? _____

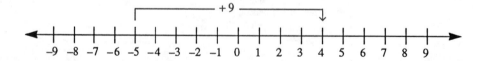

When we start at -5 and move 9 steps to the right, we end up at $+4$. From this activity we conclude that

$$-5 + 9 = 4$$

Why is the sum of -5 and 9 the same as the sum of 9 and -5?

_____

Addition is commutative, that is, $-5 + 9 = 9 + (-5)$, and the sum in each case is 4.

Add: $9 + (-5)$, using the number line.

    What starting point did you use? _____

    At what point did you end? _____

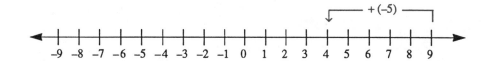

We start at 9 and end at 4 since $9 + (-5) = -5 + 9 = 4$.

    When we add a negative number, in which direction do we move on the number line? _____

We go to the right when we add a positive number, and to the left when we add a negative number.

Add: $-9 + 5$, using the number line.

    Starting point: _____

    Direction: _____

    End point: _____

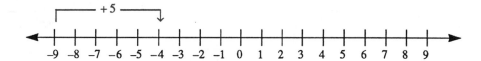

Add: $5 + (-9)$, using the number line.

    Starting point: _____

    Direction: _____

    End point: _____

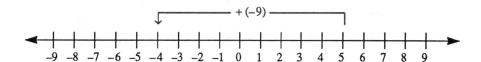

The answer to both of these problems is -4.

To summarize, we found that

$$9 + (-5) = 4 \quad \text{and} \quad (-9) + 5 = -4$$

Go over the following examples in detail so that you learn how to add a positive and a negative number.

Example 1 $6 + (-3) = ?$

Solution.

Determine the operation: $+$
Determine the signs of the numbers: $+$ and $-$
Find the absolute values of the numbers 6 and -3: 6 and 3
Subtract the smaller absolute value from the larger one: $6 - 3 = 3$
Apply the sign of the number with the larger absolute value: $+$

$$6 + (-3) = 3$$ ■

Example 2 $-9 + 7 = ?$

Solution.

Determine the operation: $+$
Determine the signs of the numbers: $-$ and $+$
Find the absolute values of the numbers -9 and 7: 9 and 7
Subtract the smaller absolute value from the larger: $9 - 7 = 2$
Apply the sign of the number with the larger absolute value: $-$

$$-9 + 7 = -2$$ ■

Example 3 $2 + (-1) + (-5) = ?$

Solution.

Method 1: Perform the additions from left to right.

$$2 + (-1) = 1$$
$$1 + (-5) = -4$$

Method 2: Add the two numbers with the same sign first:

$$-1 + (-5) = -6$$
$$\text{Add:} \quad 2 + (-6) = -4$$

By either method,

$$2 + (-1) + (-5) = -4$$ ■

MINI-PRACTICE   Add:
(Use this space for your work.)

(a) $(-8) + 9$ (b) $8 + (-9)$ (c) $(+8) + (-9)$

(d) $7 + (-12)$ (e) $(-7) + 12$ (f) $(-7) + (+12)$

Answers: (a) _____   (b) _____   (c) _____

(d) _____   (e) _____   (f) _____

Answers to Mini-Practice:
(a) 1   (b) −1   (c) −1   (d) −5   (e) 5   (f) 5

IN YOUR OWN WORDS   write a rule for addition of signed numbers with different signs: _____

_____

_____

Answer:

    To **add** numbers with different signs: (a) **subtract** their absolute values; (b) the sum gets the sign of the number with the larger absolute value.

Practice Exercises

In Exercises 1–64, add.

1. $(-9) + 13$ *Answer:* _____
2. $28 + (-79)$ *Answer:* _____
3. $3 + (-5)$ *Answer:* _____
4. $-4 + 2$ *Answer:* _____
5. $6 + (-9)$ *Answer:* _____
6. $-9 + 4$ *Answer:* _____
7. $2 + (-2)$ *Answer:* _____
8. $11 + (-7)$ *Answer:* _____
9. $-2 + 8$ *Answer:* _____
10. $-8 + 3$ *Answer:* _____
11. $-13 + 9$ *Answer:* _____
12. $18 + (-27)$ *Answer:* _____
13. $14 + (-3) + (-9)$ *Answer:* _____
14. $-2 + (-3) + 4$ *Answer:* _____
15. $-5 + (-3) + (-28)$ *Answer:* _____
16. $18 + (-21) + 6$ *Answer:* _____
17. $15 + (-9) + (-20)$ *Answer:* _____
18. $2 + (-3) + (-4)$ *Answer:* _____
19. $7 + (-2) + (-8)$ *Answer:* _____
20. $9 + (-6) + (-3)$ *Answer:* _____
21. $-3 + (-8) + 12$ *Answer:* _____
22. $13 + 62 + (-38)$ *Answer:* _____
23. $9 + (-6) + (-16)$ *Answer:* _____
24. $31 + (-23) + (-9)$ *Answer:* _____
25. $-15 + 21 + (-18)$ *Answer:* _____
26. $17 + 5 + (-33)$ *Answer:* _____
27. $-13 + (-9) + 17$ *Answer:* _____
28. $22 + 10 + 2 + (-18)$ *Answer:* _____
29. $-6 + (-8) + 13 + (-4)$ *Answer:* _____
30. $-25 + 31 + 24 + 19$ *Answer:* _____
31. $-17 + (-3) + 29$ *Answer:* _____
32. $-22 + 10 + 12 + (-18)$ *Answer:* _____
33. $-14 + (-3) + 7 + (-6)$ *Answer:* _____
34. $19 + (-9) + (-2) + 3$ *Answer:* _____
35. $-41 + (-13) + 38 + (-9)$ *Answer:* _____
36. $-12 + 30 + 3 + (-17)$ *Answer:* _____

37.
```
    12
  − 9
    15
  −17
```
Answer:

38.
```
  −34
  −18
   19
  − 7
```
Answer:

39.
```
    52
  −28
  −37
  −16
```
Answer:

40.
```
    13
    18
  −15
  − 5
```
Answer:

41.
```
  −29
  −41
   14
  − 2
```
Answer:

42.
```
  − 5
  −17
  −23
   41
```
Answer:

43. $0.3 + (-0.4) + 5.2$ *Answer:* _____

44. $4.3 + (-8.2) + 0.75$ *Answer:* _____

45. $0.45 + 3.26 + (-7.2)$ *Answer:* _____

46. $5.5 + 7.2 + (-10)$ *Answer:* _____

47. $3.2 + (-0.12) + 3.14$ *Answer:* _____

48. $(-4.8) + 8.4 + (-9.18)$ *Answer:* _____

49. $-0.2 + (-0.21) + 5.2$ *Answer:* _____

50. $5.2 + (-7.01) + (-9)$ *Answer:* _____

51. $11 + 12.36 + (-17.01)$ *Answer:* _____

52. $(-31.2) + (-3) + 42.8$ *Answer:* _____

53. $\dfrac{3}{5} + \left(\dfrac{-1}{5}\right)$ *Answer:* _____

54. $\left(-\dfrac{1}{10}\right) + \dfrac{9}{10}$ *Answer:* _____

55. $\dfrac{5}{6} + \left(\dfrac{-1}{6}\right)$ *Answer:* _____

56. $\dfrac{5}{12} + \left(\dfrac{-7}{12}\right)$ *Answer:* _____

57. $1 + \left(\dfrac{-1}{4}\right)$ *Answer:* _____

58. $\dfrac{-3}{4} + \dfrac{1}{6}$ *Answer:* _____

59. $\dfrac{1}{100} + \left(\dfrac{-7}{100}\right) + \dfrac{3}{100}$ *Answer:* _____

60. $\dfrac{1}{3} + \dfrac{5}{3} + \left(\dfrac{-2}{3}\right)$ *Answer:* _____

61. $\dfrac{-3}{13} + \left(\dfrac{-10}{13}\right) + \dfrac{12}{13}$ *Answer:* _____

62. $\dfrac{3}{4} + \left(-\dfrac{1}{4}\right) + \dfrac{3}{8}$ *Answer:* _____

63. $\dfrac{5}{18} + \left(\dfrac{-2}{9}\right) + \left(\dfrac{-5}{9}\right)$ *Answer:* _____

64. $\dfrac{1}{12} + \left(\dfrac{-1}{4}\right) + \dfrac{3}{8}$ *Answer:* _____

1.6 SUBTRACTION OF POSITIVE NUMBERS

We will now compare subtraction of signed numbers with addition. We first consider subtraction of positive numbers and then subtraction of negative numbers.

$$9 + (-5) = 4$$
$$9 - 5 = 4$$

How do you check your answer in a subtraction problem? _____

You probably remember that to check, you add 4 and 5. Thus, $4 + 5 = 9$, so the subtraction is correct.
    Now try these.

$3 + (-8) =$ _____

$3 - 8 =$ _____

Check by adding your answer to 8.

Is the sum 3? _____

If not, make a number line and do the subtraction in the following way:

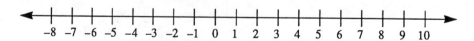

Start at 3; move 8 units to the *left*. Where is your end point? _____

Here are two more related problems.

$$-2 + (-4) = \underline{\hspace{2cm}}$$

$$-2 - 4 = \underline{\hspace{2cm}}$$

Do this second problem on the preceding number line. Check by adding 4 to your answer.

Is the sum equal to –2? _____

In the preceding activities you saw that

$$9 + (-5) = 4 \quad \text{and} \quad 9 - 5 = 4$$
$$3 + (-8) = -5 \quad \text{and} \quad 3 - 8 = -5$$
$$-2 + (-4) = -6 \quad \text{and} \quad -2 - 4 = -6$$

Go over the following examples carefully so that you learn how to subtract a positive number.

Example 1 $4 - 2 = ?$

Solution.

Determine the operation: –
Determine the signs of the numbers: + and +
Change the operation from subtraction to addition: +
Write the opposite of the second number: –2
Rewrite the problem: $4 + (-2)$
Use the rule for addition of signed numbers: $4 + (-2) = 2$

$$4 - 2 = 2$$

∎

Example 2 $-3 - 6 = ?$

Solution.

Determine the operation: –
Determine the signs of the numbers: – and +
Change the operation from subtraction to addition: +
Write the opposite of the second number: –6
Rewrite the problem: $-3 + (-6)$
Use the rule for addition of signed numbers: $-3 + (-6) = -9$

$$-3 - 6 = -9$$

∎

Example 3 $\dfrac{2}{5} - \dfrac{9}{10}$

Solution. First find the least common denominator.

The l.c.d. = 10

$$\frac{2}{5} = \frac{4}{10}$$

Rewrite the problem: $\dfrac{4}{10} - \dfrac{9}{10}$

The operation is subtraction. Both numbers are positive. Rewrite the problem as addition of the opposite: $\dfrac{4}{10} + \left(-\dfrac{9}{10}\right)$

Use the rule for addition:

$$\frac{4}{10} + \left(-\frac{9}{10}\right) = -\frac{5}{10} = -\frac{1}{2} \qquad\blacksquare$$

MINI-PRACTICE   Subtract:
(Use this space for your work.)

    (a) $4 - 10$ (b) $-7 - 3$ (c) $4 - 9$ (d) $7 - 6$

    (e) $-7 - 6$ (f) $10 - 15$ (g) $-10 - 15$

Answers: (a) _____   (b) _____   (c) _____   (d) _____

          (e) _____   (f) _____   (g) _____

IN YOUR OWN WORDS   explain how to subtract a positive number: _____

_____

_____

Answer:

    When subtracting a positive number, we change the subtraction to "addition of the opposite," and then apply the rules for addition of signed numbers.

    In symbols:

$$a - b = a + (-b)$$

Practice Exercises

In Exercises 1–42, subtract.

1. $5 - 17$ *Answer:* _____   **2.** $-5 - 17$ *Answer:* _____   **3.** $23 - 18$ *Answer:* _____

4. $-23 - 18$ *Answer:* _____   **5.** $16 - 8$ *Answer:* _____   **6.** $6 - 9$ *Answer:* _____

7. $-2 - 5$ *Answer:* _____   **8.** $2 - 7$ *Answer:* _____   **9.** $-4 - 4$ *Answer:* _____

Answers to Mini-Practice:

(a) -6 (b) -10 (c) -5 (d) 1 (e) -13 (f) -5 (g) -25

10. $-4 - 5$ *Answer:* _____   **11.** $3 - 8$ *Answer:* _____   **12.** $6 - 7$ *Answer:* _____

13. $-6 - 7$ *Answer:* _____   **14.** $3 - 9$ *Answer:* _____   **15.** $-3 - 9$ *Answer:* _____

16. $2 - 10$ *Answer:* _____   **17.** $-2 - 10$ *Answer:* _____   **18.** $5 - 15$ *Answer:* _____

19. $-5 - 15$ *Answer:* _____   **20.** $-100 - 1000$ *Answer:* _____

21. $29 - 33$ *Answer:* _____   **22.** $-45 - 23$ *Answer:* _____

23. $5.3 - 3.5$ *Answer:* _____   **24.** $4.8 - 10.2$ *Answer:* _____

25. $-9.5 - 11.25$ *Answer:* _____   **26.** $-10.1 - 5.75$ *Answer:* _____

27. $0.15 - 1.95$ *Answer:* _____   **28.** $17.21 - 14.2$ *Answer:* _____

29. $59.7 - 67.05$ *Answer:* _____   **30.** $-12.7 - 14.81$ *Answer:* _____

31. $-68.5 - 49$ *Answer:* _____   **32.** $-0.78 - 2.18$ *Answer:* _____

33. $\dfrac{3}{5} - \dfrac{4}{5}$ *Answer:* _____   **34.** $-\dfrac{3}{4} - \dfrac{5}{4}$ *Answer:* _____

35. $\dfrac{7}{3} - \dfrac{4}{3}$ *Answer:* _____   **36.** $\dfrac{7}{9} - \dfrac{8}{9}$ *Answer:* _____

37. $\dfrac{-21}{23} - \dfrac{3}{23}$ *Answer:* _____   **38.** $\dfrac{5}{18} - \dfrac{7}{18}$ *Answer:* _____

39. $-\dfrac{1}{2} - \dfrac{1}{4}$ *Answer:* _____   **40.** $\dfrac{1}{10} - \dfrac{7}{100}$ *Answer:* _____

41. $\dfrac{-1}{3} - \dfrac{1}{2}$ *Answer:* _____   **42.** $\dfrac{5}{6} - \dfrac{1}{3}$ *Answer:* _____

1.7 A SHORTCUT FOR SUBTRACTING

We have found that

$$-5 - 3 = -5 + (-3) = -8$$

Try to find a shortcut to go directly from $-5 - 3$ to -8.

Example 1 $-1 - 7 = ?$

Solution.

$$-1 + (-7) = -8$$
$$1 + 7 = 8$$
$$-1 - 7 = -8$$

∎

Example 2 $-8 - 10 = ?$

Solution.

$$8 + 10 = 18$$
$$-8 - 10 = -18$$

∎

MINI-PRACTICE   Use your shortcut in the following problems:

(Use this space for your work.)

(a) $-5-9$	(b) $-6-11$	(c) $-10-2$
(d) $-15-10$	(e) $-2-12$	

Answers: (a) _____   (b) _____   (c) _____

(d) _____   (e) _____

IN YOUR OWN WORDS   explain the shortcut you can use when you subtract a positive number from a negative number: _____

_____

_____

Answer:

When we subtract a positive number from a negative number, we simply add the absolute values of the numbers and give the answer a minus sign.

Practice Exercises

In Exercises 1–42, use the shortcut method to subtract.

1. $-2-11$ *Answer:* _____
2. $-25-25$ *Answer:* _____
3. $-13-14$ *Answer:* _____
4. $-23-25$ *Answer:* _____
5. $-10-10$ *Answer:* _____
6. $-2-67$ *Answer:* _____
7. $-11-11$ *Answer:* _____
8. $-2-8$ *Answer:* _____
9. $-9-19$ *Answer:* _____
10. $-30-40$ *Answer:* _____
11. $-45-55$ *Answer:* _____
12. $-6-6$ *Answer:* _____
13. $-7-4$ *Answer:* _____
14. $-57-74$ *Answer:* _____
15. $-115-45$ *Answer:* _____
16. $-200-500$ *Answer:* _____
17. $-938-721$ *Answer:* _____
18. $-328-421$ *Answer:* _____
19. $-0.45-0.55$ *Answer:* _____
20. $-1.23-.47$ *Answer:* _____
21. $-4.06-5.34$ *Answer:* _____
22. $-0.13-0.29$ *Answer:* _____
23. $-0.89-1.38$ *Answer:* _____
24. $-3.14-0.27$ *Answer:* _____
25. $-63.01-0.99$ *Answer:* _____
26. $-81.2-33.28$ *Answer:* _____
27. $-0.05-0.51$ *Answer:* _____
28. $-0.32-0.9$ *Answer:* _____
29. $-0.54-68.31$ *Answer:* _____
30. $-0.91-12$ *Answer:* _____
31. $\dfrac{-5}{6}-\dfrac{5}{6}$ *Answer:* _____
32. $-\dfrac{3}{10}-\dfrac{1}{10}$ *Answer:* _____
33. $-\dfrac{1}{3}-\dfrac{4}{3}$ *Answer:* _____
34. $\dfrac{-3}{4}-\dfrac{7}{4}$ *Answer:* _____
35. $\dfrac{-1}{2}-\dfrac{5}{2}$ *Answer:* _____

Answers to Mini-Practice:

(a) -14 (b) -17 (c) -12 (d) -25 (e) -14

36. $\dfrac{-3}{29} - \dfrac{17}{29}$ *Answer:* _____   **37.** $\dfrac{-4}{35} - \dfrac{1}{35}$ *Answer:* _____

38. $\dfrac{-5}{8} - \dfrac{7}{16}$ *Answer:* _____   **39.** $\dfrac{-11}{36} - \dfrac{5}{18}$ *Answer:* _____

40. $\dfrac{-3}{14} - \dfrac{3}{28}$ *Answer:* _____   **41.** $\dfrac{-3}{4} - \dfrac{1}{6}$ *Answer:* _____

42. $\dfrac{-7}{8} - \dfrac{1}{10}$ *Answer:* _____

1.8 SUBTRACTION OF NEGATIVE NUMBERS

Rewrite the following subtraction problems as addition of the opposite:

$$-7 - (-4) = -7 + \underline{\hphantom{xxx}} = \underline{\hphantom{xxx}}$$ Check your answer.

$$2 - (-6) = 2 + \underline{\hphantom{xxx}} = \underline{\hphantom{xxx}}$$ Check your answer.

You probably realized that the opposite of a negative number is positive, so that

$$-7 - (-4) = -7 + 4 = -3$$

This is correct because $-4 + (-3) = -7$. Furthermore,

$$2 - (-6) = 2 + 6 = 8$$

This is also correct because $-6 + 8 = 2$.

Example 1 $5 - (-3) = ?$

Solution.

Determine the operation: $-$
Determine the signs of the numbers: $+$ and $-$
Change the operation to addition: $+$
Write the opposite of the second number: $+3$
Rewrite the problem: $5 + 3 = ?$
Use the rule for addition of signed numbers: $5 + 3 = 8$

$$5 - (-3) = 5 + 3 = 8$$ ■

Example 2 $-5 - (-3) = ?$

Solution.

Determine the operation: $-$
Determine the signs of the numbers: $-$ and $-$
Change the operation to addition: $+$
Write the opposite of the second number: 3
Rewrite the problem: $-5 + 3 = ?$
Use the rule for addition of signed numbers: $-5 + 3 = -2$

$$-5 - (-3) = -5 + 3 = -2$$ ■

Example 3 $-4 - (-6) - (-9) = ?$

Solution. There are two subtractions. Change each of them to addition of the opposite:

$$-4 + 6 + 9$$

Method 1: Go from left to right:

$$-4 + 6 = 2$$
$$2 + 9 = 11$$

Method 2: Add the positive numbers:

$$6 + 9 = 15$$
$$-4 + 15 = 11$$ ■

MINI-PRACTICE   Perform the indicated operations:
(Use this space for your work.)

(a) $-3 - (-4)$ (b) $-2 - (-8)$ (c) $4 - (-3)$

(d) $2 - (-5)$ (e) $-3 - (-6) - (-4)$

Answers: (a) _____   (b) _____   (c) _____

(d) _____   (e) _____

IN YOUR OWN WORDS   explain how to subtract a negative number: _____

_____

_____

When subtracting a negative number, we change the subtraction to "addition of the opposite" and then apply the rules for addition of signed numbers.

In symbols:

$$a - b = a + (-b)$$

Compare this rule with the rule you found for subtraction of a positive number. In general, we can state:

When subtracting a signed number, change subtraction to "addition of the opposite," and then apply the rules for addition of signed numbers.

Practice Exercises

In Exercises 1–48, perform the indicated operations:

1. $-4 - (-3)$ *Answer:* _____   2. $12 - (-7)$ *Answer:* _____   3. $-12 - (-8)$ *Answer:* _____

4. $-6 - (-8)$ *Answer:* _____   5. $-30 - (-65)$ *Answer:* _____   6. $-8 - (-1)$ *Answer:* _____

7. $-12 - (-3)$ *Answer:* _____   8. $13 - (-7)$ *Answer:* _____   9. $42 - (-82)$ *Answer:* _____

10. $42 - (-30)$ *Answer:* _____   11. $-65 - (-11)$ *Answer:* _____   12. $-84 - (-27)$ *Answer:* _____

Answers to Mini-Practice:
(a) 1   (b) 6   (c) 7   (d) 7   (e) 7

13. 13 – (–5)   Answer: _____   14. –13 – (–20)   Answer: _____

15. –5 – (–7)   Answer: _____   16. –13 – (–5)   Answer: _____

17. –13 – (–5) – (–20)   Answer: _____   18. –8 – (–5) – (–15)   Answer: _____

19. –12 – (–5) – (–4)   Answer: _____   20. 25 – (–25) – (–25)   Answer: _____

21. –14 – (–17) – (–9)   Answer: _____   22. 72 – (–83) – (–4)   Answer: _____

23. –5 – (–12) – 13   Answer: _____   24. 62 – 14 – (–35)   Answer: _____

25. –13 – 17 – (–2)   Answer: _____   26. –19 – (–18) – 48   Answer: _____

27. –18 – (–61) – 85   Answer: _____   28. 51 – (–33) – 76   Answer: _____

29. 1.2 – (–3.5)   Answer: _____   30. (–5.3) – (–3.85)   Answer: _____

31. 0.25 – (–3.25)   Answer: _____   32. 5.3 – (–0.5)   Answer: _____

33. (–3.2) – (–0.25)   Answer: _____   34. 3.9 – (–7.4)   Answer: _____

35. (–12.7) – (–8.68)   Answer: _____   36. 0.57 – (–2.18)   Answer: _____

37. 62.1 – (–0.9)   Answer: _____   38. –24.5 – 2.34   Answer: _____

39. $\frac{5}{8} - \left(-\frac{3}{8}\right)$ Answer: _____   40. $-\frac{2}{13} - \left(\frac{-8}{13}\right)$ Answer: _____

41. $\left(-\frac{3}{4}\right) - \left(-\frac{3}{4}\right)$ Answer: _____   42. $\frac{2}{5} - \left(-\frac{1}{5}\right)$ Answer: _____

43. $\frac{1}{2} - \left(-\frac{1}{4}\right)$ Answer: _____   44. $\frac{5}{10} - \left(-\frac{1}{2}\right)$ Answer: _____

45. $\frac{1}{4} - \left(\frac{-3}{8}\right)$ Answer: _____   46. $\frac{3}{5} - \frac{1}{5} - \frac{2}{5}$ Answer: _____

47. $\frac{5}{11} - \left(\frac{-7}{11}\right) - \left(\frac{-4}{11}\right)$ Answer: _____   48. $\frac{3}{8} - \left(-\frac{1}{8}\right) - \frac{5}{8}$ Answer: _____

1.9 MULTIPLICATION OF NUMBERS WITH UNLIKE SIGNS

We know that

$$-3 + (-3) + (-3) + (-3) = -12$$

Rewrite: $-3 + (-3) + (-3) + (-3)$ in terms of multiplication: _____ (–3)

Therefore, 4(–3) = _____

Why is –3(4) also equal to –12? _____

You probably answered "because multiplication is commutative."

Example 1   5(–6) = ?

Solution.   5(–6) = –6 + (–6) + (–6) + (–6) + (–6) = 30

                     5(–6) = –30   ∎

Example 2 $-10(2) = ?$

Solution. $-10(2) = 2(-10) = -10 + (-10) = -20$

$$-10(2) = -20 \qquad \blacksquare$$

MINI-PRACTICE   Multiply:

(Use this space for your work.)

      (a) $6(-6)$ (b) $-6(6)$ (c) $5(-2)$ (d) $2(-5)$ (e) $-1(7)$

Answers: (a) _____   (b) _____   (c) _____   (d) _____   (e) _____

IN YOUR OWN WORDS   explain the rule of the sign of the product when a positive number and a negative number are multiplied: _____

_____

_____

Answer:

    When a positive number and a negative number are multiplied, the *sign* of the answer is negative.

WATCH OUT!   Why do we need the parentheses when writing $4(-3)$?_____

_____

Answer:

    The parentheses indicate multiplication. We want the *product* of 4 and –3. Thus,

$$4(-3) = -12$$

Without parentheses, $4 - 3$ would indicate subtraction. That is,

$$4 - 3 = 1$$

and *not* –12. We *cannot* drop the parentheses when writing the product of $4(-3)$.

Practice Exercises

In Exercises 1–44, multiply.

1. $-3(8)$ *Answer:* _____   2. $7(-4)$ *Answer:* _____   3. $12(-5)$ *Answer:* _____

4. $-4(2)$ *Answer:* _____   5. $-5(13)$ *Answer:* _____   6. $17(-5)$ *Answer:* _____

7. $-24(17)$ *Answer:* _____   8. $37(-2)$ *Answer:* _____   9. $-56(3)$ *Answer:* _____

10. $-8(72)$ *Answer:* _____   11. $47(-29)$ *Answer:* _____   12. $-69(28)$ *Answer:* _____

13. $16(-42)$ *Answer:* _____   14. $-92(61)$ *Answer:* _____   15. $3(-4)(5)$ *Answer:* _____

Answers to Mini-Practice:

  (a) –36   (b) –36   (c) –10   (d) –10   (e) –7

16. $-6(7)(11)$ *Answer:* _____   17. $-4(2)(7)$ *Answer:* _____   18. $3(-5)(0)$ *Answer:* _____

19. $3(-5)(1)$ *Answer:* _____   20. $9(-7)(8)$ *Answer:* _____   21. $-5(6)(13)$ *Answer:* _____

22. $-12(4)(5)$ *Answer:* _____   23. $-2(0)(98)$ *Answer:* _____   24. $14(-42)(0)$ *Answer:* _____

25. $12(-2)(1)$ *Answer:* _____   26. $7(25)(-1)$ *Answer:* _____   27. $0.5(-1.2)$ *Answer:* _____

28. $-4.2(3.1)$ *Answer:* _____   29. $0.25(-0.51)$ *Answer:* _____   30. $5.1(-0.2)$ *Answer:* _____

31. $3.6(-0.3)$ *Answer:* _____   32. $0.01(-0.25)$ *Answer:* _____   33. $0.7(-2.9)$ *Answer:* _____

34. $-0.4(12.1)$ *Answer:* _____   35. $-7.2(5.8)$ *Answer:* _____   36. $12.1(-0.3)$ *Answer:* _____

37. $\left(\frac{3}{4}\right)\left(-\frac{1}{2}\right)$ *Answer:* _____   38. $\left(-\frac{2}{5}\right)\left(\frac{5}{2}\right)$ *Answer:* _____   39. $\left(\frac{1}{2}\right)\left(-\frac{3}{4}\right)$ *Answer:* _____

40. $\left(\frac{-1}{5}\right)\left(\frac{5}{3}\right)$ *Answer:* _____   41. $\left(\frac{3}{100}\right)\left(\frac{-1}{10}\right)$ *Answer:* _____   42. $\left(\frac{4}{17}\right)\left(\frac{-5}{12}\right)$ *Answer:* _____

43. $\left(\frac{-3}{16}\right)\left(\frac{8}{15}\right)$ *Answer:* _____   44. $\left(\frac{-1}{10}\right)\left(\frac{1}{250}\right)$ *Answer:* _____

1.10 MULTIPLICATION OF TWO NEGATIVE NUMBERS

Complete the table:

$4(-3)$	=	-12
$3(-3)$	=	-9
$2(-3)$	=	-6
$1(-3)$	=	-3
$0(-3)$	=	0
$-1(-3)$	=	___
$-2(-3)$	=	___
___(-3)	=	9
___(-3)	=	___

The table is completed as follows:

$4(-3)$	=	-12
$3(-3)$	=	-9
$2(-3)$	=	-6
$1(-3)$	=	-3
$0(-3)$	=	0
$-1(-3)$	=	3
$-2(-3)$	=	6
$-3(-3)$	=	9
$-4(-3)$	=	12

When a negative number is multiplied by a negative number, what is the *sign* of the product? _____

According to this pattern, a negative number times a negative number is a positive number. Another way to come to the conclusion that the product of two negative numbers is positive is the following:

$$-3(-2) = -[3(-2)] = \text{the opposite of } [3(-2)]$$

The opposite of a number is the negative of the number.

The opposite of $[-6]$ = _____

The opposite of -6 is 6. Therefore, $-3(-2) = 6$ (*positive*).

Example 1 $-4(-5) = ?$

Solution. $-[4(-5)] = -[-20] = \text{the opposite of } -20 = 20$
$$-4(-5) = 20 \qquad \blacksquare$$

Example 2 $(-2)(-7) = ?$

Solution:
$$(-2)(-7) = -[2(-7)] = -[-14] = \text{the opposite of } -14 = 14$$
$$(-2)(-7) = 14 \qquad \blacksquare$$

Example 3 $(-1)(-2)(-3) = ?$

Solution. $(-1)(-2) = 2$
$$2(-3) = -6 \qquad \blacksquare$$

MINI-PRACTICE
(Use this space for your work.)

Multiply.

(a) $3(-7)$ (b) $(-3)7$ (c) $(-3)(-7)$

(d) $(3)(7)$ (e) $2(-5)(-6)$

Answers: (a) _____   (b) _____   (c) _____

(d) _____   (e) _____

IN YOUR OWN WORDS   state the rule for the sign of the product of two negative numbers: _____

_____

_____

Answer:

The product of two negative numbers is a positive number. The numbers that are multiplied are called *factors*. In general, we can say:

If the signs of the two factors are the same, the sign of the product is positive.

If the signs of the two factors are different, the sign of the product is negative.

_____

Answers to Mini-Practice:

(a) -21 (b) -21 (c) 21 (d) 21 (e) 60

In symbols, the rule for multiplication of signed numbers can be expressed in this way:

$$\left.\begin{array}{l}(+)(+) \\ (-)(-)\end{array}\right\} = + \qquad \left.\begin{array}{l}(+)(-) \\ (-)(+)\end{array}\right\} = -$$

More generally:

If there are an even number of negative factors, the product is positive.

If there are an odd number of negative factors, the product is negative.

For example,

$$(+)(-)(+)(-)(-)(-) = + \quad \text{(4 negative factors)}$$
$$(-)(+)(+)(-)(+)(-) = - \quad \text{(3 negative factors)}$$

Practice Exercises

In Exercises 1–44, multiply.

1. $-6(-6)$ *Answer:* _____   **2.** $-2(-4)$ *Answer:* _____   **3.** $-3(-8)$ *Answer:* _____

4. $-5(-7)$ *Answer:* _____   **5.** $(-5)(-2)$ *Answer:* _____   **6.** $0(-9)$ *Answer:* _____

7. $-7(5)$ *Answer:* _____   **8.** $-28(0)$ *Answer:* _____   **9.** $-32(-5)$ *Answer:* _____

10. $-1(735)$ *Answer:* _____   **11.** $(-2)(-2)(-2)$ *Answer:* _____

12. $-5(-2)(3)$ *Answer:* _____   **13.** $-5(-2)(-3)$ *Answer:* _____

14. $4(-1)(-5)$ *Answer:* _____   **15.** $4(-1)(5)$ *Answer:* _____

16. $-4(-1)(-5)$ *Answer:* _____   **17.** $-4(-15)(-20)$ *Answer:* _____

18. $(-5)(-5)(-5)(-5)$ *Answer:* _____   **19.** $(-3)(1)(-2)(-2)$ *Answer:* _____

20. $-3(-2)(7)(-4)$ *Answer:* _____   **21.** $3(-2)(-7)(4)$ *Answer:* _____

22. $3(-2)(-7)(-4)$ *Answer:* _____   **23.** $-3(-2)(-7)(-4)$ *Answer:* _____

24. $2(-1)(-2)(-1)(-2)$ *Answer:* _____   **25.** $-2(1)(-2)(-1)(2)$ *Answer:* _____

26. $-2(-1)(-2)(-1)(-2)$ *Answer:* _____   **27.** $(-3.8)(4.2)$ *Answer:* _____

28. $(-0.25)(-0.10)$ *Answer:* _____   **29.** $(-5.1)(-0.2)$ *Answer:* _____

30. $(-10.2)(-1.5)$ *Answer:* _____   **31.** $(-5.8)(-2.0)$ *Answer:* _____

32. $(-2.9)(7.5)$ *Answer:* _____   **33.** $(-8.17)(4.3)$ *Answer:* _____

34. $(-0.19)(-0.2)$ *Answer:* _____   **35.** $\left(\dfrac{-1}{10}\right)\left(\dfrac{-7}{10}\right)$ *Answer:* _____

36. $\left(\dfrac{-3}{8}\right)\left(\dfrac{-1}{4}\right)$ *Answer:* _____   **37.** $\left(\dfrac{-3}{5}\right)\left(\dfrac{-5}{2}\right)$ *Answer:* _____   **38.** $\left(\dfrac{-1}{2}\right)\left(\dfrac{-3}{5}\right)$ *Answer:* _____

39. $\left(-\dfrac{9}{10}\right)\left(-\dfrac{20}{9}\right)$ *Answer:* _____   **40.** $\left(\dfrac{-7}{20}\right)\left(\dfrac{-10}{21}\right)$ *Answer:* _____

41. $\left(\dfrac{-1}{3}\right)\left(\dfrac{5}{3}\right)$ *Answer:* _____   **42.** $\dfrac{39}{32}\left(\dfrac{-64}{13}\right)$ *Answer:* _____

43. $\dfrac{4}{5}\left(\dfrac{-3}{2}\right)\left(\dfrac{-10}{3}\right)$ *Answer:* _____   **44.** $\dfrac{-1}{2}\left(\dfrac{-2}{5}\right)\left(\dfrac{-10}{3}\right)$ *Answer:* _____

1.11 DIVISION

You probably remember that the reason why $9 - 5 = 4$ is because $5 + 4 = 9$.

Why is $\dfrac{27}{3} = 9$? _____

Your answer was most likely "because $3(9) = 27$." Complete the following table:

Division Problem		Reason	
$\dfrac{27}{-3} =$ _____	because	$-3(\_\_\_)$	$= 27$
$\dfrac{-27}{3} =$ _____	because	_____	$= -27$
$\dfrac{-27}{-3} =$ _____	because	_____	$=$ _____

The table is completed as follows:

Division Problem		Reason
$\dfrac{27}{-3} = -9$	because	$-3(-9) = 27$
$\dfrac{-27}{3} = -9$	because	$3(-9) = -27$
$\dfrac{-27}{-3} = 9$	because	$-3(9) = -27$

Example 1 $\dfrac{15}{-3} = ?$

Solution.

15 divided by 3 is 5;
-3 times $5 = -15$;
-3 times $-5 = 15$.

$$\dfrac{15}{-3} = -5$$

■

Example 2 $\dfrac{-15}{-3} = ?$

Solution.

15 divided by 3 is 5;
−3 times 5 = −15.

$$\frac{-15}{-3} = 5$$

■

MINI-PRACTICE
(Use this space for your work.)

Divide:

(a) $\dfrac{12}{6}$ (b) $\dfrac{-12}{6}$ (c) $\dfrac{12}{-6}$ (d) $\dfrac{-12}{-6}$

Answers: (a) _____   (b) _____   (c) _____   (d) _____

IN YOUR OWN WORDS

state the rule for the sign of the quotient in division of signed numbers. _____

_____

_____

Answer:

If the signs of the two numbers to be divided are the same, the sign of the quotient is positive.

If the signs of the two numbers to be divided are different, the sign of the quotient is negative.

In symbols, these rules can be summarized as follows:

$$\frac{+}{+} = + \qquad \frac{-}{-} = + \qquad \frac{+}{-} = - \qquad \frac{-}{+} = -$$

Practice Exercises

In Exercises 1–32, divide.

1. $\dfrac{35}{-5}$ *Answer:* _____   2. $\dfrac{-35}{5}$ *Answer:* _____   3. $\dfrac{-35}{-5}$ *Answer:* _____

4. $\dfrac{26}{2}$ *Answer:* _____   5. $\dfrac{-56}{-8}$ *Answer:* _____   6. $\dfrac{56}{-7}$ *Answer:* _____

7. $-\dfrac{63}{-9}$ *Answer:* _____   8. $\dfrac{49}{-7}$ *Answer:* _____   9. $\dfrac{-36}{12}$ *Answer:* _____

10. $\dfrac{-32}{-16}$ *Answer:* _____   11. $-\dfrac{48}{-8}$ *Answer:* _____   12. $-\dfrac{-63}{-9}$ *Answer:* _____

13. $\dfrac{45}{-15}$ *Answer:* _____   14. $-\dfrac{60}{-12}$ *Answer:* _____   15. $\dfrac{-0.8}{0.2}$ *Answer:* _____

16. $\dfrac{-0.12}{0.4}$ *Answer:* _____   17. $\dfrac{0.2}{-0.4}$ *Answer:* _____   18. $\dfrac{-0.04}{0.2}$ *Answer:* _____

19. $-\dfrac{0.0016}{-0.04}$ *Answer:* _____   **20.** $\dfrac{0.0027}{-.09}$ *Answer:* _____   **21.** $\dfrac{-1.2}{0.6}$ *Answer:* _____

22. $\dfrac{10.2}{-3}$ *Answer:* _____   **23.** $\dfrac{15.6}{5.2}$ *Answer:* _____   **24.** $\dfrac{0.36}{-6}$ *Answer:* _____

25. $\dfrac{6.012}{-0.36}$ *Answer:* _____   **26.** $\dfrac{0}{0.31}$ *Answer:* _____   **27.** $\dfrac{0.01}{-0.1}$ *Answer:* _____

28. $\dfrac{4.9}{-0.07}$ *Answer:* _____   **29.** $\dfrac{-3.24}{-1.62}$ *Answer:* _____   **30.** $\dfrac{-800}{0.04}$ *Answer:* _____

31. $\dfrac{-10.5}{-0.21}$ *Answer:* _____   **32.** $\dfrac{0.02}{-0.004}$ *Answer:* _____

SUMMARY

Addition: When the signs are the same, add the absolute values and keep the common sign.

When the signs are opposite, subtract the absolute values; keep the sign of the number with the larger absolute value.

Subtraction: Change subtraction to addition of the opposite, and follow the rules for addition.

Multiplication and Division: Two like signs yield plus, two unlike signs yield minus.

The Rules in Action are illustrated in the flowchart on page 51.

Review Practice Exercises

Exercises 1–60 include all the basic operations as well as the grouping symbols you were introduced to in the introductory chapter. Perform the indicated operations.

1. $-2 + (-6)$ *Answer:* _____   **2.** $7 + (-5)$ *Answer:* _____

3. $1 - (-2)$ *Answer:* _____   **4.** $-4 - 4$ *Answer:* _____

5. $8 - 16$ *Answer:* _____   **6.** $-11 + 19$ *Answer:* _____

7. $-8 + 3 - 1$ *Answer:* _____   **8.** $-6 - (-3) + 1$ *Answer:* _____

9. $-3 + 2 + (-4)$ *Answer:* _____   **10.** $-2 - (-3 + 1) - 5$ *Answer:* _____

11. $-4 - (-2)$ *Answer:* _____   **12.** $3 + (-2) - (-1) + (-4)$ *Answer:* _____

13. $-2 - (-3) + (-4) - (-5)$ *Answer:* _____   **14.** $5 + (-7) - (-8) + (-3) - (-4)$ *Answer:* _____

15. $6(-8)$ *Answer:* _____   **16.** $-4(-15)$ *Answer:* _____

17. $(-10)(-2)(-5)(11)$ *Answer:* _____   **18.** $(-3)(5)$ *Answer:* _____

19. $(-1)(-1)(-1)(-1)$ *Answer:* _____   **20.** $(-2)(-2)(-2)$ *Answer:* _____

Operations with Two Signed Numbers

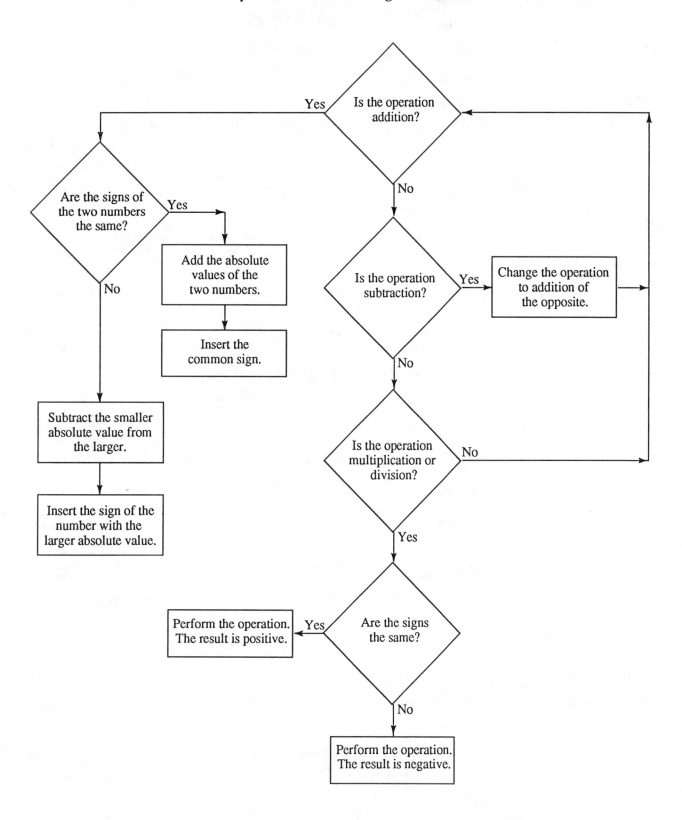

21. $(2)(-3)(-2)(5)(2)$ *Answer:* _____   **22.** $(-35) \div 7$ *Answer:* _____

23. $(-72) \div (-12)$ *Answer:* _____   **24.** $(-16) \div 4$ *Answer:* _____

25. $0 \div (-9)$ *Answer:* _____   **26.** $(-9) \div 0$ *Answer:* _____

27. $(-60) \div (-10)$ *Answer:* _____   **28.** $(-10) \div (-60)$ *Answer:* _____

29. $\dfrac{32}{-8}$ *Answer:* _____   **30.** $\dfrac{-25}{-5}$ *Answer:* _____

31. $-\dfrac{-28}{-7}$ *Answer:* _____   **32.** $\dfrac{900}{-30}$ *Answer:* _____

33. $\dfrac{-844}{-4}$ *Answer:* _____   **34.** $-3\,|-4\,|$ *Answer:* _____

35. $-3 + |-4\,|$ *Answer:* _____   **36.** $|-6\,| - |7\,|$ *Answer:* _____

37. $|-6\,| \cdot |7\,|$ *Answer:* _____   **38.** $|-1\,| \cdot |-2\,| \cdot |-3\,|$ *Answer:* _____

39. $(-2) \cdot |-2\,| \cdot |-2\,|$ *Answer:* _____   **40.** $\dfrac{1}{2} + \left(-\dfrac{1}{2}\right)$ *Answer:* _____

41. $\dfrac{7}{10} + \left(\dfrac{-3}{10}\right)$ *Answer:* _____   **42.** $\dfrac{7}{9} + \left(-\dfrac{2}{9}\right)$ *Answer:* _____

43. $\dfrac{5}{8} - \dfrac{1}{8}$ *Answer:* _____   **44.** $\dfrac{5}{12} - \left(\dfrac{-1}{12}\right)$ *Answer:* _____

45. $\dfrac{3}{4} + \left(\dfrac{-1}{2}\right)$ *Answer:* _____   **46.** $\dfrac{-1}{10} + \dfrac{3}{5}$ *Answer:* _____

47. $\dfrac{1}{12} - \dfrac{3}{8}$ *Answer:* _____   **48.** $\dfrac{9}{10} - \left(\dfrac{-1}{5}\right)$ *Answer:* _____

49. $\left(\dfrac{1}{2}\right)\left(\dfrac{3}{4}\right)$ *Answer:* _____   **50.** $\left(\dfrac{-3}{5}\right)\left(\dfrac{10}{21}\right)$ *Answer:* _____

51. $\left(\dfrac{7}{12}\right)\left(\dfrac{-3}{14}\right)$ *Answer:* _____   **52.** $\left(\dfrac{-4}{5}\right)\left(\dfrac{-15}{8}\right)$ *Answer:* _____

53. $-0.25 + 0.75$ *Answer:* _____   **54.** $1.2 + (-0.3)$ *Answer:* _____

55. $-0.43 - (-2.7)$ *Answer:* _____   **56.** $-12.8 + 2.1$ *Answer:* _____

57. $(0.02)(-1.2)$ *Answer:* _____   **58.** $(-3.1)(-.04)$ *Answer:* _____

59. $(-12.2)(-.005)$ *Answer:* _____   **60.** $(48.2)(-0.002)$ *Answer:* _____

If you have questions about this chapter, write them here:

_____

_____

_____

_____

2 Exponents

Remember to

* *STUDY EVERY STEP OF THE WORKED-OUT EXAMPLES*
* *MEMORIZE ALL OF THE RULES*
* *LEARN THE LESSONS FROM THE WATCH OUT! SUGGESTIONS TO AVOID COMMON MISTAKES*
* *REVIEW THE MATERIAL IN THE CHAPTER SUMMARIES*

In this chapter you will learn about powers, exponential notation, and operations with numbers in exponential notation. You will also explore exponents that are fractions or negative numbers.

2.1 POWERS

Suppose you start a new job in which you get paid $2 the first day and your salary is doubled every subsequent day. How much money do you earn on the tenth day? Fill out the column labeled "Earnings ($)" in the table below to find out:

Day Number	Earnings ($)	Power of 2
1	2	_____
2	4	_____
3	8	_____
4	16	_____
5	_____	_____
6	_____	_____
7	_____	_____
8	_____	_____
9	_____	_____
10	_____	_____

The Worked-out Examples

1. **Read every example worked out in the text.**
    • They are like problems you will have for homework.

2. **Go over every step.**
    • If there is even one step that is unclear, ask for help.

3. **When you understand all the examples of a section:**
    • Try the **MINI-PRACTICE**.
    • Then try the **Practice Exercises** at the end of the section.

4. **If you are stumped on a homework exercise:**
    • Try to find a worked-out example like it.
    • Read over any rules that might tell you what to do.

This way of paying you would lead the boss to bankruptcy. $1024 for the 10th day! In fact, at this rate, on the 20th day, you would earn more than $1,000,000!

This was an example of *exponential growth* or the *power of powers*. In our daily lives we hear about population explosion, radioactive decay, and compound interest. These topics can be discussed in terms of numbers expressed as **powers**, that is, in terms of **exponential notation**.

Translate into mathematics: The third power of two. _____

You probably remembered from the introductory chapter that 2^3 is called "two to the third power" or "the third power of two."

Go back to the activity on the preceding page in which you calculated your daily salary. Express each earning as a power of 2 ($2 = 2^1$; $4 = 2^2$; $8 = 2^3$, etc.). The salary for the tenth day is 2^{10}. You can calculate that number by multiplying it out on a calculator ($2 \times 2 \times 2 \ldots$), or, if your calculator has a key marked x^y, do the following:

$$2 \quad x^y \quad 10 = \text{_____}$$

The answer should again be 1024.

Example 1   Translate "the fifth power of three" into mathematics.

Solution.   The fifth power of three

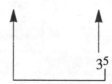

$$3^5$$

■

Example 2   Write 64 as a number to the second power.

Solution.

What number squared is 64?

64 is the square of 8, so

$$64 = 8^2$$ ∎

MINI-PRACTICE

(Use this space for your work.)

In parts (a) and (b), translate into mathematics:

(a) the fifth power of four   *Answer:* _____

(b) the ninth power of six   *Answer:* _____

(c) What power of two equals thirty-two?   *Answer:* _____

(d) Twenty-five is five to what power: *Answer:* _____

(e) Write 64 as a power of 4.   *Answer:* _____

IN YOUR OWN WORDS   explain how to find the power of a number: _____

_____

_____

Answer:

   The power of a number tells us how many times the number should be used as a factor.

Practice Exercises

In Exercises 1–12 translate into mathematics. (Do not calculate the result.)

1. The fourth power of nine   *Answer:* _____   2. The ninth power of four   *Answer:* _____

3. The fifth power of eight   *Answer:* _____   4. The forty-fourth power of nine   *Answer:* _____

5. The hundredth power of one   *Answer:* _____   6. The third power of seven   *Answer:* _____

7. The sixth power of ten   *Answer:* _____   8. The seventh power of five   *Answer:* _____

9. The fifteenth power of two   *Answer:* _____   10. The eleventh power of three   *Answer:* _____

11. The nineteenth power of one   *Answer:* _____   12. The twenty-second power of one   *Answer:* _____

13. What positive number to the second power equals sixty-four?   *Answer:* _____

14. What power of two equals one hundred twenty-eight?   *Answer:* _____

15. What number raised to the third power equals twenty-seven?   *Answer:* _____

16. What positive number to the second power equals forty-nine?   *Answer:* _____

17. What number to the third power equals eight?   *Answer:* _____

Answers to Mini-Practice:

(a) 4^5 (b) 6^9 (c) 2^5 (d) 5^2 (e) 4^3

18. What number to the third power equals sixty-four?   *Answer:* _____

19. What positive number to the fourth power equals one?   *Answer:* _____

20. What positive number to the fourth power equals sixteen?   *Answer:* _____

21. What number raised to the fifth power equals thirty-two?   *Answer:* _____

22. What number raised to the fifth power equals one?   *Answer:* _____

23. What power of three equals nine?   *Answer:* _____

24. What power of ten equals one hundred?   *Answer:* _____

25. What power of ten equals one thousand?   *Answer:* _____

26. What power of five equals five?   *Answer:* _____

27. What power of seven equals seven?   *Answer:* _____

28. Eighty-one can be written as a positive number to the (a) second power; (b) fourth power. Find these numbers.

    Answer: (a) _____   (b) _____

29. Sixteen can be written as a positive number to the (a) second power; (b) fourth power. Find these numbers.

    Answer: (a) _____   (b) _____

30. Nine can be written as a positive number to the (a) second power; (b) first power. Find these numbers.

    Answer: (a) _____   (b) _____

31. Forty-nine can be written as a positive number to the (a) second power; (b) first power. Find these numbers.

    Answer: (a) _____   (b) _____

32. Sixty-four can be written as a positive number to the (a) third power; (b) second power; (c) sixth power. Find these numbers.

    Answer: (a) _____   (b) _____   (c) _____

33. One hundred twenty-five can be written as a number to the (a) third power; (b) first power. Find these numbers.

    Answer: (a) _____   (b) _____

34. One hundred can be written as a positive number to the (a) second power; (b) first power. Find these numbers.

    Answer: (a) _____   (b) _____

35. One million can be written as a number to the (a) second power; (b) third power; (c) sixth power. Find these numbers.

    Answer: (a) _____   (b) _____   (c) _____

2.2 EXPONENTIAL NOTATION

Write 16 as a product of two numbers: _____

Write 16 as a product of the smallest whole numbers you can find:

_____

When a number is written as a product, such as $16 = 2 \times 8$, we say that the number is **factored**. Here, 2 and 8 are the factors of the product 16.

$16 = 2 \times 2 \times 2 \times 2$ is written in factored form with the smallest possible whole numbers (other than 1). We say that **16 is factored into prime factors.** You probably know that prime numbers are numbers that cannot be written in factored form (unless we use 1 and the number itself). Examples of prime numbers are 2, 3, 5, 7, 11, and 13.

As you saw in the introductory chapter, $2 \times 2 \times 2 \times 2$ can be written in exponential notation.

Write this product in exponential notation: _____

What is the base? _____   What is the exponent? _____

Put a line under the base and a circle around the exponent.

In the expression

$$16 = 2 \times 2 \times 2 \times 2 = 2^4$$

the base is equal to two and the exponent is equal to four. We can also say that sixteen is the fourth power of two or that two is raised to the fourth power.

Example 1   When writing 4^3, what is the base; what is the exponent?

Solution. $4^3 \leftarrow$ exponent
$\uparrow$
base

4 is the base; 3 is the exponent.   ■

Example 2   Find the value of 5^3.

Solution. $5^3 = \underbrace{5 \times 5 \times 5}_{\text{3 factors of 5}} = \underset{\text{value}}{\overset{\uparrow}{125}}$

The value of 5^3 is 125.   ■

MINI-PRACTICE   (a) Write in exponential notation: $3 \times 3 \times 3 \times 3$. (Do *not* calculate the result.)
(Use this space for your work.)

Answer: _____

(b) Find the value of 6^3.   *Answer:* _____

(c) Rewrite in exponential form: four squared.   *Answer:* _____

IN YOUR OWN WORDS   explain how multiplication is related to exponential notation: _____

_____

_____

_____
Answers to Mini-Practice:
(a) 3^4 (b) 216   (c) 4^2

Answer:

Exponential notation is a shorthand form of multiplication when a number is multiplied by itself several times. The base indicates the number that is multiplied and the exponent tells us how many times that number is used as a factor.

Practice Exercises

In Exercises 1–4, identify the base and the exponent.

1. 5^3 Answer: _____

2. 6^2 Answer: _____

3. $(-3)^4$ Answer: _____

4. $(-10)^2$ Answer: _____

In Exercises 5–18 write in exponential notation. Do NOT calculate the result.

5. 1×1 Answer: _____

6. $2 \times 2 \times 2 \times 2 \times 2 \times 2$ Answer: _____

7. $4 \times 4 \times 4 \times 4 \times 4 \times 4 \times 4$ Answer: _____

8. $15 \times 15 \times 15 \times 15$ Answer: _____

9. $(-3)(-3)(-3)(-3)(-3)$ Answer: _____

10. $(0.1)(0.1)(0.1)$ Answer: _____

11. $(2.5)(2.5)$ Answer: _____

12. $(-3.8)(-3.8)(-3.8)$ Answer: _____

13. $(7.6)(7.6)(7.6)(7.6)$ Answer: _____

14. $(0.25)(0.25)$ Answer: _____

15. $\left(\frac{1}{2}\right)\left(\frac{1}{2}\right)$ Answer: _____

16. $\left(\frac{1}{4}\right)\left(\frac{1}{4}\right)$ Answer: _____

17. $\left(-\frac{1}{2}\right)\left(-\frac{1}{2}\right)$ Answer: _____

18. $\left(-\frac{1}{4}\right)\left(-\frac{1}{4}\right)$ Answer: _____

In Exercises 19–40, find each value.

19. 3^6 Answer: _____

20. 1^5 Answer: _____

21. 2^5 Answer: _____

22. 4^3 Answer: _____

23. 10^5 Answer: _____

24. 5^3 Answer: _____

25. 9^2 Answer: _____

26. 7^2 Answer: _____

27. 8^3 Answer: _____

28. 20^1 Answer: _____

29. 0^2 Answer: _____

30. 0^5 Answer: _____

31. $(0.5)^2$ Answer: _____

32. $(0.01)^3$ Answer: _____

33. $(3.2)^2$ Answer: _____

34. $(0.2)^4$ Answer: _____

35. $(0.1)^2$ Answer: _____

36. $(0.3)^2$ Answer: _____

37. $(2.0)^3$ Answer: _____

38. $(0.4)^2$ Answer: _____

39. $(0.25)^1$ Answer: _____

40. $(0.001)^1$ Answer: _____

In Exercises 41–47, rewrite in exponential form.

41. The cube of 6   Answer:_____

42. The fourth power of 3   Answer:_____

43. Five to the third power   Answer:_____

44. The cube of 5.1   Answer:_____

45. 3.5 to the fourth power   Answer:_____

46. The fifth power of $\frac{1}{2}$ Answer:_____

47. $\frac{1}{8}$ to the second power   Answer:_____

2.3 EXPONENTS VERSUS FACTORS

Translate 2^3 in English: _____

Evaluate 2^3: _____

Translate 2(3) into English: _____

Evaluate 2(3): _____

Translate "Use two as a factor three times" into mathematical symbols: _____

Translate "Multiply two by three" into mathematical symbols:

_____

In the expression $2^3 = 2 \times 2 \times 2 = 8$, we use two as a factor three times. 3 is the exponent.

In the expression $2(3) = 6$, we multiply two by three. 3 is a factor.

From this activity we can draw the conclusion that

$$2^3 \neq 2(3)$$

Example 1   Does $5^2 = 5(2)$?

Solution.
$$5^2 = (5)(5) = 25$$
$$\uparrow$$
two equal factors of 5

$$5(2) = 10$$
$$\uparrow$$
two unequal factors

Thus, $5^2 \neq 5(2)$.   ∎

It is of interest to note that
$$1^1 = 1 \quad \text{and} \quad 1(1) = 1$$
so that
$$1^1 = 1(1)$$
Also,
$$2^2 = 4 \quad \text{and} \quad 2(2) = 4$$
so that
$$2^2 = 2(2)$$

But for numbers larger than 2, this is never the case.

Example 2   Does $3^3 = (3(3)$?

Solution.
$$3^3 = (3)(3)(3) = 27$$
$$\uparrow$$
three equal factors of 3

$$3(3) = 9$$

↑

two factors   (Each factor is equal to 3.)

Thus, $3^3 \neq 3(3)$.   ■

Example 3   Does $(3.2)^2 = 2(3.2)$?

Solution. $(3.2)^2 = (3.2)(3.2) = 10.24$

↑

Two equal factors of (3.2)

$$2(3.2) = 6.4$$

Thus, $(3.2)^2 \neq 2(3.2)$.   ■

Example 4   Does $\left(\frac{2}{3}\right)^2 = 2\left(\frac{2}{3}\right)$?

Solution. $\left(\frac{2}{3}\right)^2 = \left(\frac{2}{3}\right)\left(\frac{2}{3}\right) = \frac{2(2)}{[(3)(3)]} = \frac{4}{9}$

↑

Two equal factors of $\left(\frac{2}{3}\right)$

$$2\left(\frac{2}{3}\right) = \left(\frac{2}{1}\right)\left(\frac{2}{3}\right) = \frac{2(2)}{1(3)} = \frac{4}{3}$$

Thus, $\left(\frac{2}{3}\right)^2 \neq 2\left(\frac{2}{3}\right)$ ■

MINI-PRACTICE
(Use this space for your work.)

Evaluate:

(a) 2^5 (b) $2(5)$ (c) $(5)^2$ (d) $(5)2$

Answers: (a) _____   (b) _____   (c) _____   (d) _____

IN YOUR OWN WORDS

explain how exponential notation is related to the factors of a product: _____

_____

_____

Answer:

When the factors are the same, we can write multiplication in exponential form.

Practice Exercises

In Exercises 1–38, find each value.

1. 2^4 *Answer:* _____   2. $2(4)$ *Answer:* _____   3. 3^2 *Answer:* _____

Answers to Mini-Practice:

(a) 32   (b) 10   (c) 25   (d) 10

4. $3(2)$ *Answer:* _____   **5.** 6^3 *Answer:* _____   **6.** $6(3)$ *Answer:* _____

7. $(1)^3$ *Answer:* _____   **8.** $(1)(3)$ *Answer:* _____   **9.** 8^3 *Answer:* _____

10. $8(3)$ *Answer:* _____   **11.** 5^2 *Answer:* _____   **12.** $5(2)$ *Answer:* _____

13. 7^3 *Answer:* _____   **14.** $7(3)$ *Answer:* _____   **15.** 9^4 *Answer:* _____

16. $9(4)$ *Answer:* _____   **17.** 4^5 *Answer:* _____   **18.** $4(5)$ *Answer:* _____

19. 11^2 *Answer:* _____   **20.** $11(2)$ *Answer:* _____   **21.** 10^3 *Answer:* _____

22. $10(3)$ *Answer:* _____   **23.** $2(5.1)$ *Answer:* _____   **24.** $(5.1)^2$ *Answer:* _____

25. $(0.02)^3$ *Answer:* _____   **26.** $3(0.02)$ *Answer:* _____   **27.** $(0.1)^3$ *Answer:* _____

28. $3(0.1)$ *Answer:* _____   **29.** $(0.5)^2$ *Answer:* _____   **30.** $2(0.5)$ *Answer:* _____

31. $(0.3)^2$ *Answer:* _____   **32.** $2(0.3)$ *Answer:* _____   **33.** $\left(\frac{1}{2}\right)^2$ *Answer:* _____

34. $2\left(\frac{1}{2}\right)$ *Answer:* _____   **35.** $\left(\frac{1}{3}\right)^2$ *Answer:* _____   **36.** $2\left(\frac{1}{3}\right)$ *Answer:* _____

37. $\left(\frac{3}{2}\right)^3$ *Answer:* _____   **38.** $3\left(\frac{3}{2}\right)$ *Answer:* _____

2.4 NEGATIVE BASES

Complete the table:

	Base	*Exponent*	*Value*
$(-2)^2 = (-2)(-2) =$			
$(-3)^4 =$			
$(-2)^3 =$			
$(-1)^5 =$			

From this activity determine the following:

When a negative base is raised to an *even* power, what is the sign of the result? _____

When a negative base is raised to an *odd* power, what is the sign of the result? _____

Example 1   Evaluate: (a) $(-1)^2$ (b) $(-1)^3$

Solution.

(a)
$$(-1)^2 = (-1)(-1) = +1 = 1$$

negative even   positive
base   exponent   answer

(b)
$$(-1)^3 = (-1)(-1)(-1) = -1$$

negative odd   negative
base   exponent   answer   ∎

Example 2   Evaluate: (a) $(-3)^3$ (b) $(-3)^4$

Solution.

(a)
$$(-3)^3 = (-3)(-3)(-3) = -27$$

(b)
$$(-3)^4 = (-3)(-3)(-3)(-3) = +81 = 81$$ ∎

MINI-PRACTICE   Find the value of the following:
(Use this space for your work.)

(a) $(-1)^4$ (b) $(-1)^3$ (c) $(-4)^2$ (d) $(-2)^5$

Answers: (a) _____   (b) _____   (c) _____   (d) _____

IN YOUR OWN WORDS   state the rules for the signs of the results when negative bases are raised to powers:

_____

_____

_____

Answer:

When a negative base is raised to an even power, the result is positive. When a negative base is raised to an odd power, the result is negative.

RULE:

$$(\text{NEGATIVE BASE})^{\text{EVEN EXPONENT}} = \text{POSITIVE RESULT}$$

$$(\text{NEGATIVE BASE})^{\text{ODD EXPONENT}} = \text{NEGATIVE RESULT}$$

Answers to Mini-Practice:
(a) 1   (b) −1   (c) 16   (d) −32

Practice Exercises

In Exercises 1–36, find each value.

1. $(-2)^6$ *Answer:* _____
2. $(-3)^5$ *Answer:* _____
3. $(-5)^2$ *Answer:* _____
4. $(-12)^2$ *Answer:* _____
5. $(-6)^3$ *Answer:* _____
6. $(-7)^3$ *Answer:* _____
7. $(-5)^3$ *Answer:* _____
8. $(-1)^{20}$ *Answer:* _____
9. $(-9)^2$ *Answer:* _____
10. $(-4)^3$ *Answer:* _____
11. $(-8)^2$ *Answer:* _____
12. $(-10)^2$ *Answer:* _____
13. $(-100)^1$ *Answer:* _____
14. $(-1)^4$ *Answer:* _____
15. $(-11)^2$ *Answer:* _____
16. $(-12)^1$ *Answer:* _____
17. $(-0.1)^2$ *Answer:* _____
18. $(-0.5)^3$ *Answer:* _____
19. $(-0.25)^1$ *Answer:* _____
20. $(-1.5)^1$ *Answer:* _____
21. $(-0.9)^2$ *Answer:* _____
22. $(-3.1)^2$ *Answer:* _____
23. $(-5.5)^3$ *Answer:* _____
24. $(-0.3)^2$ *Answer:* _____
25. $(-0.04)^2$ *Answer:* _____
26. $(-7.1)^2$ *Answer:* _____
27. $\left(\frac{-1}{2}\right)^2$ *Answer:* _____
28. $\left(\frac{-1}{2}\right)^3$ *Answer:* _____
29. $\left(\frac{-1}{5}\right)^2$ *Answer:* _____
30. $\left(\frac{-1}{4}\right)^3$ *Answer:* _____
31. $\left(\frac{-1}{10}\right)^2$ *Answer:* _____
32. $\left(\frac{-2}{5}\right)^3$ *Answer:* _____
33. $\left(\frac{-3}{2}\right)^1$ *Answer:* _____
34. $\left(\frac{-4}{5}\right)^2$ *Answer:* _____
35. $\left(\frac{-2}{3}\right)^2$ *Answer:* _____
36. $\left(\frac{-9}{10}\right)^3$ *Answer:* _____

2.5 THE NEGATIVE SIGN IN EXPONENTIAL NOTATION

Complete the following table:

Number	Base	Exponent	Value
$(-2)^3$			
-2^3			
$(-2)^4$			
-2^4			

Why is $(-2)^3 = -2^3$, whereas $(-2)^4 \neq -2^4$? _____

When -2 is raised to an *odd* power the result is negative. *Minus* 2^3 is also negative. When we raise (-2) to an *even* power, our answer is positive. But *minus* 2^4 is still negative.

Example 1 Evaluate: (a) $(-4)^2$ (b) -4^2

Solution.

(a)

$$(-4)^2 = (-4)(-4) = 16$$
$$\uparrow$$
-4 is the base.

(b)

$$-4^2 = -(4)(4) = -16$$
$$\uparrow$$
4 is the base. ■

Example 2 Evaluate: (a) $(-4)^3$ (b) -4^3

Solution.

(a)

$$(-4)^3 = (-4)(-4)(-4) = -64$$

(b)

$$-4^3 = -(4)(4)(4) = -64$$ ■

MINI-PRACTICE
(Use this space for your work.)

Find each value:

(a) $(-3)^2$ (b) -3^2 (c) $(-3)^3$ (d) -3^3

Answers: (a) _____ (b) _____ (c) _____ (d) _____

IN YOUR OWN WORDS

explain how you can tell whether or not the minus sign belongs to the base:

_____

_____

Answer:

   If the minus sign is ENCLOSED in parentheses together with a number, it belongs to the base. If the minus sign is written BEFORE the parentheses or if parentheses are not used, it does NOT belong to the base.

Practice Exercises

In Exercises 1–36, evaluate the following expressions.

1. -2^2 *Answer:* _____ **2.** $(-2)^2$ *Answer:* _____ **3.** $(-2)^4$ *Answer:* _____

4. $-(2)^4$ *Answer:* _____ **5.** -3^4 *Answer:* _____ **6.** $(-3)^4$ *Answer:* _____

Answers to Mini-Practice:
(a) 9   (b) –9   (c) –27   (d) –27

 7. $(-6)^3$ *Answer:* _____   8. -6^3 *Answer:* _____   9. $-(-2)^3$ *Answer:* _____

10. $-(-2)^4$ *Answer:* _____   11. -4^3 *Answer:* _____   12. $(-4)^3$ *Answer:* _____

13. -7^2 *Answer:* _____   14. $(-7)^2$ *Answer:* _____   15. $-(-9)^2$ *Answer:* _____

16. -9^2 *Answer:* _____   17. -5^3 *Answer:* _____   18. $-(-5)^3$ *Answer:* _____

19. $-(-10)^2$ *Answer:* _____   20. -10^2 *Answer:* _____   21. $-(0.1)^3$ *Answer:* _____

22. $(-0.1)^3$ *Answer:* _____   23. $-(0.5)^2$ *Answer:* _____   24. $(-0.5)^2$ *Answer:* _____

25. $(-0.4)^3$ *Answer:* _____   26. $-(0.4)^3$ *Answer:* _____   27. $(-0.2)^4$ *Answer:* _____

28. $-(0.2)^4$ *Answer:* _____   29. $-(-0.25)^2$ *Answer:* _____   30. $-(0.25)^2$ *Answer:* _____

31. $\left(\dfrac{-1}{2}\right)^3$ *Answer:* _____   32. $\left(\dfrac{-1}{2}\right)^4$ *Answer:* _____   33. $\left(\dfrac{-1}{4}\right)^2$ *Answer:* _____

34. $-\left(\dfrac{1}{4}\right)^2$ *Answer:* _____   35. $-\left(\dfrac{1}{5}\right)^3$ *Answer:* _____   36. $\left(\dfrac{-1}{5}\right)^3$ *Answer:* _____

2.6 THE NUMBER ONE AS EXPONENT

Complete the following table:

Number	Number of Factors	Exponent
2^3		
3^5		
3		

In the last example, we have a single factor of 3. We consider the exponent to be equal to 1, even if it is not written out. It is not wrong to write the exponent 1, but it is usually not done.

Example 1   Identify the base and the exponent of 6.

Solution   6
                                            ↑
                                           base

The base is 6 and the exponent is understood to be 1.   ■

Example 2   Evaluate 5^1.

Solution. 5^1 is the same as 5, that is,

$$5^1 = 5$$ ∎

MINI-PRACTICE   Find each value.
(Use this space for your work.)

(a) 4^1 (b) 0^1 (c) $(-2)^1$

Answers: (a) _____   (b) _____   (c) _____

IN YOUR OWN WORDS   explain why a number without a specific exponent is understood to have the exponent 1: _____

_____

_____

Answer:

A number raised to the first power is equal to itself; thus, a number without a specific exponent equals that number with exponent 1.

Practice Exercises

In Exercises 1 and 2, identify the base and the exponent.

1. 5^1 *Answer:* _____   2. (-7) *Answer:* _____

In Exercises 3–16, find each value.

3. 10^1 *Answer:*_____   4. 1000^1 *Answer:* _____   5. $(-35)^1$ *Answer:*_____

6. 0^1 *Answer:*_____   7. $(-9)^1$ *Answer:* _____   8. $(-4)^1$ *Answer:*_____

9. $(-2)^1(-3)^1$ *Answer:*_____   10. $(-6)^1(-2)^1$ *Answer:* _____   11. $(-3)^1(-8)^1$ *Answer:*_____

12. $(-9)^1(-4)^1$ *Answer:*_____   13. $(0.5)^1$ *Answer:* _____   14. $(-0.1)^1$ *Answer:*_____

15. $\left(\dfrac{2}{3}\right)^1$ *Answer:*_____   16. $\left(\dfrac{-4}{5}\right)^1$ *Answer:* _____

Answers to Mini-Practice:
(a) 4   (b) 0   (c) –2

2.7 ADDITION AND SUBTRACTION IN EXPONENTIAL NOTATION

Complete the following:

Number	Value
2^3	_____
2^2	_____
Sum: $2^3 + 2^2 =$	_____

We now have

$$2^3 + 2^2 = 8 + 4 = 12$$

The numbers we add or subtract are called **terms**. 2^3 and 2^2 are terms of the expression $2^3 + 2^2$; also, 8 and 4 are terms of the expression $8 + 4$.

Example 1 Add the terms 3^3 and 3^2.

Solution. $3^3 = 27$; $3^2 = 9$

$3^3 + 3^2 = 27 + 9 = 36$ ∎

Example 2 Subtract the terms 2^3 and 2^2 in this order.

Solution. $2^3 - 2^2 = 8 - 4 = 4$ ∎

MINI-PRACTICE
(Use this space for your work.)

Find each value.

(a) $3^2 + 3^2$ (b) $4^2 - 2^2$ (c) $2^3 - 2$

(d) $3^2 - 2^2$ (e) $1^5 + 1^6$

Answers: (a) _____   (b) _____   (c) _____

(d) _____   (e) _____

IN YOUR OWN WORDS explain what we have to do when we perform addition or subtraction with terms in exponential notation: _____

_____

_____

Answers to Mini-Practice:
(a) 18   (b) 12   (c) 6   (d) 5   (e) 2

Answer:

You probably said that we have to **evaluate the power of each term individually and then add or subtract the numbers.**

Practice Exercises

In Exercises 1–38, find each value.

1. $6^3 + 10$ *Answer:* _____

2. $5^2 - 2^3$ *Answer:* _____

3. $6^2 + 5^2$ *Answer:* _____

4. $2^2 + 2^4$ *Answer:* _____

5. $4^3 + 4^2$ *Answer:* _____

6. $5^2 - 5$ *Answer* _____

7. $3^3 - 3^4$ *Answer:* _____

8. $(-2)^3 - (-2)^2$ *Answer:* _____

9. $(-5)^2 + (-5)^1$ *Answer:* _____

10. $6^2 - 4^2$ *Answer:* _____

11. $10^4 - 10^3$ *Answer:* _____

12. $1^3 + 10^2$ *Answer:* _____

13. $-2^2 + 4^2$ *Answer:* _____

14. $7^2 - (-2)^3$ *Answer:* _____

15. $-3^2 - 9$ *Answer:* _____

16. $8^2 + (-6)^2$ *Answer:* _____

17. $(-10)^3 + 5^3$ *Answer:* _____

18. $9^2 - (-2)^3$ *Answer:* _____

19. $-7^2 + (-3)^3$ *Answer:* _____

20. $-4^3 - (-5)^2$ *Answer:* _____

21. $(-0.2)^2 + 0.1$ *Answer:* _____

22. $(0.3) - 5^2$ *Answer:* _____

23. $(0.5)^2 - (0.5)^3$ *Answer:* _____

24. $(0.1)^3 - (0.1)^2$ *Answer:* _____

25. $(-5.1)^2 - (-5.1)$ *Answer:* _____

26. $(0.7)^3 + 0.7$ *Answer:* _____

27. $(6.1)^2 - (6.1)^2$ *Answer:* _____

28. $(3.8)^2 - (3.8)^3$ *Answer:* _____

29. $(-0.25)^2 + 3.5$ *Answer:* _____

30. $1^3 - (-0.4)^2$ *Answer:* _____

31. $\left(\frac{1}{2}\right)^2 - \left(\frac{1}{2}\right)^3$ *Answer:* _____

32. $\frac{1}{10} + \left(\frac{1}{10}\right)^3$ *Answer:* _____

33. $\left(\frac{3}{10}\right)^2 - \left(\frac{3}{10}\right)^4$ *Answer:* _____

34. $\left(\frac{1}{3}\right)^2 + \left(\frac{-4}{3}\right)$ *Answer:* _____

35. $\left(\frac{1}{4}\right) - \left(\frac{-2}{3}\right)^2$ *Answer:* _____

36. $\left(\frac{-2}{5}\right)^2 + \left(\frac{1}{5}\right)$ *Answer:* _____

37. $\left(\frac{2}{3}\right)^2 - \left(\frac{3}{2}\right)^2$ *Answer:* _____

38. $\left(\frac{-4}{5}\right)^2 - \left(\frac{-5}{2}\right)^2$ *Answer:* _____

2.8 MULTIPLICATION WHEN THE BASES ARE THE SAME

Write 2^3 in factored form: _____

Write 2^4 in factored form: _____

What is $(2 \times 2 \times 2)(2 \times 2 \times 2 \times 2)$ in exponential form? _____

What is $2^3(2)^4$ in exponential form? _____

Your answer should be $2^3(2)^4 = 2^7$.

Example 1   Multiply $5^2 \cdot 5$. Leave your answer in exponential form.

Solution. $5^2 \cdot 5 = (5 \times 5) \times 5 = 5^3$

2 factors of 5 multiplied by 1 factor of 5 equal 3 factors of 5.   ■

Example 2   Multiply: $3^2 \times 3^3$. Leave your answer in exponential form.

Solution. $3^2 \times 3^3 = (3 \times 3) \times (3 \times 3 \times 3) = 3^5$

2 factors of 3 multiplied by 3 factors of 3 equal 5 factors of 3.   ■

MINI-PRACTICE   Multiply and leave the answer in exponential form:
(Use this space for your work.)

(a) $3^2 \cdot 3^3$ (b) $2 \cdot 2^2$ (c) $5^6 \cdot 5^7$

(d) $(-2)^2 \cdot (-2)^3$ (e) $(0.5)^4 \cdot (0.5)^5$

Answers: (a) _____   (b) _____   (c) _____

(d) _____   (e) _____

IN YOUR OWN WORDS   explain what shortcut we can use for multiplication of numbers written in exponential notation when the bases are the same? _____

_____

_____

Answer:

When multiplying powers of the same base, keep the common base and add the exponents.

RULE:

$4^2 \cdot 4^4 = 4^6$
$\downarrow \quad \downarrow$
Same base 4
Keep the common base. Add the exponents.

Thus, for example

$$8^4 \cdot 8^6 = 8^{4+6} = 8^{10}$$

When multiplying three or more powers of the same base, this rule also applies. Thus,

$$5^4 \cdot 5^2 \cdot 5^3 = 5^{4+2+3} = 5^9$$

Answers to Mini-Practice:

(a) 3^5 (b) 2^3 (c) 5^{13} (d) $(-2)^5$ (e) $(0.5)^9$

Practice Exercises

In Exercises 1–34, multiply and leave the answer in exponential form.

1. $3^2 \cdot 3^3$ Answer: _____   2. $2^4 \cdot 2^2$ Answer: _____

3. $5^2 \cdot 5^2$ Answer: _____   4. $7^4 \cdot 7^5$ Answer: _____

5. $3 \cdot 3^3$ Answer: _____   6. $4^2 \cdot 4^3$ Answer: _____

7. $9^4 \cdot 9^5$ Answer: _____   8. $8^3 \cdot 8^6$ Answer: _____

9. $2^1 \cdot 2^1 \cdot 2^1$ Answer: _____   10. $2 \cdot 2^2 \cdot 2^3 \cdot 2^4$ Answer: _____

11. $3 \cdot 3^4 \cdot 3^3$ Answer: _____   12. $(-10)^3 \cdot (-10)^4$ Answer: _____

13. $(-2)^3 \cdot (-2)^6$ Answer: _____   14. $(-4)^2 \cdot (-4) \cdot (-4)^3$ Answer: _____

15. $(-9)^5 \cdot (-9)^4 \cdot (-9)^3$ Answer: _____   16. $5^2 \cdot 5^4 \cdot 5 \cdot 5^3$ Answer: _____

17. $(-6)^3 \cdot (-6)^4 \cdot (-6)^2 \cdot (-6)$ Answer: _____   18. $(-3)^2 \cdot (-3)^3 \cdot (-3)^4 \cdot (-3)^5$ Answer: _____

19. $7^2 \cdot 7 \cdot 7^3 \cdot 7 \cdot 7^4$ Answer: _____   20. $(-8)^2 \cdot (-8) \cdot (-8)^4 \cdot (-8)^7 \cdot (-8)$ Answer: _____

21. $11^2 \cdot 11^4 \cdot 11^6 \cdot 11 \cdot 11^3$ Answer: _____   22. $100^4 \cdot 100^2 \cdot 100^3 \cdot 100^5 \cdot 100$ Answer: _____

23. $(0.1)^3 \cdot (0.1)^5$ Answer: _____   24. $(2.5)^2 \cdot (2.5)^4$ Answer: _____

25. $(-7.1)^3 \cdot (-7.1)$ Answer: _____   26. $(5.8)^2 \cdot (5.8)^5$ Answer _____

27. $(-9.2) \cdot (-9.2)^6$ Answer: _____   28. $(4.6)^3 \cdot (4.6)^7$ Answer: _____

29. $\left(\frac{1}{2}\right)^3 \cdot \left(\frac{1}{2}\right)^2$ Answer: _____   30. $\left(\frac{2}{5}\right)^2 \cdot \left(\frac{2}{5}\right)^2$ Answer: _____

31. $\left(\frac{3}{10}\right)^2 \cdot \left(\frac{3}{10}\right)^3$ Answer: _____   32. $\left(\frac{-1}{5}\right)^2 \cdot \left(\frac{-1}{5}\right)^3 \cdot \left(\frac{-1}{5}\right)^5$ Answer: _____

33. $\left(\frac{2}{3}\right)^4 \cdot \left(\frac{2}{3}\right) \cdot \left(\frac{2}{3}\right)^3 \cdot \left(\frac{2}{3}\right)^2$ Answer: _____   34. $\left(\frac{-3}{4}\right)^3 \cdot \left(\frac{-3}{4}\right)^5 \cdot \left(\frac{-3}{4}\right) \cdot \left(\frac{-3}{4}\right)^{10}$ Answer: _____

2.9 MULTIPLICATION WHEN THE BASES ARE DIFFERENT

Write 2^3 in factored form: _____

Write 3^4 in factored form: _____

Find the value of 2^3: _____

Find the value of 3^4: _____

Find the value of $2^3 \cdot 3^4$: _____

Example 1   Is there a shortcut when multiplying $4^2 \cdot 5^2$?

Solution. $4^2 \cdot 5^2 = 16(25) = 400$
            ↑   ↑
          base 4 base 5

The bases are NOT equal; there is NO shortcut.   ■

Example 2   Is there a shortcut when multiplying $2 \cdot 5 \cdot 2^3 \cdot 5^2$?

Solution.

$$2 \cdot 5 \cdot 2^3 \cdot 5^2 = 2^1 \cdot 2^3 \cdot 5^1 \cdot 5^2 = 2^{1+3} \cdot 5^{1+2} = 2^4 \cdot 5^3$$

bases   2   5   2   5

The bases are 2 and 5; for equal bases, add the exponents. There is no shortcut for unequal bases.   ∎

MINI-PRACTICE
(Use this space for your work.)

1. Find the value of the products:

   (a) $2^3 \cdot 3^2$ *Answer:* _____   (b) $6 \cdot 7^2$ *Answer:* _____

2. Multiply and leave the answer in exponential notation:

   (a) $5^3 \cdot 6^3 \cdot 5^4 \cdot 6^7$ *Answer:* _____   (b) $9 \cdot 10^2 \cdot 10^4$ *Answer:* _____

IN YOUR OWN WORDS   explain how to perform multiplication when not all bases are the same: _____

_____

_____

Answer:

   We need only add the exponents when the bases are the same. In other cases we have to evaluate each exponential form before we multiply the factors.

WATCH OUT!   Explain why $2^3 \cdot 3^4$ is not equal to 6^7. _____

_____

_____

Answer:

   In multiplication of numbers in exponential notation, we can add the exponents *only when the bases are the same*. Here,

$$2^3 \cdot 3^4 = 8 \cdot 81 = 648, \quad \text{but} \quad 6^7 = 279{,}936$$

Practice Exercises

In Exercises 1–13, find each product.

1. $4^2 \cdot 3^2$ *Answer:* _____   2. $3^3 \cdot 2^3$ *Answer:* _____

3. $5^3 \cdot 4^2$ *Answer:* _____   4. $(-1)^2 \cdot 3^4$ *Answer:* _____

Answers to Mini-Practice:

1. (a) 72   (b) 294   **2.** (a) $5^7 \cdot 6^{10}$ (b) $9 \cdot 10^6$

5. $(-2)^3 \cdot (-3)^2$ *Answer:* _____

6. $(0.1)^2 \cdot (0.5)^3$ *Answer:* _____

7. $(-5.1)^2 \cdot (0.1)$ *Answer:* _____

8. $(2.3)^3 \cdot (-3.2)^2$ *Answer:* _____

9. $(0.02)^3 \cdot (-7.8)$ *Answer:* _____

10. $(2.9)^2 \cdot (9.2)^3$ *Answer:* _____

11. $\left(\frac{1}{2}\right)^2 \cdot \left(\frac{1}{3}\right)^2$ *Answer:* _____

12. $\left(\frac{3}{4}\right)^3 \cdot \left(\frac{-1}{2}\right)^2$ *Answer:* _____

13. $\left(\frac{1}{10}\right)^5 \cdot \left(\frac{5}{8}\right)^2$ *Answer:* _____

In Exercises 14–34, multiply and leave the answer in exponential notation.

14. $3 \cdot 4^2 \cdot 3^5 \cdot 4^6$ *Answer:* _____

15. $7^2 \cdot 9^4 \cdot 7^5 \cdot 9^7$ *Answer:* _____

16. $(-3)^3 \cdot 5^2 \cdot (-3)^3 \cdot 5^6$ *Answer:* _____

17. $2 \cdot 5 \cdot 7 \cdot 2^2 \cdot 5^3 \cdot 7^4$ *Answer:* _____

18. $5 \cdot 7^2 \cdot 5^3 \cdot 7^2$ *Answer:* _____

19. $2^2 \cdot 3^2 \cdot 5^3 \cdot 2 \cdot 3^3$ *Answer:* _____

20. $(-6)^3 \cdot 9^5 \cdot (-6)^2 \cdot 9^4$ *Answer:* _____

21. $10^2 \cdot 11^3 \cdot 13^5 \cdot 10^3 \cdot 11^2 \cdot 13$ *Answer:* _____

22. $5^2 \cdot 6^4 \cdot 7^5 \cdot 5^9 \cdot 6^2 \cdot 7$ *Answer:* _____

23. $(0.2)(0.3)^2$ *Answer:* _____

24. $(0.1)^2(0.01)^2$ *Answer:* _____

25. $(0.2)^3(0.5)^2$ *Answer:* _____

26. $(0.1)^2 \cdot (0.5)^3 \cdot (0.1)^6 \cdot (0.5)^4$ *Answer:* _____

27. $(-3.1)^2 \cdot (0.1)^3 \cdot (-3.1)$ *Answer:* _____

28. $(5.2) \cdot (0.2)^3 \cdot (0.3)^2 \cdot (0.2)^5$ *Answer:* _____

29. $(9.2)^3 \cdot (9.2)^4 \cdot (0.25)^2 \cdot (9.2)^3 \cdot (0.25)^3$ *Answer:* _____

30. $(5.7)^3 \cdot (5.3)^4 \cdot (5.7)^2$ *Answer:* _____

31. $(4.1)^2 \cdot (1.4)^3 \cdot (1.4)^5 \cdot (4.1)^6$ *Answer:* _____

32. $\left(\frac{1}{4}\right)^2 \cdot \left(\frac{1}{2}\right)$ *Answer:* _____

33. $\left(\frac{2}{3}\right)^2 \cdot \left(\frac{1}{3}\right)^2$ *Answer:* _____

34. $\left(\frac{1}{10}\right)^2 \cdot \left(\frac{3}{10}\right)^2$ *Answer:* _____

2.10 POWERS OF EXPONENTIAL EXPRESSIONS

Write $(4^2)^3$ as a product of three equal factors: _____

Use the rule for multiplication of equal bases to obtain the product of $4^2 \cdot 4^2 \cdot 4^2$:

$$(4^2)^3 = 4^6$$

Example 1   Write $(2^3)^4$ as a power of 2.

Solution. Here 2^3 is the base and 4 is the exponent.

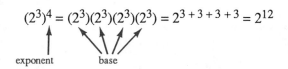

$$(2^3)^4 = (2^3)(2^3)(2^3)(2^3) = 2^{3+3+3+3} = 2^{12}$$

exponent   base

Example 2   Write $(5^2)^3$ as a power of 5.

$$(5^2)^3 = (5^2)(5^2)(5^2) = 5^{2+2+2} = 5^6$$

MINI-PRACTICE
(Use this space for your work.)

(a) Write $(2^2)^2$ as a power of 2.

    Answer: _____

(b) Write $(4^2)^3$ as a power of 4.

    Answer: _____

(c) Write $(5^3)^4$ as a power of 5.

    Answer: _____

(d) Write $(5^4)^3$ as a power of 5.

    Answer: _____

(e) Write $(3^4)^3$ as a power of 3.

    Answer: _____

IN YOUR OWN WORDS   explain the shortcut you could use to go directly from $(4^2)^3$ to 4^6: _____

_____

_____

Answer:

When we raise an exponential expression to a power, we leave the base as it is and multiply the exponents. Thus,

$$(4^2)^3 = 4^{2 \cdot 3} = 4^6$$

Practice Exercises

1. Write $(6^6)^6$ as a power of 6.   *Answer:* _____
2. Write $(2^5)^7$ as a power of 2.   *Answer:* _____
3. Write $(10^5)^8$ as a power of 10.   *Answer:* _____
4. Write $(25^2)^4$ as a power of 25.   *Answer:* _____
5. Write $(0.5^3)^6$ as a power of 0.5.   *Answer:* _____
6. Write $(1.2^3)^2$ as a power of 1.2.   *Answer:* _____
7. Write $(9.5^4)^8$ as a power of 9.5.   *Answer:* _____
8. Write $(0.25^7)^3$ as a power of 0.25.   *Answer:* _____
9. Write $(3^4)^2$ as a power of 3.   *Answer:* _____
10. Write $(3^2)^4$ as a power of 3.   *Answer:* _____
11. Write $(10^3)^5$ as a power of 10.   *Answer:* _____
12. Write $(5^5)^5$ as a power of 5.   *Answer:* _____
13. Write $(11^3)^4$ as a power of 11.   *Answer:* _____
14. Write $(7^4)^6$ as a power of 7.   *Answer:* _____

Answers to Mini-Practice:

(a) 2^4 (b) 4^6 (c) 5^{12} (d) 5^{12} (e) 3^{12}

15. Write $\left[\left(\frac{1}{2}\right)^2\right]^3$ as a power of $\frac{1}{2}$.   *Answer:* _____   16. Write $\left[\left(\frac{2}{3}\right)^3\right]^4$ as a power of $\frac{2}{3}$.   *Answer:* _____

17. Write $\left[\left(\frac{4}{5}\right)^4\right]^2$ as a power of $\frac{4}{5}$.   *Answer:* _____   18. Write $\left[\left(\frac{7}{9}\right)^5\right]^3$ as a power of $\frac{7}{9}$.   *Answer:* _____

2.11 POWERS OF PRODUCTS

Write in factored form: $5 \cdot 2^3 =$ _____

Write in factored form: $(5 \cdot 2)^3 =$ _____

Each exponent belongs to the base in front of it. If the base is enclosed in parentheses, everything inside the parentheses should be raised to that particular power.

$$5 \cdot 2^3 = 5(2^3) = 5 \cdot 8 = 40$$

$$(5 \cdot 2)^3 = (5 \cdot 2)(5 \cdot 2)(5 \cdot 2) = 10 \cdot 10 \cdot 10 = 1000$$

$(5 \cdot 2)(5 \cdot 2)(5 \cdot 2)$ can also be written as $5^3 \cdot 2^3 = 125 \cdot 8 = 1000$

Write $(2^3 \cdot 3^2)^4$ as a repeated multiplication of $2^3 \cdot 3^2$:

(   )(   )(   )(   )

Write all the 2^3 factors first and then the 3^2 factors:

_____

Simplify the final result by using the rules for multiplication:

$2^?3^? =$ _____

Your answer should be $2^{12} \cdot 3^8$.

In raising $3 \cdot 5^2 \cdot 7^3$ to the fifth power, that is, $(3 \cdot 5^2 \cdot 7^3)^5$, we raise each factor, in turn, to the fifth power. Fill in the blanks in the following table.

3	3^1	$(3^1)^5$	3^5
5	5^2		
7	7^3		

$(3 \cdot 5^2 \cdot 7^3)^5 =$ _____

You probably answered $3^5 \cdot 5^{10} \cdot 7^{15}$.

Example 1   Write $(4^2 \cdot 5^3)^3$ as a product in exponential form.

Solution. $(4^2 \cdot 5^3)^3 = \underbrace{(4^2 \cdot 5^3)}(4^2 \cdot 5^3)(4^2 \cdot 5^3) = 4^6 \cdot 5^9$ ■

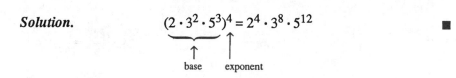

base   exponent

Example 2   Write $(2 \cdot 3^2 \cdot 5^3)^4$ as a product in exponential form.

Solution. $(2 \cdot 3^2 \cdot 5^3)^4 = 2^4 \cdot 3^8 \cdot 5^{12}$ ■

base   exponent

MINI-PRACTICE   (a) Write $(3^2 \cdot 4^3)^2$ as a product in exponential form.   *Answer:* _____

(Use this space for your work.)   (b) Write $(2^5 \cdot 5^3)^3$ as a product in exponential form.   *Answer:* _____

(c) Write $(2 \cdot 3^2)^5$ as a product in exponential form.   *Answer:* _____

(d) Write $(2 \cdot 3)^2$ as a product in exponential form.   *Answer:* _____

(e) Write $(3 \cdot 4 \cdot 5)^3$ as a product in exponential form.   *Answer:* _____

IN YOUR OWN WORDS   explain how you raise a product in exponential form to a power: _____

_____

_____

Answer:

When you have to raise a product in exponential form to a power, you multiply each exponent in the product by the outside exponent.

Practice Exercises

In Exercises 1–24, raise each product to the indicated power. (Do not multiply out.)

1. $(2 \cdot 3 \cdot 5)^4$ *Answer:* _____   **2.** $(5 \cdot 6^2 \cdot 8)^4$ *Answer:* _____

3. $(3^3 \cdot 7^5)^6$ *Answer:* _____   **4.** $(4^2 \cdot 5^4 \cdot 6^5)^7$ *Answer:* _____

5. $(3 \cdot 4^3 \cdot 5^6 \cdot 7^2)^5$ *Answer:* _____   **6.** $(2 \cdot 3^2 \cdot 5^3)^2$ *Answer:* _____

7. $(5^3 \cdot 6^2 \cdot 7^4)^3$ *Answer:* _____   **8.** $(3^{11} \cdot 5^6 \cdot 9^2)^3$ *Answer:* _____

9. $(2^5 \cdot 8^2 \cdot 11^3)^3$ *Answer:* _____   **10.** $(7^4 \cdot 12^5 \cdot 13^3)^2$ *Answer:* _____

11. $(2 \cdot 5^4 \cdot 7^2 \cdot 9^3)^5$ *Answer:* _____   **12.** $[(0.1)^5 (0.25)^3]^2$ *Answer:* _____

Answers to Mini-Practice:
(a) $3^4 \cdot 3^6$ (b) $2^{15} \cdot 5^9$ (c) $2^5 \cdot 3^{10}$ (d) $2^2 \cdot 3^2$ (e) $3^3 \cdot 4^3 \cdot 5^3$

13. $[(3.2)^2(4.5)^3]^5$ *Answer:* _____   **14.** $[(-0.2)^4(0.3)^2]^4$ *Answer:* _____

15. $[(2.9)^3(3.5)^2]^5$ *Answer:* _____   **16.** $[(5.7)^2(-7.5)^3(0.1)^5]^6$ *Answer:* _____

17. $[(0.1)^2(0.01)^3(0.2)]^7$ *Answer:* _____   **18.** $[(3.6)^3(4.7)^2(5.9)^4]^2$ *Answer:* _____

19. $\left[\left(\frac{1}{4}\right)^2\left(\frac{1}{2}\right)^3\right]^2$ *Answer:* _____   **20.** $\left[\left(\frac{1}{2}\right)^3\left(\frac{2}{3}\right)^4\right]^7$ *Answer:* _____

21. $\left[\left(\frac{4}{7}\right)^2\left(\frac{1}{5}\right)^3\right]^4$ *Answer:* _____   **22.** $\left[\left(\frac{1}{8}\right)^2\left(\frac{2}{9}\right)^4\left(\frac{3}{11}\right)^7\right]^3$ *Answer:* _____

23. $\left[\left(\frac{3}{5}\right)^4\left(\frac{-2}{3}\right)^3\left(\frac{1}{8}\right)^5\right]^2$ *Answer:* _____   **24.** $\left[\left(\frac{3}{5}\right)^4\left(\frac{1}{12}\right)^2\left(\frac{2}{7}\right)^6\right]^9$ *Answer:* _____

2.12 DIVISION

In multiplication using exponential notation, we add the exponents whenever the bases are the same. For example, $2^3 \cdot 2^6 = 2^9$.

What would you do with the exponents in division? _____

$$3^5 \div 3^3 = \frac{3 \times 3 \times 3 \times 3 \times 3}{3 \times 3 \times 3} = 3 \times 3 = 3^2$$

Example 1   Divide 4^3 by 4^2.

Solution.   3 factors of 4 → $\dfrac{4^3}{4^2} = 4$ ←1 factor of 4
2 factors of 4 →

$$4^3 \div 4^2 = 4$$ ∎

Example 2   Divide 4^5 by 4. Leave your answer in exponential form.

Solution.   5 factors of 4 → $\dfrac{4^5}{4} = 4^4$ ← 4 factors of 4
1 factor of 4 →

$$4^5 \div 4 = 4^4$$ ∎

MINI-PRACTICE   Divide. (Leave the answer in exponential form.)
(Use this space for your work.)

(a) $\dfrac{2^{10}}{2^3}$ (b) $\dfrac{4^{12}}{4^5}$ (c) $\dfrac{(-5)^3}{(-5)^2}$ (d) $\dfrac{(4 \cdot 3)^5}{(4 \cdot 3)^2}$ (e) $\dfrac{4^4}{4^3}$

Answers: (a) _____   (b) _____   (c) _____   (d) _____   (e) _____

Answers to Mini-Practice:

(a) 2^7 (b) 4^7 (c) -5 (d) $(4 \cdot 3)^3$ (e) 4^1

IN YOUR OWN WORDS   explain what shortcut we use for division of numbers written in exponential notation: _____

_____

_____

Answer:

When dividing powers of the same base, keep the common base and subtract the exponents.

$$\frac{4^6}{4^2} = 4^4$$

Keep the common base.
Subtract the exponents.

As another example,

$$\frac{5^8}{5^6} = 5^2$$

WATCH OUT!   Why is $\dfrac{6^9}{3^3}$ unequal to 2^6? _____

_____

Answer:

We can use the method of subtracting the exponents *only when the bases are the same*!

Practice Exercises

In Exercises 1-30, divide. (Leave the answer in exponential form.)

1. $\dfrac{5^6}{5^4}$ Answer: _____

2. $\dfrac{7^8}{7^5}$ Answer: _____

3. $\dfrac{8^{10}}{8^9}$ Answer: _____

4. $\dfrac{2^5}{2^3}$ Answer: _____

5. $\dfrac{3^9}{3^8}$ Answer: _____

6. $\dfrac{(-4)^5}{(-4)^3}$ Answer _____

7. $\dfrac{(-6)^9}{(-6)^7}$ Answer: _____

8. $\dfrac{9^{11}}{9^8}$ Answer: _____

9. $\dfrac{11^5}{11^2}$ Answer: _____

10. $\dfrac{10^{14}}{10^7}$ Answer: _____

11. $\dfrac{(-12)^5}{(-12)^3}$ Answer: _____

12. $\dfrac{153^4}{153^3}$ Answer: _____

13. $\dfrac{(-45)^6}{(-45)^4}$ Answer: _____

14. $\dfrac{(-93)^8}{(-93)^7}$ Answer: _____

15. $\dfrac{(0.7)^6}{(0.7)^3}$ Answer: _____

16. $\dfrac{(0.1)^5}{(0.1)^3}$ Answer: _____

17. $\dfrac{(0.3)^4}{(0.3)^3}$ Answer: _____

18. $\dfrac{(10.5)^{11}}{(10.5)^7}$ Answer: _____

19. $\dfrac{(-5.2)^4}{(-5.2)}$ *Answer:* _____

20. $\dfrac{(9.6)^5}{(9.6)^2}$ *Answer:* _____

21. $\dfrac{(0.02)^6}{(0.02)^3}$ *Answer:* _____

22. $\dfrac{(-7.1)^5}{(-7.1)^2}$ *Answer:* _____

23. $\dfrac{(2\cdot 5)^7}{(2\cdot 5)^4}$ *Answer:* _____

24. $\dfrac{(5\cdot 9)^4}{(5\cdot 9)^3}$ *Answer:* _____

25. $\dfrac{\left(\frac{1}{2}\right)^3}{\left(\frac{1}{2}\right)^2}$ *Answer:* _____

26. $\dfrac{\left(-\frac{2}{3}\right)^4}{\left(-\frac{2}{3}\right)}$ *Answer:* _____

27. $\dfrac{\left(\frac{4}{11}\right)^8}{\left(\frac{4}{11}\right)^5}$ *Answer:* _____

28. $\dfrac{\left(\frac{5}{7}\right)^6}{\left(\frac{5}{7}\right)^2}$ *Answer:* _____

29. $\dfrac{\left(\frac{-9}{8}\right)^3}{\left(\frac{-9}{8}\right)}$ *Answer:* _____

30. $\dfrac{\left(\frac{6}{7}\right)^8}{\left(\frac{6}{7}\right)^5}$ *Answer:* _____

2.13 POWERS OF QUOTIENTS

In the expression $\left(\frac{2}{3}\right)^3$, what is the base? _____

What is the exponent? _____

Here, $\frac{2}{3}$ is the base and 3 the exponent.

Rewrite the expression in terms of division. _____

What is the answer? _____

$$\left(\frac{2}{3}\right)^3 = \frac{2^3}{3^3} = \frac{2\times 2\times 2}{3\times 3\times 3} = \frac{8}{27}$$

Example 1   Raise $\frac{3}{4}$ to the fourth power. Leave the numerator and denominator in exponential form.

Solution. $$\left(\frac{3}{4}\right)^4 = \frac{3\times 3\times 3\times 3}{4\times 4\times 4\times 4} = \frac{3^4}{4^4}$$

$$\left(\frac{3}{4}\right)^4 = \frac{3^4}{4^4}$$ ∎

Example 2   Raise $\frac{7}{11}$ to the fifth power. Leave the numerator and denominator in exponential form.

Solution.
$$\left(\frac{7}{11}\right)^5 = \frac{7 \times 7 \times 7 \times 7 \times 7}{11 \times 11 \times 11 \times 11 \times 11} = \left(\frac{7}{11}\right)^5$$

$$\left(\frac{7}{11}\right)^5 = \frac{7^5}{11^5}$$ ∎

MINI-PRACTICE
(Use this space for your work.)

Evaluate the following expressions:

(a) $\left(\frac{1}{2}\right)^2$ 　　　　　(b) $\left(\frac{3}{4}\right)^3$ 　　　　　(c) $\left(\frac{1}{4}\right)^3$

Answers:　(a) _____ 　　　(b) _____ 　　　(c) _____

IN YOUR OWN WORDS　state the rule for simplifying fractions raised to a power: _____

_____

_____

Answer:

When we raise a fraction to a power, we must raise both the numerator and the denominator to that power.

WATCH OUT!　In the case of $3 \div 3^4$ we have

$$\frac{3}{3 \times 3 \times 3 \times 3} = \frac{1}{3 \times 3 \times 3} = \frac{1}{3^3} \quad \text{or} \quad \frac{1}{27}$$

Notice that we divided both numerator and denominator by three. Why must we leave our answer with a "1" in the numerator? _____

_____

_____

Answer:

First notice that when we divide the numerator 3 by 3, we obtain 1, *not* 0. Also, you might recall from arithmetic that $27 = \frac{27}{1}$; but $\frac{27}{1}$ and $\frac{1}{27}$ are not the same numbers.

Compare this with money: 1 quarter, or $\$\frac{25}{100}$, is \$0.25, and that is certainly not the same as $\$\frac{100}{25}$, or \$4!

Answers to Mini-Practice:

(a) $\frac{1}{2^2}$ or $\frac{1}{4}$ 　　　(b) $\frac{3^3}{4^3}$ or $\frac{27}{64}$ 　　　(c) $\frac{1}{4^3}$ or $\frac{1}{64}$

Practice Exercises

In Exercises 1–12, evaluate the following expressions.

1. $\left(\dfrac{3}{10}\right)^3$ Answer: _____

2. $\left(\dfrac{3}{7}\right)^2$ Answer: _____

3. $\left(\dfrac{2}{5}\right)^3$ Answer: _____

4. $\left(\dfrac{1}{2}\right)^5$ Answer: _____

5. $\left(\dfrac{5}{9}\right)^3$ Answer: _____

6. $\left(\dfrac{2}{3}\right)^3$ Answer: _____

7. $\left(\dfrac{1}{5}\right)^2$ Answer: _____

8. $\left(\dfrac{2}{11}\right)^2$ Answer: _____

9. $\left(\dfrac{1}{2}\right)^4$ Answer: _____

10. $\left(\dfrac{3}{5}\right)^3$ Answer: _____

11. $\left(\dfrac{1}{3}\right)^5$ Answer: _____

12. $\left(\dfrac{3}{4}\right)^3$ Answer: _____

In Exercises 13–20, evaluate. Leave each numerator and denominator in exponential form.

13. $\left(\dfrac{7}{3}\right)^4$ Answer: _____

14. $\left(\dfrac{5}{2}\right)^7$ Answer: _____

15. $\left(\dfrac{9}{7}\right)^3$ Answer: _____

16. $\left(\dfrac{4}{9}\right)^5$ Answer: _____

17. $\left(\dfrac{6}{7}\right)^8$ Answer: _____

18. $\left(\dfrac{11}{12}\right)^5$ Answer: _____

19. $\left(\dfrac{4}{9}\right)^7$ Answer: _____

20. $\left(\dfrac{3}{13}\right)^{11}$ Answer: _____

2.14 THE NUMBER ZERO AS EXPONENT

Below is a table showing the number three raised to different powers. Some of the products are calculated. Complete the table.

3^4	3^3	3^2	3^1	3^0
81	27			

When we move from left to right in the top line of the table, we decrease each exponent by 1.

When we move from left to right in the bottom line of the table, we divide 81 by 3 to get 27, 27 by 3 to get 9, and so on.

When we divide 3 by 3, what is the quotient? _____

As you recall from the previous section on division, division implies subtraction of exponents. So, when we divide 3^1 by 3^1, we get:

$$3^1 \div 3^1 = 3^{1-1} = 3^0$$

What is 3^1 divided by 3^1? _____

What is a number divided by itself? _____

Any number divided by itself equals one. The only exception is $0 \div 0$, which is undefined. With this in mind, we define

$$3^0 = 3^1 \div 3^1 = 1$$

and we complete the table as follows:

3^4	3^3	3^2	3^1	3^0
81	27	9	3	1

Example 1   Divide 3^2 by 3^2.

Solution.   Subtract the exponents!

$$\frac{3^2}{3^2} = 3^{2-2} = 3^0$$

but

$$\frac{3^2}{3^2} = \frac{9}{9} = 1$$

Therefore,

$$3^2 \div 3^2 = 3^0 = 1$$

■

Example 2   Evaluate 5^0.

Solution.   Since $0 = 1 - 1$, we have

$$5^0 = 5^{1-1} = \frac{5^1}{5^1} = 1$$

Therefore,

$$5^0 = 1$$

■

MINI-PRACTICE
(Use this space for your work.)

Evaluate:

(a) 6^0 (b) 100^0 (c) $(-7)^0$ (d) $\dfrac{5^4}{5^4}$

Answers: (a) _____   (b) _____   (c) _____   (d) _____

Answers to Mini-Practice:

(a) 1   (b) 1   (c) 1   (d) 1 or 5^0

IN YOUR OWN WORDS   explain what answer we get when we raise a number (other than 0) to the zeroth power: _____

_____

_____

Answer:

   A number raised to the zeroth power is one. This statement, which actually is a definition, holds for all number bases with the exception of 0 itself.

Practice Exercises

1. Divide 5^4 by 5^4. *Answer:* _____
2. Divide 11^2 by 11^2. *Answer:* _____
3. Divide 10^5 by 10^5. *Answer:* _____
4. Divide $(-4)^3$ by $(-4)^3$. *Answer:* _____

In Exercises 5–32, evaluate each expression.

5. $\dfrac{7^2}{7^2}$ *Answer:* _____
6. $\dfrac{5^6}{5^6}$ *Answer:* _____

7. 4^0 *Answer:* _____
8. 1000^0 *Answer:* _____

9. $(1,000,000)^0$ *Answer:* _____
10. -6^0 *Answer:* _____

11. $5^0(4)$ *Answer:* _____
12. $3 \cdot 7^0$ *Answer:* _____

13. $5^0 + 9$ *Answer:* _____
14. $4 + 9^0$ *Answer:* _____

15. $4^2 \cdot 4^1 \cdot 4^0$ *Answer:* _____
16. $4^2 + 4^1 + 4^0$ *Answer:* _____

17. $3^3 + 3^2 + 3^1 + 3^0$ *Answer:* _____
18. $3^3 \cdot 3^2 \cdot 3^1 \cdot 3^0$ *Answer:* _____

19. $5.6 + (3.35)^0$ *Answer:* _____
20. $3.1 - (2.4)^0$ *Answer:* _____

21. $(5.3)^0 + (3.2) - (1.7)^0$ *Answer:* _____
22. $(4.6)^0[0.05 \div 0.5]$ *Answer:* _____

23. $10.8 \div (51.38)^0$ *Answer:* _____
24. $7.5 + (5.7)^0 + (3.8)^0(3.8)$ *Answer:* _____

25. $(-7.36)^0 + 6.5 \div (3.9)^0$ *Answer:* _____
26. $2 \cdot (6.9)^0 + 3 \div (7.5)^0$ *Answer:* _____

27. $\left(\dfrac{1}{2}\right)^0 + \left(\dfrac{1}{2}\right) + \left(\dfrac{1}{2}\right)^2$ *Answer:* _____
28. $\left(\dfrac{3}{8}\right)\left(\dfrac{8}{3}\right)^0$ *Answer:* _____

29. $\left(\dfrac{10}{3}\right)^0 \div \dfrac{9}{8}$ *Answer:* _____
30. $\left(-\dfrac{3}{4}\right)^0 - \dfrac{5}{6} + \left(\dfrac{9}{2}\right)^0$ *Answer:* _____

31. $\left(\dfrac{7}{5}\right)^0 + 1$ *Answer:* _____
32. $\left(\dfrac{2}{3}\right)^2 + \left(\dfrac{2}{3}\right)^1 + \left(\dfrac{2}{3}\right)^0$ *Answer:* _____

2.15 POWERS WITH NEGATIVE EXPONENTS

Divide $\dfrac{3^5}{3^3}$ using the rule for division. _____

Divide $\dfrac{3^3}{3^5}$ using the rule for division. _____

The form of your answer depends on how you did the subtraction:

$$\frac{3^3}{3^5} = \frac{1}{3^{5-3}} = \frac{1}{3^2} \qquad \text{or} \qquad \frac{3^3}{3^5} = 3^{3-5} = 3^{-2}$$

But *these answers must be equal*, so that

$$3^{-2} = \frac{1}{3^2}$$

$$\frac{2}{2^4} = \underline{\hspace{3cm}} \qquad \text{or} \qquad \underline{\hspace{3cm}}$$

Note that 2^{-3} is the same as $\frac{1}{2^3}$ or $\frac{1}{8}$.

Example 1   Rewrite 4^{-1} with a positive exponent.

Solution. $\qquad\qquad 4^{-1} = \frac{1}{4^1} = \frac{1}{4}$

Thus,

$$4^{-1} = \frac{1}{4} \qquad\qquad\qquad \blacksquare$$

Example 2   Evaluate $\frac{1}{5^{-2}}$.

Solution. $\qquad\qquad\qquad \frac{1}{5^{-2}} = \frac{1}{\frac{1}{5^2}}$

The fraction bar is replaced with ÷, and division of fractions is performed. Thus,

$$\frac{1}{5^{-2}} = \frac{1}{\frac{1}{5^2}} = 1 \div \frac{1}{5^2} = 1 \times \frac{5^2}{1} = 5^2 = 25 \qquad\qquad \blacksquare$$

MINI-PRACTICE   Express with positive exponents.

(Use this space for your work.)

(a) 2^{-1} (b) $\left(\frac{1}{2}\right)^{-1}$ (c) $\frac{1}{2^{-2}}$ (d) $\left(\frac{1}{2}\right)^{-2}$ (e) $(-2)^{-2}$

Answers: (a) _____ (b) _____ (c) _____ (d) _____ (e) _____

Answers to Mini-Practice:

(a) $\frac{1}{2}$ (b) 2   (c) 4   (d) 2^2 or 4   (e) $\frac{1}{(-2)^2}$ or $\frac{1}{4}$

IN YOUR OWN WORDS   explain how you can rewrite a number with a negative exponent so that the exponent becomes positive: _____

_____

_____

Answer:

In order to change a negative exponent to a positive exponent, we invert the number. To **invert** (or to take the **reciprocal**) is to exchange the numerator and denominator. Thus

$$\left(\frac{2}{3}\right)^{-4} = \left(\frac{3}{2}\right)^{4}$$

RULE:

POWERS WITH NEGATIVE EXPONENTS

$$(\text{BASE})^{\text{NEGATIVE EXPONENT}} = \frac{1}{(\text{BASE})^{\text{POSITIVE EXPONENT}}}$$

$$\frac{1}{(\text{BASE})^{\text{NEGATIVE EXPONENT}}} = (\text{BASE})^{\text{POSITIVE EXPONENT}}$$

Practice Exercises

In Exercises 1–12, rewrite the following problems with positive exponents. Do not evaluate:

1. 3^{-3} *Answer:* _____
2. 4^{-5} *Answer:* _____
3. 6^{-1} *Answer:* _____
4. 7^{-3} *Answer:* _____
5. 11^{-4} *Answer:* _____
6. 16^{-6} *Answer:* _____
7. 20^{-9} *Answer:* _____
8. $\left(\frac{1}{4}\right)^{-3}$ *Answer:* _____
9. $\frac{1}{15^{-1}}$ *Answer:* _____
10. $\frac{1}{18^{-1}}$ *Answer:* _____
11. $\frac{1}{9^{-1}}$ *Answer:* _____
12. $\left(\frac{1}{3}\right)^{-1}$ *Answer:* _____

In Exercises 13–58, write with positive exponents. Simplify whenever possible.

13. 4^{-2} *Answer:* _____
14. $(-4)^{-2}$ *Answer:* _____
15. $(-2)^{-3}$ *Answer:* _____
16. 5^{-2} *Answer:* _____
17. 3^{-3} *Answer:* _____
18. 7^{-1} *Answer:* _____
19. 12^{-1} *Answer:* _____
20. 6^{-3} *Answer:* _____
21. $(-5)^{-3}$ *Answer:* _____
22. $(-3)^{-2}$ *Answer:* _____
23. $(-4)^{-3}$ *Answer:* _____
24. 3^{-4} *Answer:* _____
25. $(-2)^{-5}$ *Answer:* _____
26. $(-4)^{-3}$ *Answer:* _____
27. $(-10)^{-2}$ *Answer:* _____
28. $\left(\frac{-1}{5}\right)^{-3}$ *Answer:* _____
29. $\left(\frac{-1}{3}\right)^{-4}$ *Answer:* _____
30. $\frac{1}{5^{-2}}$ *Answer:* _____

31. $\dfrac{1}{3^{-4}}$ Answer: _____

32. $\left(-\dfrac{1}{4}\right)^{-3}$ Answer: _____

33. $\left(-\dfrac{1}{7}\right)^{-2}$ Answer: _____

34. $\dfrac{1}{7^{-2}}$ Answer: _____

35. $\left(\dfrac{1}{6}\right)^{-3}$ Answer: _____

36. $(0.1)^{-3}$ Answer: _____

37. $(0.2)^{-4}$ Answer: _____

38. $(0.4)^{-2}$ Answer: _____

39. $(0.5)^{-2}$ Answer: _____

40. $(-3)^{-4}$ Answer: _____

41. $\dfrac{1}{(-2)^{-4}}$ Answer: _____

42. $\dfrac{3^{-2}}{3}$ Answer: _____

43. $\dfrac{5^{-3}}{5}$ Answer: _____

44. $\dfrac{2^{-3}}{2^3}$ Answer: _____

45. $\dfrac{3^{-2}}{3^2}$ Answer: _____

46. $\dfrac{(0.32)^{-2}}{(0.32)^{-3}}$ Answer: _____

47. $(-0.1)^{-5}(10)^2$ Answer: _____

48. $3^{-2}(9)$ Answer: _____

49. $\left(\dfrac{3}{4}\right)^{-2}$ Answer: _____

50. $\left(\dfrac{5}{8}\right)^{-3}$ Answer: _____

51. $\dfrac{5^{-2}}{8^3}$ Answer: _____

52. $1 + \left(\dfrac{7}{9}\right)^{-1}$ Answer: _____

53. $\dfrac{4}{3^{-2}}$ Answer: _____

54. $\dfrac{1}{(-5)^{-3}}$ Answer: _____

55. $\dfrac{4^{-1}}{4}$ Answer: _____

56. $\dfrac{6^{-2}}{6}$ Answer: _____

57. $\dfrac{10^{-1}}{10^2}$ Answer: _____

58. $\dfrac{8}{8^{-2}}$ Answer: _____

2.16 SQUARE ROOTS

What number squared is 1? _____

What number squared is 4? _____

What number squared is 16? _____

What number squared is 25? _____

What number squared is 36? _____

You probably gave 1, 2, 4, 5, and 6 as answers. For example,

$$1^2 = 1 \times 1 = 1 \quad \text{and} \quad 2^2 = 2 \times 2 = 4$$

But note that

$$(-1)^2 = (-1)(-1) = 1 \qquad (-2)^2 = (-2)(-2) = 4$$

Instead of using the expression "what number squared is ____, " we can say, "the **square root** of ____ is." Thus, both 1 and –1 are the square roots of 1, and both 2 and –2 are the square roots of 4.

What is the symbol for square root? _____

The symbol for "square root" is $\sqrt{}$. This symbol is understood to indicate the *positive* square root or 0. Thus,

$$\sqrt{36} = 6 \text{ and } -\sqrt{36} = -6$$

Also,

$$\sqrt{0} = 0$$

Find the following square roots. Show that when each square root is multiplied by itself, the product is the number under the square root sign:

Number	Square Root	Product
4	$\sqrt{4} = 2$	$2 \times 2 = 4$
9	$\sqrt{9} =$	$3 \times 3 = 9$
25		
49		

Example 1 $\sqrt{64} = \sqrt{8 \times 8} = 8$ ■

Example 2 $\sqrt{144} = \sqrt{12 \times 12} = 12$ ■

Example 3 $\sqrt{2.25} = \sqrt{1.5 \times 1.5} = 1.5$ ■

MINI-PRACTICE   Find the following square roots:
(Use this space for your work.)

(a) $\sqrt{100}$ (b) $\sqrt{0.01}$ (c) $\sqrt{121}$ (d) $\sqrt{\dfrac{4}{9}}$

Answers: (a) _____   (b) _____   (c) _____   (d) _____

IN YOUR OWN WORDS   give a definition of square root. _____

_____

_____

Answer:

A square root of a number can be defined as a number that when multiplied by itself yields the original number. We could also say: A square root of a number is one of the two equal factors of the number.

_____

Answers to Mini-Practice:

(a) 10   (b) 0.1   (c) 11   (d) $\dfrac{2}{3}$

Practice Exercises

In Exercises 1–20, find the square roots.

1. $\sqrt{81}$ *Answer:* _____
2. $\sqrt{144}$ *Answer:* _____
3. $\sqrt{10,000}$ *Answer:* _____

4. $\sqrt{169}$ *Answer:* _____
5. $\sqrt{22,500}$ *Answer:* _____
6. $\sqrt{256}$ *Answer:* _____

7. $\sqrt{289}$ *Answer:* _____
8. $\sqrt{324}$ *Answer:* _____
9. $\sqrt{196}$ *Answer:* _____

10. $\sqrt{361}$ *Answer:* _____
11. $\sqrt{400}$ *Answer:* _____
12. $\sqrt{625}$ *Answer:* _____

13. $\sqrt{484}$ *Answer:* _____
14. $\sqrt{40,000}$ *Answer:* _____
15. $\sqrt{1,000,000}$ *Answer:* _____

16. $\sqrt{0.09}$ *Answer:* _____
17. $\sqrt{0.0016}$ *Answer:* _____
18. $\sqrt{1.69}$ *Answer:* _____

19. $\sqrt{\dfrac{16}{49}}$ *Answer:* _____
20. $\sqrt{\dfrac{64}{169}}$ *Answer:* _____

2.17 CUBE ROOTS

Evaluate:
$2^3 =$ _____ $3^3 =$ _____

$4^3 =$ _____ $5^3 =$ _____

What number cubed equals 8? _____

What number cubed equals 64? _____

What number cubed equals 125? _____

Instead of saying "what number cubed is _____," we can say "the cube root of _____ is."

Therefore, we can say that the **cube root** of 8 is 2.

The symbol for cube root is $\sqrt[3]{}$, and we write $\sqrt[3]{8} = 2$.

The 3 in the symbol $\sqrt[3]{}$ is called the **index**.

Example 1 $\sqrt[3]{125} = \sqrt[3]{5 \times 5 \times 5} = 5$ ■

Example 2 $\sqrt[3]{1000} = \sqrt[3]{10 \times 10 \times 10} = 10$ ■

MINI-PRACTICE
(Use this space for your work.)

Find the following cube roots:

(a) $\sqrt[3]{216}$ (b) $\sqrt[3]{343}$ (c) $\sqrt[3]{8000}$ (d) $\sqrt[3]{0.125}$ (e) $\sqrt[3]{\dfrac{8}{125}}$

Answers: (a) _____   (b) _____   (c) _____   (d) _____   (e) _____

Answers to Mini-Practice:

(a) 6   (b) 7   (c) 20   (d) 0.5   (e) $\dfrac{2}{5}$

IN YOUR OWN WORDS   write a definition of cube root. _____

_____

_____

Answer:

A cube root of a number is one of the three equal factors of that number.

Practice Exercises

In Exercises 1–15, find the cube roots of the numbers.

1. 512   Answer: _____   2. 729   Answer: _____   3. 27,000   Answer: _____

4. 0.001   Answer: _____   5. 0.000001   Answer: _____   6. 0.512   Answer: _____

7. 0.008   Answer: _____   8. $\dfrac{1}{27}$ Answer: _____   9. $\dfrac{8}{125}$ Answer: _____

10. $\dfrac{27}{1000}$ Answer: _____   11. 1,000,000   Answer: _____   12. 8,000,000   Answer: _____

13. $\dfrac{8}{27}$ Answer: _____   14. $\dfrac{27}{1,000,000}$ Answer: _____   15. $\dfrac{125}{512}$ Answer: _____

2.18 FRACTIONAL EXPONENTS

Square roots can be multiplied. Thus,

$$\sqrt{4} \times \sqrt{9} = 2 \times 3 = 6 = \sqrt{36} = \sqrt{4 \times 9}$$

It is also true that

$$\sqrt{2} \times \sqrt{3} = \sqrt{2 \times 3} = \sqrt{6}$$

Similarly,

$$\sqrt{2} \times \sqrt{2} = \sqrt{4}$$

But since $\sqrt{4} = 2$, we can write

$$\sqrt{2} \times \sqrt{2} = 2$$

Now consider this question: What number multiplied by itself gives a product of 2? Recall that $2 = 2^1$, and we have

$$2^? \cdot 2^? = 2^1$$

Now when multiplying with equal bases, we add the exponents; thus, our problem is: What number added to itself is 1?

$$\underline{\hspace{1cm}} + \underline{\hspace{1cm}} = 1$$

Did you think of fractions?

$$\frac{1}{2} + \frac{1}{2} = 1$$

Since $2^{1/2} \cdot 2^{1/2} = 2$, we can draw the conclusion that

$$2^{1/2} \text{ is the same as } \sqrt{2}$$

Example 1

$$\sqrt{5} \times \sqrt{5} = 5$$
$$5^{1/2} \times 5^{1/2} = 5^1 = 5$$
$$\sqrt{5} = 5^{1/2}$$ ∎

Cube roots can also be multiplied. Thus, with 3 equal factors:

or
$$\sqrt[3]{4} \times \sqrt[3]{4} \times \sqrt[3]{4} = \sqrt[3]{4 \times 4 \times 4} = \sqrt[3]{64} = 4$$

$$\underbrace{4^{1/3} \times 4^{1/3} \times 4^{1/3}}_{3 \text{ equal factors}} = 4^{1/3+1/3+1/3} = 4^1 = 4$$

MINI-PRACTICE

(Use this space for your work.)

1. Write in exponential form:

   (a) $\sqrt[5]{36}$ *Answer:* _____   (b) $\sqrt[7]{100}$ *Answer:* _____

2. Perform the following operations:

   (a) $\sqrt{5} \times \sqrt{7}$ *Answer:* _____   (b) $3^{1/2} \times 3^{1/2}$ *Answer:* _____

IN YOUR OWN WORDS   Explain the relationship between fractional exponents and the index of the roots.

_____

_____

_____

Answer:

   The index of the root is the same as the *denominator* of the fractional exponent; here, the numerator is one.

Practice Exercises

In Exercises 1–12, write in exponential form.

1. $\sqrt{123}$ *Answer:* _____
2. $\sqrt[3]{56}$ *Answer:* _____
3. $\sqrt[6]{216}$ *Answer:* _____
4. $\sqrt[8]{895}$ *Answer:* _____
5. $\sqrt[10]{10,000}$ *Answer:* _____
6. $\sqrt{45}$ *Answer:* _____
7. $\sqrt[3]{39}$ *Answer:* _____
8. $\sqrt[4]{64}$ *Answer:* _____
9. $\sqrt[5]{75}$ *Answer:* _____
10. $\sqrt[7]{113}$ *Answer:* _____
11. $\sqrt[9]{81}$ *Answer:* _____
12. $\sqrt{94}$ *Answer:* _____

Answers to Mini-Practice:
1. (a) $36^{1/5}$ (b) $100^{1/7}$ 2. (a) $\sqrt{35}$ (b) 3

In Exercises 13–24, write in root notation.

13. $3^{1/2}$ *Answer:* _____   **14.** $7^{1/3}$ *Answer:* _____   **15.** $8^{1/3}$ *Answer:* _____

16. $10^{1/4}$ *Answer:* _____   **17.** $64^{1/6}$ *Answer:* _____   **18.** $25^{1/2}$ *Answer:* _____

19. $93^{1/3}$ *Answer:* _____   **20.** $11^{1/2}$ *Answer:* _____   **21.** $15^{1/4}$ *Answer:* _____

22. $100^{1/5}$ *Answer:* _____   **23.** $1000^{1/10}$ *Answer:* _____   **24.** $863^{1/8}$ *Answer:* _____

In Exercises 25–30, multiply.

25. $4^{1/2} \cdot 4^{1/2}$ *Answer:* _____   **26.** $5^{1/3} \cdot 5^{1/3} \cdot 5^{1/3}$ *Answer:* _____

27. $10^{1/5} \cdot 10^{1/5} \cdot 10^{1/5} \cdot 10^{1/5} \cdot 10^{1/5}$ *Answer:* _____   **28.** $2^{1/2} \cdot 2^{1/2} \cdot 2^{1/2} \cdot 2^{1/2}$ *Answer:* _____

29. $3^{1/6} \cdot 3^{1/6} \cdot 3^{1/6} \cdot 3^{1/6} \cdot 3^{1/6} \cdot 3^{1/6}$ *Answer:* _____   **30.** $9^{1/4} \cdot 9^{1/4}$ *Answer:* _____

2.19 EXPONENTS AND CALCULATORS

Many calculators have keys for exponents and roots. If you have an "x^y" key on yours, try the following: Press the following keys:

$$5 \quad x^y \quad 2 \quad =$$

What answer did you get? _____

It should have been 25 because $5^2 = 25$. Now try: $25^{1/2}$. On the calculator press

$$25 \quad x^y \quad 0.5 \quad =$$

This should give you an answer of 5 because $\sqrt{25} = 5$ and because 0.5 is the same as 1/2. Calculators cannot use fractional exponents—only whole numbers and decimals.

Now try:

$$25 \quad x^y \quad (\quad 1 \quad \div \quad 2 \quad) \quad =$$

The calculator is programmed to perform the operations inside parentheses first; it changes 1/2 to 0.5. The square-root key gives you the answer immediately if you press

$$25 \quad \sqrt{}$$

In all three cases we find that the square root of 25 is 5.

Example 1   Find $49^{1/2}$ with a calculator.

Solution. $49 \quad x^y \quad (\quad 1 \quad \div \quad 2 \quad) \quad =$

$49^{1/2}$ is equal to $\sqrt{49}$, which equals 7.

■

Example 2   Find $\sqrt[3]{64}$ with a calculator.

Solution. $\sqrt[3]{64}$ is the same as $64^{1/3}$. Press:

$$64 \quad x^y \quad (\quad 1 \quad \div \quad 3 \quad) \quad =$$

The cube root of 64 is 4.   ∎

MINI-PRACTICE
(Use this space for your work.)

Use the calculator to find the following:

(a) $\sqrt{36}$ (b) $\sqrt[3]{27}$ (c) $64^{1/3}$ (d) $81^{1/2}$ (e) $81^{1/4}$

Answers: (a) _____ (b) _____ (c) _____ (d) _____ (e) _____

IN YOUR OWN WORDS

explain, when using a calculator, how to find the values of numbers in exponential notation, where the exponent is a fraction of the form $1/N$. _____

_____

_____

Answer:

You must change the fraction to a decimal or put parentheses around the fraction. The x^y key is used to find the value.

Practice Exercises

In Exercises 1–20, use a calculator to find the following roots. Express your answers to one decimal place.

1. $\sqrt{35}$ *Answer:* _____
2. $\sqrt{187}$ *Answer:* _____
3. $\sqrt{3590}$ *Answer:* _____
4. $\sqrt[3]{0.008}$ *Answer:* _____
5. $\sqrt[5]{32}$ *Answer:* _____
6. $\sqrt{125}$ *Answer:* _____
7. $\sqrt{390}$ *Answer:* _____
8. $\sqrt{8649}$ *Answer:* _____
9. $\sqrt[3]{19,683}$ *Answer:* _____
10. $\sqrt[3]{450}$ *Answer:* _____
11. $\sqrt[4]{99}$ *Answer:* _____
12. $\sqrt[4]{50,625}$ *Answer:* _____
13. $\sqrt[5]{1025}$ *Answer:* _____
14. $\sqrt[5]{6900}$ *Answer:* _____
15. $\sqrt[6]{4562}$ *Answer:* _____
16. $\sqrt[6]{10,928}$ *Answer:* _____
17. $\sqrt[7]{2187}$ *Answer:* _____
18. $\sqrt[7]{1100}$ *Answer:* _____
19. $\sqrt[8]{256}$ *Answer:* _____
20. $\sqrt[8]{4592}$ *Answer:* _____

In Exercises 21–40, use a calculator to find the following values. Express your answers to one decimal place.

21. $45^{1/2}$ *Answer:* _____
22. $1331^{1/3}$ *Answer:* _____
23. $7776^{1/5}$ *Answer:* _____
24. $1024^{1/10}$ *Answer:* _____
25. $0.0016^{1/4}$ *Answer:* _____
26. $1521^{1/2}$ *Answer:* _____
27. $494^{1/8}$ *Answer:* _____
28. $1024^{1/5}$ *Answer:* _____
29. $3933^{1/3}$ *Answer:* _____
30. $625^{1/4}$ *Answer:* _____
31. $7200^{1/5}$ *Answer:* _____
32. $16,384^{1/7}$ *Answer:* _____

Answers to Mini-Practice:

(a) 6   (b) 3   (c) 4   (d) 9   (e) 3

33. $625^{1/2}$ *Answer:* _____

34. $999,999^{1/6}$ *Answer:* _____

35. $729^{1/6}$ *Answer:* _____

36. $1,000,000^{1/6}$ *Answer:* _____

37. $1331^{1/3}$ *Answer:* _____

38. $68^{1/2}$ *Answer:* _____

39. $192^{1/4}$ *Answer:* _____

40. $83^{1/5}$ *Answer:* _____

SUMMARY

1. A whole-number exponent of a number indicates how many factors there are of that number. Thus,

$$2^5 \text{ is the same as } \underbrace{2 \times 2 \times 2 \times 2 \times 2}_{5 \text{ factors}}$$

2. When the exponent is a whole number, the base is the number that is multiplied. Thus,

$$\text{in } 2^5, 2 \text{ is the base}$$

3. To find the product in exponential form of two or more numbers having the same base, keep the common base and add the exponents.

$$5^4 \cdot 5^7 = 5^{4+7} = 5^{11}$$

4. To simplify the power of a product, raise each factor of the product to the given exponent.

$$(2 \cdot 3)^5 = 2^5 \cdot 3^5$$

5. To simplify the power of a power, multiply the exponents.

$$(2^3)^4 = 2^{3 \cdot 4} = 2^{12}$$

6. To divide two numbers with the same base, keep the base and subtract the exponents.

$$\frac{5^6}{5^2} = 5^{6-2} = 5^4$$

7. A number raised to a negative exponent is the same as the reciprocal of that number raised to the opposite of the exponent.

$$5^{-2} = \frac{1}{5^2}$$

8. The square of a number is the product of the number multiplied by itself.

$$8^2 = 8 \times 8$$

9. A square root of a number is one of two equal factors of the number.

$$\sqrt{64} = 8 \quad \text{and} \quad -\sqrt{64} = -8$$

10. The cube of a number is the product of the number used as a factor three different times.

$$4^3 = 4 \times 4 \times 4 = 64$$

11. The cube root of a number is one of three equal factors of the number.

$$\sqrt[3]{64} = \sqrt[3]{4(4)(4)} = \sqrt[3]{4} \times \sqrt[3]{4} \times \sqrt[3]{4} = 4$$

12. If x represents any number other than zero, then $x^0 = 1$.

13. A number raised to a fractional exponent, $1/N$, can be written as a root in which the denominator, N, of the fraction is the index of the root.

$$5^{1/3} = \sqrt[3]{5}$$

Review Practice Exercises

In Exercises 1–10, identify the base and the exponent. (In some exercises there may be other operations involved.)

1. 5^2 Answer: _____   **2.** $(-3)^4$ Answer: _____   **3.** $5 \cdot 2^3$ Answer: _____

4. $(5.2)^3$ Answer: _____   **5.** -3.7^3 Answer: _____   **6.** $(-3.7)^3$ Answer: _____

7. $\dfrac{3^2}{4}$ Answer: _____   **8.** $\left(\dfrac{3}{4}\right)^5$ Answer: _____   **9.** $\left(-\dfrac{5}{8}\right)^3$ Answer: _____

10. $\dfrac{(-5)^3}{8}$ Answer: _____

In Exercises 11–17, write each expression using exponents.

11. $3 \cdot 3 \cdot 3 \cdot 3 \cdot 3$ Answer: _____   **12.** $4 \cdot 4 \cdot 4$ Answer: _____

13. $5 \cdot 5 \cdot 5 \cdot 5 \cdot 5 \cdot 5 \cdot 5$ Answer: _____   **14.** $(5.1)(5.1)(5.1)$ Answer: _____

15. $(0.05)(0.05)(0.05)(0.05)$ Answer: _____   **16.** $\left(\dfrac{3}{8}\right)\left(\dfrac{3}{8}\right)\left(\dfrac{3}{8}\right)$ Answer: _____

17. $\left(-\dfrac{1}{2}\right)\left(-\dfrac{1}{2}\right)\left(-\dfrac{1}{2}\right)\left(-\dfrac{1}{2}\right)\left(-\dfrac{1}{2}\right)$ Answer: _____

In Exercises 18–33, evaluate.

18. 2^6 Answer: _____   **19.** 7^1 Answer: _____   **20.** 5^3 Answer: _____

21. 6^2 Answer: _____   **22.** $\left(\dfrac{3}{4}\right)^2$ Answer: _____   **23.** $\left(\dfrac{2}{3}\right)^3$ Answer: _____

24. $3^2 + 3^4$ Answer: _____   **25.** $2^8 - 2^6$ Answer: _____   **26.** $4^2 + 4^3$ Answer: _____

27. $4^0 + 5^0$ Answer: _____   **28.** $2^3 \cdot 3^2$ Answer: _____   **29.** $(0.01)^2$ Answer: _____

30. $(2.5)^2$ Answer: _____   **31.** $\left(\dfrac{3}{4}\right)^3$ Answer: _____   **32.** $\left(\dfrac{-5}{8}\right)^2$ Answer: _____

33. $\left(\dfrac{1}{2}\right)^2 + \left(\dfrac{1}{2}\right)^3$ Answer: _____

In Exercises 34–42, multiply. Leave your answer in exponential form.

34. $4^3 \cdot 4^2$ Answer: _____   35. $2^3 \cdot 2^2$ Answer: _____   36. $5^3 \cdot 5^4$ Answer: _____

37. $3^4 \cdot 3^2$ Answer: _____   38. $4 \cdot 4^5$ Answer: _____   39. $(3.1)^2 (3.1)^3$ Answer: _____

40. $(0.05)^5 (0.05)^{-3}$ Answer: _____   41. $\left(\dfrac{3}{8}\right)^2 \left(\dfrac{3}{8}\right)^5$ Answer: _____

42. $\left(\dfrac{6}{7}\right)^{-3} \left(\dfrac{6}{7}\right)^9$ Answer: _____

43. Write $(3^2)^4$ as a power of 3.   Answer: _____   44. Write $(5^3)^4$ as a power of 5.   Answer: _____

45. Write $(4^2)^3$ as a power of 4.   Answer: _____   46. Write $(2^5)^3$ as a power of 2.   Answer: _____

47. Write $(6^3)^2$ as a power of 6.   Answer: _____

In Exercises 48–52, raise each product to the indicated power. Leave the answer in exponential form.

48. $(2 \cdot 7)^2$ Answer: _____   49. $(5 \cdot 2)^3$ Answer: _____   50. $(5 \cdot 3)^0$ Answer: _____

51. $(3 \cdot 5^2 \cdot 7^4)^2$ Answer: _____   52. $(2 \cdot 3^3 \cdot 5^6)^3$ Answer: _____

In Exercises 53–63, evaluate.

53. $(-10)^4$ Answer: _____   54. $-(-4)^2$ Answer: _____   55. $(-3)^3$ Answer: _____

56. $-(-5)^3$ Answer: _____   57. $(-2 \cdot 3^2)^2$ Answer: _____   58. $(-3.5)^2$ Answer: _____

59. $-(-5.01)^2$ Answer: _____   60. $(-0.01)^3$ Answer: _____   61. $\left(-\dfrac{3}{8}\right)^3$ Answer: _____

62. $-\left(-\dfrac{5}{7}\right)^2$ Answer: _____   63. $\dfrac{(-3)^2}{5}$ Answer: _____

In Exercises 64–72, divide. Leave the answer in exponential form.

64. $\dfrac{2^7}{2^4}$ Answer: _____   65. $\dfrac{8^9}{8^3}$ Answer: _____   66. $\dfrac{8^3}{8^9}$ Answer: _____

67. $\dfrac{5^{10}}{5^4}$ Answer: _____   68. $\dfrac{5^7}{5^7}$ Answer: _____   69. $\dfrac{(5.3)^7}{(5.3)^3}$ Answer: _____

70. $\dfrac{(6.8)^{-2}}{(6.8)^5}$ Answer: _____   71. $\dfrac{\left(\dfrac{7}{5}\right)^3}{\left(\dfrac{7}{5}\right)}$ Answer: _____   72. $\dfrac{\left(\dfrac{9}{4}\right)^5}{\left(\dfrac{9}{4}\right)^2}$ Answer: _____

In Exercises 73–80, first simplify; then evaluate.

73. $\dfrac{5^6 \cdot 5^4}{5^7}$ Answer: _____   74. $\dfrac{3^3 \cdot 3^5}{3^2 \cdot 3^6}$ Answer: _____

75. $\dfrac{(3 + 1)^2 (7 - 3)^3}{4^2}$ Answer: _____   76. $\dfrac{(4 + 2)^7 (8 - 2)^5}{(6^2)^5}$ Answer: _____

77. $\dfrac{(5.3)^2 (0.01)^2}{(5.3)(0.01)}$ Answer: _____   78. $\dfrac{(0.25)^3 (0.25)^4}{(0.25)^5}$ Answer: _____

79. $\dfrac{\left(\frac{2}{5}\right)^2 \left(\frac{2}{5}\right)^3}{\left(\frac{2}{5}\right)^3}$ Answer: _____

80. $\dfrac{\left(\frac{1}{2}\right)^3 \left(\frac{1}{2}\right)^5}{\left(\frac{1}{2}\right)^2}$ Answer: _____

In Exercises 81–84, find the square roots.

81. $\sqrt{441}$ Answer: _____

82. $\sqrt{225}$ Answer: _____

83. $\sqrt{0.04}$ Answer: _____

84. $\sqrt{\dfrac{25}{36}}$ Answer: _____

85. Find the cube root of 125.   Answer: _____

86. Find the cube root of 0.027.   Answer: _____

In Exercises 87 and 88, use a calculator to find the following.

87. $\sqrt[3]{0.008}$ Answer: _____

88. $\sqrt[4]{28,561}$ Answer: _____

If you have questions about this chapter, write them here. _____

_____

_____

_____

_____

_____

3 Order of Operations

Remember to

  ✱ *REVIEW FRACTIONS AND DECIMALS IN THE APPENDIX*
  ✱ *MEMORIZE THE ORDER OF OPERATIONS IN THE SUMMARY*
  ✱ *ANSWER ALL QUESTIONS IN WRITING*
  ✱ *DO ALL THE PRACTICE EXERCISES*

Now you will learn about the order to follow when you find several operations and grouping symbols in one problem. In the introductory chapter you saw that braces, { }, brackets, [ ], and parentheses, ( ), serve as grouping symbols. You are also familiar with the four basic operations as well as with exponential notation.

3.1 GROUPING SYMBOLS

Here is a list of things that are happening at a supermarket. They are not in the correct order. Rearrange the sentences so that they are correct.

1. Mary hands the cashier a $5 bill.
2. The cashier subtracts the total of $1.05 + $0.60 + $0.90 from $5.
3. The cashier adds all of Mary's purchases: $1.05 + $0.60 + $0.90.
4. Mary buys a loaf of bread for $1.05, butter for $0.60 and a carton of milk for $0.90.
5. Mary enters the supermarket.
6. The cashier hands Mary the change.

   The correct order of these sentences is:

   ——, ——, ——, ——, ——, ——

We need a symbol to show that we want to add *all* of Mary's purchases before subtracting.

Which symbol could we use? _____

Let's use parentheses to group all the purchases together. Then we could represent Step 2 of the supermarket transactions in the following way:

$$5 - (1.05 + 0.60 + 0.90)$$

As an example of the use of several grouping symbols, consider

$$\{2 + [5 - (3 + 1)]\} + 4$$

Perform the operation inside the parentheses, the *innermost* grouping symbol, first: _____

Next, simplify inside the brackets: _____

Now simplify inside the braces, the *outermost* grouping symbol:

_____

You should now have

$$3 + 4 = 7$$

Example 1   Evaluate $10 - [2 + (65 - 59)]$.

Solution. $10 - [2 + (65 - 59)]$ Do inside ( ) first.
$10 - [2 + (6)]$ Drop ( ); add 2 and 6
$10 - [8] = 2$ Drop [ ]; subtract.   ■

Example 2   Evaluate $14 - \{22 - [10 + (14 - 8)]\}$

Solution. $14 - \{22 - [10 + (14 - 8)]\}$ Do ( ) first.
$14 - \{22 - [10 + (6)]\}$ Do [ ] next.
$14 - \{22 - [16]\}$ Do { } next.
$14 - \{6\} = 8$ ■

MINI-PRACTICE   Evaluate:
(Use this space for your work.)

(a) $16 - \{20 - [8 + (15 - 12)]\}$ *Answer:* _____

(b) $23 - \{27 - [15 + (22 - 14)]\}$ *Answer:* _____

(c) $10 - \{14 - [2 + (9 - 6)]\}$ *Answer:* _____

(d) $4 + [3 + (-5 + 4)]$ *Answer:* _____

(e) $15 - \{-4 + [-2 - (3 + 4)] - 5\}$ *Answer:* _____

Answers to Mini-Practice:

(a) 7   (b) 19   (c) 1   (d) 6   (e) 33

Why Study Math?

1. **Math is useful in many careers.**
   - You don't have to be a scientist or engineer to need math.

2. **Here are some fields that use the math in this book.**
   - nursing
   - psychology
   - business
   - teaching
   - sociology
   - economics
   - dental technology
   - anthropology
   - computer science
   - health sciences
   - drafting
   - law
   - machine technology
   - architecture
   - medicine
   - dentistry

3. **Math helps you to think.**
   - You learn simple, direct ways to solve problems.
   - These problems arise in the above fields.
   - You learn to think logically.

4. **Who cares if you take math?**
   - Your math skills will be valued by employers.
   - They want to hire people who can cope with difficult material.

IN YOUR OWN WORDS explain the rule we follow when we have a problem involving more than one grouping symbol: _____

_____

_____

Answer:

To simplify a problem with several grouping symbols, we simplify inside the innermost parentheses first. We then work outward.

Practice Exercises

In Exercises 1–30, evaluate.

1. $14 - [3 + (8 - 4)]$ *Answer:* _____

2. $13 - [5 + (9 - 7)]$ *Answer:* _____

3. $44 - [10 + (26 - 12)]$ *Answer:* _____

4. $14 - \{22 - [10 + (14 - 8)]\}$ *Answer:* _____

5. $16 - \{20 - [8 + (15 - 12)]\}$ *Answer:* _____

6. $23 - \{27 - [15 + (22 - 14)]\}$ *Answer:* _____

7. $10 - \{14 - [2 + (9 - 6)]\}$ *Answer:* _____

8. $11 - \{11 - [5 + (10 - 7)]\}$ *Answer:* _____

9. $5 + \{10 + [8 - (6 - 4)]\} + 4$ *Answer:* _____

10. $4 - [3 + (-10)]$ *Answer:* _____

11. $(4 - 2 - 1) - (3 + 5 - 2)$ *Answer:* _____

12. $[4 - (2 - 1)] - [3 - (5 - 2)]$ *Answer:* _____

13. $[16 + (-17)] + [(-8) + 10]$ *Answer:* _____

14. $42 - [(-30) - 65 - (-11)]$ *Answer:* _____

15. $-30 - [(-65) - (29 - 4)]$ *Answer:* _____

16. $[-18 - 49 - (-84)] - 27$ *Answer:* _____

17. $[32 - (-82)] - (65 - 7)$ *Answer:* _____

18. $-8 - [30 - (-12) - 7 - (-14)]$ *Answer:* _____

19. $[(4 - (-3) - 12) - (-7)] - 20$ *Answer:* _____

20. $-[25 + (-31) + 24 + 19]$ *Answer:* _____

21. $5.2 - 3.2 + [6.7 - 8.1]$ *Answer:* _____

22. $3.2[5 - (4.1 - 3.5)]$ *Answer:* _____

23. $7.4 - 2.3 + [7.1 - 6.8]$ *Answer:* _____

24. $(0.8 + 6.4) - 2.3 + [7.1 - 6.8]$ *Answer:* _____

25. $-3.09 - [4.6 - 273.3 + 1.27]$ *Answer:* _____

26. $4.2 - [1.8 + (3.7 - 7.4)]$ *Answer:* _____

27. $\dfrac{1}{2} - \dfrac{3}{2} + \left[\dfrac{8}{5} + \left(\dfrac{1}{5} - 2\right)\right]$ *Answer:* _____

28. $\dfrac{4}{3}\left[\dfrac{27}{40} - \left(\dfrac{5}{8} - \dfrac{6}{5}\right)\right]$ *Answer:* _____

29. $8 - 5 - \left[\dfrac{8}{3} - \dfrac{3}{2} + \dfrac{2}{3}\right]$ *Answer:* _____

30. $\dfrac{5}{12} - \dfrac{9}{16} + 2\dfrac{3}{4}$ *Answer:* _____

3.2 THE BASIC OPERATIONS

Simplify: $6 + 2 \times 5$

What is the answer? _____

If your answer was 40, you first did the addition, $6 + 2 = 8$, and then the multiplication, $8 \times 5 = 40$.

If your answer was 16, you first did the multiplication, $2 \times 5 = 10$, and then the addition, $6 + 10 = 16$.

There is obviously only one correct answer and that is 16. *By convention*, mathematicians have agreed to do **multiplication before addition** in a problem such as this.

What is your answer to $48 \div 12 \times 4$? _____

Did you get 1 or 16? In this case 16 is again the correct answer. Mathematicians have decided to do **multiplication and division as they occur from left to right**.

When you work out a problem in arithmetic, if there are any parentheses, do all the operations inside each set of parentheses first. Then go from left to right until you have completed all operations.

Example 1   Evaluate $(2 + 5) - (1 + 4)$.

Solution. $(2 + 5) - (1 + 4) = 7 - 5$

Then start again; go from left to right:

$$7 - 5 = 2$$ ∎

Multiplication and division are performed before addition and subtraction. Multiplication and division must be done in the order in which they are encountered.

The last step is addition and subtraction. Here you have a choice: you can go from left to right and perform the operations as they occur. Alternatively, you can collect all numbers with a positive sign and add them, then collect all numbers with a negative sign and add them. Finally, you do one subtraction. Thus,

$$15 - 11 + 7 + 2 - 3 = (15 + 7 + 2) - (11 + 3) = 24 - 14 = 10$$

If people ignored these rules of order, confusion would result. For example, take the following situation: You have ordered by mail and must pay $4 postage as well as the price of the 5 books ordered, which are each $15. If you think: I have to pay

$$\$4 + 5 \times \$15 = \$4 + \$75 = \$79$$

you are right. That is,

postage plus five times the price of the book

In mathematical symbols the numerical expression would be

$$4 + 5 \times 15$$

But in this expression it is tempting to add 4 and 5 and then multiply 9 by 15. That would give you 135, which, of course, when interpreted as dollars, would be too much money to pay.

Example 2   Evaluate $9 - 3(2)$.

Solution.   Two operations: subtraction and multiplication

According to the rules, multiplication is performed first:

$$3(2) = 6$$

The problem is now seen to be:

$$9 - 6 = 3$$ ■

Example 3   Evaluate $9 \div 3(2)$.

Solution.   Two operations: division and multiplication

Going from left to right, perform division first and then multiplication:

$$9 \div 3(2)$$
$$\uparrow$$
$$3(2) = 6$$ ■

MINI-PRACTICE   Evaluate:
(Use this space for your work.)

(a) $12 + 5(4)$ (b) $12 \div (4)(3)$ (c) $12 \div (3)(4)$

(d) $12 \div [3(4)]$ (e) $-12 + 5(2 + 2)$

Answers: (a) _____   (b) _____   (c) _____

(d) _____   (e) _____

IN YOUR OWN WORDS   explain the rules we follow when we have problems involving parentheses, multiplication, division, addition, and subtraction: _____

_____

_____

Answer:

The rule states that operations are done in the following order:

1. Operations within parentheses
2. Multiplication/division from left to right (whichever comes first)
3. Addition and subtraction

WATCH OUT!   When are we allowed to drop parentheses?

Look at the following examples:

(a) $3 + (7 - 2)$
$3 + (5)$

Can these parentheses be dropped? _____

(b) $3 \cdot (7 - 2)$
$3(5)$

Can these parentheses be dropped? _____

Why can the parentheses be dropped in (a), but *not* in (b)? _____

_____

_____

Answer:

(a) $3 + (7 - 2) = 3 + (5)$
                                      $\lfloor_5\rfloor$

Here, $3 + (5)$ means the same as $3 + 5$, so the parentheses around 5 can be dropped. Therefore,

$$3 + (7 + 2) = 3 + 5 = 8$$

Without parentheses in the given example, we would perform addition and subtraction from left to right, obtaining the same final result:

$$3 + 7 - 2 = 10 - 2 = 8$$
$$\lfloor_{10}\rfloor$$

We see that parentheses could have been dropped.

Answers to Mini-Practice:

(a) 32   (b) 9   (c) 16   (d) 1   (e) 8

$$(b) \qquad\qquad 3 \cdot (7 - 2) = 3(5) = 15$$
$$\underset{5}{\underline{\quad}}$$

Here, the parentheses around the 5 is necessary to indicate that the operation is multiplication. *Without* parentheses in the given example, multiplication would be performed *before* subtraction. Thus,

$$3 \cdot 7 - 2 = 21 - 2 = 19$$
$$\underset{21}{\underline{\quad}}$$

and *not* 15, as before. Here parentheses *cannot* be dropped if we are to get the desired result.

Practice Exercises

In Exercises 1–36, evaluate.

1. $12 - 5(4)$ Answer: _____
2. $-12 + 5(-4)$ Answer: _____

3. $-12 - 5(-4)$ Answer: _____
4. $12 - 5(-4)$ Answer: _____

5. $-5 + 3(9 - 15)$ Answer: _____
6. $7 + 2(-9 - 2)$ Answer: _____

7. $7 + 3 \cdot 4$ Answer: _____
8. $15 - 5 \cdot 3$ Answer: _____

9. $6 + 4(3 + 2)$ Answer: _____
10. $6 + (4 \cdot 3 + 2)$ Answer: _____

11. $30 - 5(4 - 2)$ Answer: _____
12. $30 - (5 \cdot 4) - 2$ Answer: _____

13. $9 - 5(4 - 2)$ Answer: _____
14. $9 - (5 \cdot 4 - 2)$ Answer: _____

15. $7 - 3(1 - 5)$ Answer: _____
16. $24 - 6(-4)$ Answer: _____

17. $18 \div 6 \div 3$ Answer: _____
18. $18 \div 6 \cdot 3$ Answer: _____

19. $18 + [6(3)]$ Answer: _____

20. $3(4 + 5) \div [1 - 2(2)] \div [2(4) - (1 + 2(2))]$ Answer: _____

21. $12 \cdot \dfrac{16}{-4}$ Answer: _____
22. $1 + (-5) \cdot 2$ Answer: _____

23. $-2[5 - 2(4 - 8)]$ Answer: _____
24. $6 - 7[2 - 3(4 - 5)]$ Answer: _____

25. $5 - 2[3 - (4 + 6)]$ Answer: _____

26. $2(5 + 2)^2 - [(4 - 2 \cdot 3^2) \div (6 - 2^2)]$ Answer: _____

27. $0.4 - 4(0.2)$ Answer: _____
28. $7.3 - 5.2(1.1)$ Answer: _____

29. $-9.2(5.7 - 3.7)$ Answer: _____
30. $6.16 \div 2.8(1.2 - 3.2)$ Answer: _____

31. $(3.5 - 2.5) \div 0.25$ Answer: _____
32. $\dfrac{1}{2} - 4\left(\dfrac{3}{4}\right)$ Answer: _____

33. $\dfrac{3}{7} + \dfrac{4}{7}\left(\dfrac{7}{8} - \dfrac{5}{8}\right)$ Answer: _____
34. $\dfrac{-9}{2} \div 5\left(\dfrac{3}{10}\right)$ Answer: _____

35. $\left[\dfrac{8}{7} \div \dfrac{2}{7}\right]\left[6\left(\dfrac{3}{2}\right)\right]$ Answer: _____
36. $-\dfrac{7}{5} - \dfrac{9}{10} + \dfrac{3}{20}\left(\dfrac{5}{9}\right)$ Answer: _____

3.3 FRACTION BARS AS GROUPING SYMBOLS

What is the answer to $\dfrac{4+8}{2}$? _____

This expression could be rewritten as $(4 + 8) \div 2$, which equals $12 \div 2$, or 6.

A common mistake is to divide just 4 by 2 and get $2 + 8 = 10$. But that is wrong because the entire sum $4 + 8$ must be divided by 2. To avoid this kind of mistake, we add first and then we simplify.

An expression such as

$$\dfrac{4\,(8)}{2}$$

is the same as $4(8) \div 2$ and therefore, according to the rules, equals $32 \div 2 = 16$. However, here we could reduce by dividing 4 by 2, obtaining $2(8) = 16$, or by dividing 8 by 2, obtaining $4(4) = 16$.

Example 1   Evaluate $\dfrac{20+5}{5}$.

Solution. $\dfrac{20+5}{5} = (20 + 5) \div 5 = (25) \div 5 = 5$ ■

Example 2   Evaluate $\dfrac{20}{5} + 5$.

Solution. $\dfrac{20}{5} + 5 = (20 \div 5) + 5 = 4 + 5 = 9$

This could also be solved by addition of fractions.

$$5 = \dfrac{5}{1}$$

$$\dfrac{20}{5} + \dfrac{5}{1} = \dfrac{20+25}{5} = \dfrac{45}{5} = 9$$

Here, the "least common denominator" equals 5.   ■

MINI-PRACTICE   Evaluate.
(Use this space for your work.)

(a) $\dfrac{21 + 6}{9}$ (b) $\dfrac{20 + 7}{3}$ (c) $\dfrac{30 + 10}{5}$

(d) $\dfrac{30}{5} + 10$ (e) $30 + \dfrac{10}{5}$

Answers: (a) _____   (b) _____   (c) _____

(d) _____   (e) _____

Answers to Mini-Practice:

(a) 3   (b) 9   (c) 8   (d) 16   (e) 32

IN YOUR OWN WORDS   explain how the fraction bar can be considered a grouping symbol: _____

_____

_____

Answer:

The fraction bar groups the numbers together in the same way parentheses do, and therefore serves as a grouping symbol.

Practice Exercises

In Exercises 1–26, evaluate.

1. $\dfrac{24 + 6}{10}$ *Answer:* _____

2. $\dfrac{31 + 8}{13}$ *Answer:* _____

3. $\dfrac{4 + 2}{3}$ *Answer:* _____

4. $5 + \dfrac{10}{5}$ *Answer:* _____

5. $\dfrac{16}{4} + 3$ *Answer:* _____

6. $22 - \dfrac{26}{-2}$ *Answer:* _____

7. $22 + \dfrac{26}{-2}$ *Answer:* _____

8. $22 - \dfrac{-26}{-2}$ *Answer:* _____

9. $\dfrac{22 - 26}{-2}$ *Answer:* _____

10. $\dfrac{-30 + 20}{5}$ *Answer:* _____

11. $\dfrac{-12 - 3 + 1}{7}$ *Answer:* _____

12. $\dfrac{-11 - 9 + 2}{6}$ *Answer:* _____

13. $\dfrac{-18 - (6 - 3)}{-3}$ *Answer:* _____

14. $\dfrac{-8 - 2(-4)}{-2}$ *Answer:* _____

15. $\dfrac{6(-3) - 8}{-9 - 4}$ *Answer:* _____

16. $5 - \dfrac{3 - 5}{-2}$ *Answer:* _____

17. $4 + \dfrac{8 - 6}{7}$ *Answer:* _____

18. $\dfrac{8 - 6}{4 - 7}$ *Answer:* _____

19. $6 - \dfrac{8 - 2(-3)}{-7}$ *Answer:* _____

20. $10 + \dfrac{4 - 3(-2)}{-2}$ *Answer:* _____

21. $\dfrac{-7.3 + 5.2}{2.1}$ *Answer:* _____

22. $-7.2 + \dfrac{5.2}{\cdot 2.0}$ *Answer:* _____

23. $\dfrac{-7.2}{2.0} + 5.2$ *Answer:* _____

24. $\dfrac{8}{.2} + 12$ *Answer:* _____

25. $\dfrac{10}{.25} + \dfrac{5}{.1}$ *Answer:* _____

26. $\dfrac{-6}{1.5} + \dfrac{21.6}{-1.2}$ *Answer:* _____

3.4 POWERS

Simplify:

(a) $(3 + 5)^2$ _____   (b) $3 + 5^2$ _____

We apply operations within parentheses before applying powers, so that in (a),

$$(3 + 5)^2 = 8^2 = 64$$

We apply powers before addition, so that in (b),

$$3 + 5^2 = 3 + 25 = 28$$

Simplify:

(c) $(3 \cdot 5)^2$ _____   (d) $3 \cdot 5^2$ _____

We apply parentheses before powers, so that in (c),

$$(3 \cdot 5)^2 = (15)^2 = 225$$

We apply powers before multiplication, so that in (d),

$$3 \cdot 5^2 = 3 \cdot 25 = 75$$

Example 1   Simplify $(3 + 3^2) \div 2^2$.

Solution. Parentheses first: $3 + 3^2$ (Powers first)
$$\downarrow$$
$$3 + 9$$

$$= 12 \div 2^2 \quad \text{(Powers first)}$$
$$\downarrow$$
$$= 12 \div 4$$
$$= 3 \qquad \blacksquare$$

Example 2   Simplify $7(5 - 8)^2$.

Solution.   Parentheses first: $7(-3)^2$
Powers: $7(9) = 63 \qquad \blacksquare$

Example 3   Evaluate $10 + 2[5(2)^2 - (5 - 4)^2 + 3]$.

Solution.   The order of operations directs us to start inside the brackets:

$$5(2)^2 - (5 - 4)^2 + 3$$
$$\downarrow$$

Operations within parentheses first: $5(2)^2 - (1)^2 + 3$
$$\downarrow$$

Exponents: $5(4) - (1) + 3$
$$\downarrow$$

Multiplication: $20 - 1 + 3$
Addition/Subtraction: $19 + 3 = 22$

When we have reduced the inside of the brackets to *one* number, we continue:

$$10 + 2[22] = 10 + 44 = 54$$
$$\downarrow \qquad\qquad \downarrow$$
multiplication   addition $\blacksquare$

MINI-PRACTICE   Evaluate.
(Use this space for your work.)

(a) $5 - 2^2$ (b) $(5 - 2)^2$ (c) $(7 - 3^2)^2$

(d) $3^2(2 \cdot 3^2)^3$ (e) $10(6 - 4)^2$

Answers: (a) _____   (b) _____   (c) _____

(d) _____   (e) _____

IN YOUR OWN WORDS   explain where exponents come in the "order of operations": _____

_____

_____

Answer:

All powers are evaluated *after* operations within parentheses, brackets, and braces, but *before* multiplication/division. The order is:

1. Operations within parentheses
2. Powers
3. Multiplication/division from left to right
4. Addition/subtraction

Practice Exercises

In Exercises 1–34, evaluate.

1. $56 \div (-3 - 5^2)$ *Answer:* _____

2. $-96 \div [-5 + (-3)^3]$ *Answer:* _____

3. $5(-3)^2$ *Answer:* _____

4. $4^2 - 3(-2)$ *Answer:* _____

5. $10^2 - (-9)^2$ *Answer:* _____

6. $2(-4)^2 + 3(-6)$ *Answer:* _____

7. $5^2 - 2(-1)^2$ *Answer:* _____

8. $-5(4^2 - 1)$ *Answer:* _____

9. $-4(5) - (-4)^2$ *Answer:* _____

10. $(-3)(-2) + (-5)^2$ *Answer:* _____

11. $4[(2 - 1) - 3]$ *Answer:* _____

12. $(3 - 2)^2 + (1 - 3)^2$ *Answer:* _____

13. $\left(\dfrac{6-4}{2}\right)^2 - \dfrac{9-1}{2^2}$ *Answer:* _____

14. $\dfrac{12 - 2^2}{2^2} + 3$ *Answer:* _____

15. $\dfrac{(6-1)^2}{6-1}$ *Answer:* _____

16. $[4 - (3 - 2)^2]^2$ *Answer:* _____

17. $\dfrac{10^2 - (1-2)^2}{3^2}$ *Answer:* _____

18. $\dfrac{(12 - 4)^2}{(4 - 2)^2}$ *Answer:* _____

19. $\left(3 - \dfrac{15}{5}\right)^2$ *Answer:* _____

20. $3 - \left(\dfrac{15}{5}\right)^2$ *Answer:* _____

Answers to Mini-Practice:

(a) 1   (b) 9   (c) 4   (d) 52,488   (e) 40

21. $(7.2)(-0.1)^2 + (-2.1)^2$ *Answer:* _____   22. $-9.2(2^3 - 7.7)$ *Answer:* _____

23. $-46.8 \div [-7.6 - (-2)^3]$ *Answer:* _____   24. $6.5 - 3^2(8.2 - 9.2)^4$ *Answer:* _____

25. $[(0.2)^2 - (0.3)^3] \div [6.5 - (2.5)^2]$ *Answer:* _____   26. $(9.3 \div 18.6)^2 \div 5((3.1)^2 - 8.61)$ *Answer:* _____

27. $\dfrac{3}{2} - \dfrac{5}{2}\left[\left(\dfrac{1}{2}\right)^2 \div \left(\dfrac{1}{2}\right)^3\right]$ *Answer:* _____   28. $-\dfrac{5}{4}\left[\left(\dfrac{3}{2}\right)^2 - \dfrac{7}{4}\right]$ *Answer:* _____

29. $-\dfrac{7}{5} \div \left[-\dfrac{6}{5} + \left(\dfrac{1}{5}\right)^2\right]$ *Answer:* _____   30. $\dfrac{8}{3} - \left(\dfrac{1}{3}\right)^2\left(\dfrac{5}{2} - \dfrac{3}{2}\right)^4$ *Answer:* _____

31. $5 + 4(3)^2 - (6 - 4)^2 + 5$ *Answer:* _____   32. $20 - (7 - 4)^2 + 5(4 + 3)^2$ *Answer:* _____

33. $10[6 - (5 - 3)^2 + 2(8 - 4)^2]$ *Answer:* _____   34. $(4 + 6)^2 - 12(5 - 2)^2 + 2(9 - 4)^2$ *Answer:* _____

SUMMARY

1. Apply all operations within parentheses.
2. Apply all exponents.
3. Multiply and/or divide from left to right.
4. Add and/or subtract in any order.

Review Practice Exercises

In Exercises 1–62, evaluate.

1. $6 \div 2 + 4$ *Answer:* _____   2. $9 \div 3 + 6$ *Answer:* _____   3. $6 \div (2 + 4)$ *Answer:* _____

4. $9 \div (3 + 6)$ *Answer:* _____   5. $7 + 3 \cdot 4$ *Answer:* _____   6. $4 - 3^2 + 6 \cdot 5$ *Answer:* _____

7. $6^2 - 4^2 + 15 - 2$ *Answer:* _____   8. $3^2 - 5 + 2^4 \div 4$ *Answer:* _____

9. $3 + 44 \div 11 - 6$ *Answer:* _____   10. $2 \cdot 5^2$ *Answer:* _____

11. $5(3)^2$ *Answer:* _____   12. $(2 + 5)^2 - 1$ *Answer:* _____   13. $2 + 3 \cdot 7$ *Answer:* _____

14. $4 \cdot 3 + 2$ *Answer:* _____   15. $2 \cdot 4 + 3 \cdot 5$ *Answer:* _____   16. $17 - 5(5 - 3)$ *Answer:* _____

17. $16 \div 8 \div 2$ *Answer:* _____   18. $10 \div 2(5)$ *Answer:* _____   19. $2|3 - 7|$ *Answer:* _____

20. $\dfrac{6 + 2 \cdot 4}{2 + 5}$ *Answer:* _____   21. $\dfrac{10 - 5(-2)}{10(-2)}$ *Answer:* _____   22. $\dfrac{10 + 5(-4)}{1 + 3^2}$ *Answer:* _____

23. $\{144 - [3(2 - 1)^3]\} + 1$ *Answer:* _____

24. $8 + 4[3 + (7 - 5)^2(4 - 3)^3 + 2]$ *Answer:* _____

25. $[7 + 5(3 - 1)^2][(20 - 10) - (30 - 20)]$ *Answer:* _____

26. $[3 - (3 - 2)^4]^2 - [5 + 2(3 + 7)^2]$ *Answer:* _____

27. $10 - 2[5(2)^2 - (4 - 5) + 3]$ *Answer:* _____

28. $10 - 4[3 - (-5 + 7)^2(4 - 3)^3 + 2]$ *Answer:* _____

29. $7 - 3[4 - (7 - 6)^2]^3$ *Answer:* _____

30. $10 - 3 \cdot 5[2 \cdot 5^2 - (8 - 4)]$ *Answer:* _____

31. $8 - 4[2 \cdot 3^2 - (4 - 1)]$ *Answer:* _____

32. $7 + 3[4 + (7 - 6)^2]^3$ *Answer:* _____

33. $(3 + 1)^2[7 - 3(2 - 2)^8]$ *Answer:* _____

34. $[(10 - 9)^2 + (3 - 2 - 1)^9] + [5 + (7 - 6)^2 + (5 - 4)^2]$ *Answer:* _____

35. $\{4 + [100 - (20 - 10)^2]\}[8 - (4 - 2)^3 + 1]$ *Answer:* _____

36. $\dfrac{10 - [5(-3)]}{10(-3)}$ *Answer:* _____

37. $2 - 3[3 - (2 + 3)]$ *Answer:* _____

38. $4 - 2[4 - (3 - 5)]$ *Answer:* _____

39. $6 - [24 \div (3 \cdot 4)]$ *Answer:* _____

40. $[72 \div (-9)] \div 2$ *Answer:* _____

41. $\dfrac{[-9(-1)] - [5(3)]}{-5 - 3}$ *Answer:* _____

42. $\dfrac{[-3(7)] - 3}{-4 - 2}$ *Answer:* _____

43. $(-96) \div (12) \div (-4)$ *Answer:* _____

44. $-8 \div 2 \cdot 4$ *Answer:* _____

45. $2 + (-3)(5)$ *Answer:* _____

46. $15 - 3\{2[-3(2 - 5)] - (-4)\}$ *Answer:* _____

47. $6 - 2[4(2 - 6) - (3 - 6)]$ *Answer:* _____

48. $11 - 6[(-1)(-1)(-1) + (-2)(-2)(-3)] + (-11)$ *Answer:* _____

49. $-6(2 - 5) - (-6) \cdot (5 - 8)$ *Answer:* _____

50. $-2 + (-6 - 3) \div (-3) + (-3)[-2(-3 - 4)] - (-7)(-2)$ *Answer:* _____

51. $\dfrac{7.8 + 2.4}{5.6 - 0.5}$ *Answer:* _____

52. $6.2 \div 2.0 \div (-0.5)$ *Answer:* _____

53. $8.6|3.1 - 5.1|$ *Answer:* _____

54. $\left(\dfrac{3}{4} - \dfrac{1}{4}\right)^2 - \dfrac{9}{2}\left(\dfrac{7}{3} - \dfrac{10}{3}\right)^2$ *Answer:* _____

55. $\dfrac{6}{5} - \dfrac{9}{5}\left(\dfrac{3}{2} - \dfrac{1}{2}\right)^2 + \dfrac{1}{4}$ *Answer:* _____

56. $-4 \cdot \left(\dfrac{1}{2}\right)^2$ *Answer:* _____

57. $\dfrac{-7 - 3}{-2(-5)}$ *Answer:* _____

58. $\dfrac{-4 - 15 - 2}{1 - 4}$ *Answer:* _____

59. $\dfrac{-4(-2)(-8)}{-4(2 - 8)}$ *Answer:* _____

60. $\dfrac{-3(2) - 6}{3(-2) - 6}$ *Answer:* _____

61. $\dfrac{6 - 4(3 - 1)}{-2(-3) - 4}$ *Answer:* _____

62. $\dfrac{7 + 2(5 - 10)}{-3(-3) - 6}$ *Answer:* _____

If you have questions about this chapter, write them here: _____

_____

_____

_____

_____

_____

4 Translations

Remember to

* *READ SLOWLY AND CAREFULLY*
* *LEARN THE VOCABULARY*
* *PAY ATTENTION TO ORDER IN TRANSLATIONS*
* *KNOW WHEN TO USE PARENTHESES WHEN MORE THAN ONE OPERATION IS INVOLVED*

Now you will practice translations of the basic operations from English to mathematics and from mathematics to English.

4.1 TRANSLATIONS FROM ENGLISH INTO MATHEMATICS

Operation:
Addition
Translate the following English expressions into mathematics using numbers and addition symbols. Do not give the answers.

English	Mathematics
Three increased by two	_____
Two more than three	_____
Three plus two	_____
Add two to three.	_____

You probably translated all of these English expressions as 3 + 2 or as 2 + 3.

Does it matter if we write 3 + 2 or 2 + 3? _____

109

How to Study Math

1. **Studying math is special.**
   - Read math slowly.
   - Let every word and symbol sink in.
   - Draw pictures that help you "see" the problem.

2. **When you read the text:**
   - Answer the questions in the space provided.

3. **Take frequent breaks.**
   - When you tire, let your mind rest.
   - Then go back to work refreshed.

4. **After you finish your homework:**
   - Think about what you have learned.
   - Reread the rules you have used.

In the introductory chapter, you were introduced to "commutative" operations. This means that the order in which the numbers are given doesn't effect the answer. Since addition is commutative,

3 + 2 has the same answer as 2 + 3

What is the *answer* to an addition problem called? _____

The answer is called the *sum*. This word can be used to translate an addition problem back into English. For example, 3 + 2 can be translated as

the sum of three and two

**Operation:
Subtraction**

John had $14; he spent $5. How much does he have left? _____

Write this in mathematics: _____

John took away $5 from his $14. How much does he have left? ____

Write this in mathematics: _____

Five dollars less than the fourteen dollars John has is how much?
Write the numbers involved in mathematical symbols: _____

You probably wrote 14 − 5 in all three cases.

Translate the following expressions into mathematics using numbers and subtraction symbols. Do not give the answers. Keep in mind that subtraction is a noncommutative operation! That is, the order in which the numbers are given effects the answer.

English	*Mathematics*
Take away five from fourteen.	_____
Five less than fourteen	_____
Fourteen decreased by five	_____
Subtract five from fourteen.	_____
Fourteen minus five	_____

The correct answer is $14 - 5$.

What do we call the *answer* to a subtraction problem? _____

The answer is called the *difference*. This word can be used to translate a subtraction problem back into English. For example, $14 - 5$ can be translated as

the difference between fourteen and five (*in this order*)

WATCH OUT!   Read the following statements carefully:

Five less than fourteen	**Five is less than fourteen.**

What is different in these two statements? _____

The second statement contains the word "is," whereas the first statement does not.

Translate each statement into a mathematical expression:

Five less than fourteen: _____

Five is less than fourteen: _____

As we have said before,

"five less than fourteen" is translated as $14 - 5$

whereas the statement

"five *is* less than fourteen" is translated as $5 < 14$

The word "is" has changed the meaning from an operation into a comparison statement. Remember that the comparison statements are

$$=, \quad <, \quad \text{and} \quad >$$

Go back to the introductory chapter if you have forgotten what they stand for.

Operation: Multiplication   Translate the following phrases into mathematics using numbers and multiplication symbols. Do not give the answers.

English	Mathematics
Multiply six by three.	_____
Six times three	_____

The answer in both cases is 6 × 3. There are other ways of writing 6 × 3. Maybe you gave the answer as 6 · 3 or 6(3). If you did, you are absolutely right.

What do we call the *answer* to a multiplication problem? _____

The answer is called the *product*. This word can be used to translate a multiplication problem back into English. For example, 6 × 3 can be translated as

the product of six and three

WATCH OUT!   *Dot Notation versus the Decimal Point*

Translate into words: 6 · 3 _____

Translate into words: 6.3 _____

Do not confuse dot notation for multiplication with the decimal point!

Explain the difference in the way the two symbols are written:

_____

_____

The dot for multiplication is always centered between the factors, whereas the decimal point is placed at the bottom of the numbers. For example,

2 · 3 *indicates* 2 multiplied by 3

2.3 *indicates* the decimal two and three-tenths

Operation: Division

Translate the following expressions using numbers and division symbols. Do not give the answers.

English	Mathematics
Four divided into eight	_____
Eight divided by four	_____
How many times does four go into eight?	_____

A commonly used division symbol is ÷. If you used this symbol, all your answers should be 8 ÷ 4. However, you may have used other symbols to represent the operation of division. For example, 8 ÷ 4 may also be written as

$$4 \overline{)8} \quad \text{or} \quad \frac{8}{4} \quad \text{or} \quad 8/4$$

The horizontal line − is called a **fraction bar** and / is called a **slash**. In some countries 8 ÷ 4 is written 8:4 or 8⌊4 .

WATCH OUT!   **By** *and* **Into**

Perform the following division examples:

(a) $\dfrac{20}{5}$ = _____   (b) $\dfrac{5}{20}$ = _____

Look at the following situations:

John has $20; he divides this sum equally among his 5 children. How much does each child receive? _____

Jack has $5 and he divides this sum equally among his 20 students. How much does each student receive? _____

In the first case each child receives $4. In the second case each student receives a quarter.

To get $4 you *divided* $ _____ by _____ .

To get a quarter you *divided* $ _____ by _____ .

Observe that $\dfrac{5}{20}$ reduces to $\dfrac{1}{4}$.

From the above example you can see that there is nothing wrong in dividing a number by a larger number. In such cases the answer will be a *proper* fraction, that is, a fraction in which the numerator is smaller than the denominator.

In English we say either *divide by* or *divide into*. The prepositions "by" and "into" determine which number is the numerator and which is the denominator when expressing division in terms of fractions.

$$20 \text{ divided by 5 is written } \frac{20}{5} \quad \left(\frac{20}{5} = 4 \right)$$

$$20 \text{ into 5 is written } \frac{5}{20} \quad \left(\frac{5}{20} = \frac{1}{4} \right)$$

What do we call the *answer* in a division problem? _____

The answer is called the *quotient*. This word can be used to translate a division problem back into English. For example, 8 ÷ 4 can be translated as

the quotient of eight divided by four

WATCH OUT!   *Division Using the Calculator*

When we use the calculator, all division problems must be rewritten in terms of "divided by." For example, 20 divided by 5 is keyed on the calculator as

$$20 \div 5 =$$

The problem "6 into 42" must be rewritten as "42 divided by 6"

$$42 \div 6 =$$

before you can do the division on the calculator.

Example 1   Translate into symbols:

Five less than ten

Solution.   The words *less than* give the message to reverse the order of the numbers:

$$10 - 5$$ ∎

Example 2   Translate into symbols:

Thirty-six into six

Solution.   The word *into* gives the message to reverse the order of the numbers:

$$6 \div 36$$ ∎

MINI-PRACTICE   Translate from English to mathematical symbols:
(Use this space for your work.)

(a) Ten increased by five   *Answer:* _____

(b) Four less than seven   *Answer:* _____

(c) Find the sum of three and four.   *Answer:* _____

(d) Find the difference of ten and six (in this order).   *Answer:* _____

(e) Find the product of four and six.   *Answer:* _____

(f) Find the quotient of four divided by eight   *Answer:* _____

(g) Seventeen is less than twenty.   *Answer:* _____

(h) Seventeen less than twenty.   *Answer:* _____

IN YOUR OWN WORDS   explain about the noncommutative operations. That means you have to make sure you follow the correct order when you translate. _____

_____

_____

_____

Answers to Mini-Practice:

(a) $10 + 5$ (b) $7 - 4$ (c) $3 + 4$ (d) $10 - 6$ (e) $4(6)$ (f) $4 + 8$ (g) $17 < 20$ (h) $20 - 17$

Answer:

Subtraction and division are the basic noncommutative operations. The words *less than* and *subtracted from* give a message to reverse the order in subtraction. In division the words *divide into* give the message to reverse the order.

WATCH OUT! *The Word* And *in Translations*

Does "and" in a math problem mean addition? _____

Not necessarily! The word "and" tells us that we have more than one number: We have a number *and* also another number. "And" does not in any way instruct us *what* to do with the numbers. Actually we can use "and" for any operation, as the following examples show:

(a) The sum of 3 and 4
(b) The product of 3 and 4
(c) The difference of 3 and 4

Practice Exercises

In Exercises 1–28, translate into mathematics. Do not evaluate.

1. The sum of three and two   *Answer:* _____
2. Fourteen minus five   *Answer:* _____
3. The product of six and three   *Answer:* _____
4. Four increased by five   *Answer:* _____
5. Add five to four.   *Answer:* _____
6. Five plus four   *Answer:* _____
7. Four more than five   *Answer:* _____
8. Decrease ten by four   *Answer:* _____
9. From seventeen take away nine.   *Answer:* _____
10. Subtract six from eight.   *Answer:* _____
11. Four less than seven   *Answer:* _____
12. Twelve minus nine   *Answer:* _____
13. Ten less than fifteen   *Answer:* _____
14. Ten is less than fifteen.   *Answer:* _____
15. The product of seven and four   *Answer:* _____
16. Eight times three   *Answer:* _____
17. Ten divided by five   *Answer:* _____
18. Four into eight   *Answer:* _____
19. Five into ten   *Answer:* _____
20. Eight divided by four   *Answer:* _____
21. Ten into five   *Answer:* _____
22. Eight into four   *Answer:* _____
23. Five divided by ten   *Answer:* _____
24. Four divided by eight   *Answer:* _____
25. Subtract nine from seventeen.   *Answer:* _____
26. The quotient of thirty divided by six   *Answer:* _____
27. Six is less than seven.   *Answer:* _____
28. Six less than seven   *Answer:* _____

4.2 MORE THAN ONE OPERATION

Translate:

Subtract the sum of four and six from fifteen. _____

Here we get the message to subtract the *sum* of four and six; the sum is ten. Ten from fifteen is five.

In mathematics we must put parentheses around the sum in order to indicate that the entire sum is to be subtracted.

$$15 - (4 + 6)$$

According to the rules of the order of operations we simplify inside the parentheses first and obtain

$$15 - 10 = 15$$

Without parentheses,

$$15 - 4 + 6$$

would indicate subtracting just 4 from 15, getting 11, and then adding six to 11, obtaining 17.

Example 1   Translate: Divide twenty by eight minus three.

Solution. First subtract:

$$8 - 3 = 5$$

Then divide:

$$20 \div 5 = 4$$

or

$$20 \div (8 - 3) = 20 \div 5 = 4$$ ■

Example 2   Translate: The product of six and the sum of four and five

Solution. The sum is

$$4 + 5 = 9$$

The product is

$$6 \times 9 = 54$$

or

$$6(4 + 5) = 6 \times 9 = 54$$ ■

MINI-PRACTICE
(Use this space for your work.)

Translate into mathematics and evaluate the result:

(a) The product of seven and the sum of six and three   *Answer:* _____

(b) Twenty divided by ten less than twenty   *Answer:* _____

(c) Five added to the product of two and four   *Answer:* _____

Answers to Mini-Practice:

(a) $7(6 + 3) = 63$ (b) $\dfrac{20}{20 - 10} = 2$ (c) $2 \times 4 + 5 = 13$

IN YOUR OWN WORDS explain the use of parentheses when more than one operation has to be translated:

_____

_____

_____

Answer:

In order to apply addition and subtraction before multiplication and division, parentheses must be inserted.

Practice Exercises

In Exercises 1–20, translate the English statements into mathematics. Do not evaluate,

1. Divide fifty-four by three subtracted from twelve.   *Answer:* _____

2. The product of fifteen minus eleven and five   *Answer:* _____

3. The sum of six and four is divided by two.   *Answer:* _____

4. Ten less than the product of four and thirteen   *Answer:* _____

5. Seven subtracted from the product of six and nine   *Answer:* _____

6. Ten less than the product of three and four   *Answer:* _____

7. Ten more than one-fourth of twenty-eight   *Answer:* _____

8. Five added to the product of six and three   *Answer:* _____

9. Two less than twice five   *Answer:* _____

10. The difference of the square of three and twice three (in this order)   *Answer:* _____

11. Four less than sixteen   *Answer:* _____

12. Seven decreased by three   *Answer:* _____

13. Six times the difference between nine and seven (in this order)   *Answer:* _____

14. Ten decreased by the product of two and six   *Answer:* _____

15. Eight divided by the result of subtracting six from eight   *Answer:* _____

16. The quotient of six less than nine divided by twice nine   *Answer:* _____

17. Nineteen less than the product of five and two   *Answer:* _____

18. The ratio of nine more than twelve and twelve   *Answer:* _____

19. Four-fifths times the sum of eight and seventeen   *Answer:* _____

20. Ten increased by the quotient of four divided by eight   *Answer:* _____

4.3 TRANSLATIONS FROM MATHEMATICS INTO ENGLISH

Operation:
Addition   Translate $3 + 5$ by using each of the following expressions:

more than: _____

increased by: _____

the sum of: _____

The order in which the numbers in an addition problem are written does not effect the sum. For example, "five more than three" and "three more than five" both have a sum of eight.

Operation: Subtraction

Translate 5 – 2 by using each of the following expressions:

decreased by: _____

less than: _____

the difference of: _____

minus: _____

subtract from: _____

Which of the English translations follow the order: first 5, then 2?

_____

Hopefully, you remembered that "less than" and "subtracted from" give us a message to reverse the order:

5 – 2 is translated as "two less than five" or as "two subtracted from five."

Operation: Multiplication

Translate 5 × 6 using each of the following expressions:

the product of: _____

times: _____

"The product of five and six" and "Five times six" are the answers.

Operation: Division

Translate 10 ÷ 5 by using each of the following expressions:

the quotient of: _____

divided by: _____

divided into: _____

"The quotient of ten divided by five" is the same as "ten divided by five" and "five divided into ten."

Example 1   Translate 16 – 4 into English in three different ways.

Solution.

Direct translations:

> The difference of sixteen and four (in this order)
> Sixteen minus four

Reverse translation:

> Four less than sixteen ∎

Example 2   Translate $8 \cdot 5$ into English in two different ways.

Solution.

Multiplication is *times*:

Eight times five

The answer in a multiplication problem is the *product*:

The product of eight and five ■

MINI-PRACTICE
(Use this space for your work.)

Translate into English:

(a) $2 + 6$ *Answer:* _____

(b) $6 - 4$ *Answer:* _____

(c) $3(5)$ *Answer:* _____

(d) $10 \div 2$ *Answer:* _____

Practice Exercises

1. Translate $8 + 5$ into English in three different ways.   *Answer:* _____

   _____   _____

2. Translate $18 - 7$ into English in three different ways.   *Answer:* _____

   _____   _____

In Exercises 3–8, translate from mathematics into English.

3. $9 \cdot 3$ *Answer:* _____

4. $4(2)$ *Answer:* _____

5. $(6)(5)$ *Answer:* _____

6. $8 \div 4$ *Answer:* _____

7. $4 \div 8$ *Answer:* _____

8. $4 < 8$ *Answer:* _____

In Exercises 9–16, translate from mathematics into English in two different ways.

9. $8 + 4$ *Answer:* _____

   _____

10. $10 - 7$ *Answer:* _____

   _____

Answers to Mini-Practice:
(There are other possibilities.)
(a) The sum of two and six
(b) Four less than six
(c) The product of three and five
(d) The quotient of ten and two

11. $8 - 10$ *Answer:* _____

_____

12. $\dfrac{12}{4}$ *Answer:* _____

_____

13. 3^2 *Answer:* _____

_____

14. $2 \cdot 3$ *Answer:* _____

_____

15. $8 - 6$ *Answer:* _____

_____

16. $\dfrac{4}{5}(10)$ *Answer:* _____

_____

SUMMARY

The words
  increased by
  more than
  the sum of   give a message of " + ."
  plus
  add

The words
  decreased by
  less than
  minus   give a message of " − ."
  subtracted from
  the difference between

The words
  times
  the product of   give a message of " × ."
  multiplied by

The words
  divided by
  the quotient of   give a message of " ÷ ."
  divided into

The expressions "more than," "less than," "subtracted from," and "divided into" give us the message to reverse the order when we translate.

Review Practice Exercises

In Exercises 1–20, translate the English expressions into mathematical expressions. Do not evaluate.

1. Six more than five   Answer: _____

2. Three increased by six   Answer: _____

3. Five less than the product of four and three   Answer: _____

4. The product of three and the sum of three and five   Answer: _____

5. Twelve less than four times twenty-seven   Answer: _____

6. The sum of five times twenty-four and three times twenty-four   Answer: _____

7. Twelve more than the difference of five subtracted from nine   Answer: _____

8. Five times the sum of six and two   Answer: _____

9. The sum of five times three and two   Answer: _____

10. Sixteen decreased by the sum of twice four and eleven   Answer: _____

11. The sum of the product six times five plus the product four times five   Answer: _____

12. Four less than eight is added to six.   Answer: _____

13. Three more than seven added to twice seven   Answer: _____

14. Fifteen more than twice six   Answer: _____

15. Add three to nine and then divide the result by two.   Answer: _____

16. Twice ten is decreased by three times ten.   Answer: _____

17. Take eight and subtract three; double the result and then divide by four.   Answer: _____

18. Take away six from the square of four.   Answer: _____

19. Three times nine plus twice the square of nine   Answer: _____

20. Four times the sum of three times eleven and the cube of eleven   Answer: _____

In Exercises 21–26, translate from mathematics into English.

21. $10^2 - 8$ *Answer:* _____

22. $6(4 + 5)$ *Answer:* _____

23. $\frac{1}{2}(7 + 9)$ *Answer:* _____

24. $\frac{8}{5 - 3}$ *Answer:* _____

25. $8 + \frac{6}{3}$ *Answer:* _____

26. $\frac{7 + 5}{6 + 3}$ *Answer:* _____

27. Translate $10 - 8$ into English in three different ways.

    Answer: _____

    _____

    _____

28. Translate 9 + 7 into English in three different ways.

Answer: _____

_____

_____

If you have questions about this chapter, write them here: _____

_____

_____

_____

_____

_____

PART II

Variables and Operations

5 Operations Using Variables

Remember to

* *DISTINGUISH BETWEEN CONSTANTS AND VARIABLES*
* *USE THE INDEX TO LEARN THE USES OF NEW WORDS*
* *REVIEW THE RULES OF ARITHMETIC*
* *WRITE OUT WHAT IS CALLED FOR IN YOU OWN WORDS*

In this chapter you will learn to use basic operations involving variables. You will also apply the rules used with numbers in exponential notation to expressions with variables.

5.1 VARIABLES AND TERMS

Variables Recall that a **constant** is a symbol that represents a number, such as 5 or –7 or 0 or $\frac{1}{2}$, whose value is *fixed*. On the other hand, a **variable** is a symbol that represents *various numbers* in the course of a discussion. For example, the letter t might represent different times during the course of a chemistry experiment. In studies of the populations of cities x could represent the number of males in a city, y the number of females, and z the number of foreign-born people.

Terms In the previous chapters you have studied factors and multiplication. You have also added and subtracted numbers.

What are the numbers in an addition/subtraction problem called?

_____

Homework

1. You learn by practicing.
- Homework helps you to understand your classwork.
- To know the material, you must work the problems out for yourself.
- Understanding homework helps you on exams.

2. Work steadily!
- It is best to do some homework every day.

3. When you don't understand an exercise:
- Try to find a similar example worked out in the text.
- Go over every step of this example.
- Find similar exercises that you have already solved. Think about how you solved these.

4. Still stuck?
- Call a classmate and ask for help.
- Ask your instructor in class the next day.
- Go to your instructor's office hours.
- Go to the math lab for help.

We add or subtract **terms**. In algebra, terms can be many things. They can be number symbols (4, for example). They can be variable symbols (x, for example), or they can both be combined in multiplication ($4x$, for example, where multiplication is *understood*) or division by a *number* $\left(\dfrac{x}{4}, \text{ for example}\right)$.

Terms can be combined in various ways. $4x - 2y$ is not a term. It is an **algebraic expression** consisting of two terms, $4x$ and $2y$, that are separated by a minus sign. Also, $\dfrac{4x}{y}$ is not a term, but rather, it is the quotient of $4x$ divided by the (variable) term y. In contrast, $4x \cdot y$, or $4xy$, is a term because number symbols and variable symbols can be combined in a term by multiplication.

Encircle the terms in the following algebraic expressions:

(a) 0   (b) $3xyz$ (c) $5x + 3y$

(d) $67 - 32xy$ (e) $6xy + 5pq$ (f) $3x$

(g) 5   (h) $7ax$ (i) $5a - ab$

(a), (b), (f), (g), and (h) are expressions with one term only, whereas (c), (d), (e), and (i) are expressions with two terms.

Give an example of a term containing the variable x: _____

Give another example of a term containing the variable y: _____

Using your two terms, write an algebraic expression that is not a term: _____

Other examples of algebraic expressions are

$$a + 2b - c + 5d \quad \text{and} \quad \frac{x-1}{x+2}$$

A term, such as 8 or $3ab$, is also considered to be an algebraic expression.

IN YOUR OWN WORDS explain the relationship between terms and algebraic expressions: _____

_____

_____

Answer:

Terms can be number symbols or variable symbols, possibly combined by multiplication or by division by a number.

Algebraic expressions consist of one or more terms, combined by arithmetic operations.

Practice Exercises

In Exercises 1–10, encircle the terms in the following algebraic expressions.

1. a

2. $2a$

3. $a + 2b$

4. $ab + a$

5. -7

6. $a + b + c$

7. abc

8. $a - abcd$

9. $\dfrac{a}{2} + \dfrac{2}{a}$

10. $ab + bc + de + ef$

In Exercises 11–20, how many terms do the following expressions have?

11. $2xy + 3$ Answer: _____

12. x Answer: _____

13. $x + y + 2z$ Answer: _____

14. $3 + 2x$ Answer: _____

15. $2xy$ Answer _____

16. $\dfrac{2}{3}$ Answer: _____

17. $w + x + y + z$ Answer: _____

18. $a + b + c - d - e$ Answer: _____

19. $abcde$ Answer: _____

20. $x - xyz$ Answer: _____

5.2 VARIABLES WITH EXPONENTS

Write $a \cdot a \cdot a \cdot a \cdot a \cdot a \cdot a \cdot a \cdot a$ in exponential form: _____

Remember that each letter "a" is a factor. Here we have 9 factors by the name of "a" and we write this product as a^9.

Operations with terms containing variables and exponents are performed by applying the same rules that are used for operations with only numbers.

Review of Rules   State the rules you use in the following operations:

1. $2^3 \cdot 2^5 = $ _____

The rule is: _____

2. $\dfrac{3^8}{3^5} = $

_____

The rule is: _____

3. $(2^4)^3 = $ _____

The rule is: _____

4. $5^0 = $ _____

The rule is: _____

5. $(2 \cdot 3)^4 = $ _____

The rule is: _____

6. $\left(\dfrac{2}{3}\right)^4 = $

_____

The rule is: _____

Answers: **1.** Add the exponents.   **2.** Subtract the exponents.   **3.** Multiply the exponents.   **4.** Any nonzero number raised to the zeroth power is one. **5.** Raise each factor to the fourth power.   **6.** Raise the numerator and the denominator each to the fourth power.

Replace the numbers used as bases in the preceding review with variables. Show how the rules work. Use x or y instead of the numbers. For example, $2^3 \cdot 2^5$ becomes $x^3 \cdot x^5$.

$2^3 \cdot 2^5 = 2^8$ becomes _____

$\dfrac{3^8}{3^5} = 3^3$ becomes _____

$(2^4)^3 = 2^{12}$ becomes _____

$5^0 = 1$ becomes _____

$(2 \cdot 3)^4 = 2^4 \cdot 3^4$ becomes _____

$\left(\dfrac{2}{3}\right)^4 = \dfrac{2^4}{3^4}$ becomes _____

Here are the answers when x is the variable:

$$x^3 \cdot x^5 = x^{3+5} = x^8$$

$$\frac{x^8}{x^5} = x^{8-5} = x^3$$

$$(x^4)^3 = x^{4 \cdot 3} = x^{12}$$

$$x^0 = 1$$

This last statement will always be true for any value of x except when $x = 0$.

$$(x \cdot y)^4 = x^4 y^8$$

$$\left(\frac{x}{y}\right)^4 = \frac{x^4}{y^4}$$

Example 1   Write in exponential notation:

$$a \cdot x \cdot x$$

Solution.   There are two different variables: a and x. Note that $x \cdot x$ can be written in exponential notation: x^2

$$a \cdot x \cdot x = a \cdot x^2 \quad \text{or} \quad ax^2 \qquad \blacksquare$$

Example 2   Raise to the indicated power:

$$(3a)^2$$

Solution.   Each factor inside the parentheses, 3 and a, must be raised to the second power:

$$(3a)^2 = 3^2 a^2 = 9a^2 \qquad \blacksquare$$

Example 3   Raise to the indicated power:

$$(x \cdot y^2)^4$$

Solution.   Raise each factor to the fourth power.

$$(x \cdot y^2)^4 = x^4 \cdot (y^2)^4$$

Now raise y^2 to the fourth power.

$$(y^2)^4 = y^8$$

Therefore,

$$(x \cdot y^2)^4 = x^4 \cdot (y^2)^4 = x^4 y^8 \qquad \blacksquare$$

MINI-PRACTICE   (a) Write in exponential notation: $x \cdot x \cdot x$ *Answer:* _____

(Use this space for your work.)

(b) Multiply: $a^5 \cdot a^1$ *Answer:* _____   (c) Divide: $\dfrac{x^{11}}{x^8}$ *Answer:* _____

(d) Raise to the indicated power: $(xy)^2$ *Answer:* _____

(e) Raise to the indicated power: $(4a^2 b^3)^2$ *Answer:* _____

(f) $\left(\dfrac{2}{x}\right)^3$ Answer: _____

IN YOUR OWN WORDS   explain how the rules of exponential notation relate to variables: _____

_____

_____

Answer:

   The rules are the same as those used with numbers only.

Practice Exercises

In Exercises 1–4, write in exponential form.

1. y cubed   Answer: _____

2. $aabbbccccc$ Answer: _____

3. $4xxyyyyzz$ Answer: _____

4. $-7uuvvvvwww$ Answer: _____

In Exercises 5–14, multiply.

5. $x^4 \cdot x^4$ Answer: _____

6. $x^2 \cdot x^5$ Answer: _____

7. $x^2 \cdot x^3 \cdot x^4$ Answer: _____

8. $a^4 \cdot a^7$ Answer: _____

9. $y^3 \cdot y \cdot y^5$ Answer: _____

10. $a^3 \cdot a^2$ Answer: _____

11. $p \cdot p$ Answer: _____

12. $b^4 \cdot b^5$ Answer: _____

13. $c^5 \cdot c^4 \cdot c^9$ Answer: _____

14. $x^{10} \cdot x^{100} \cdot x^{1000}$ Answer: _____

In Exercises 15–20, divide.

15. $\dfrac{c^7}{c^3}$ Answer: _____

16. $\dfrac{x^{14}}{x^6}$ Answer: _____

17. $\dfrac{b^{15}}{b^{10}}$ Answer: _____

18. $\dfrac{x^{14}}{x^{11}}$ Answer: _____

19. $\dfrac{a^{12}}{a^4}$ Answer: _____

20. $\dfrac{b^{12}}{b^{12}}$ Answer: _____

In Exercises 21–42, raise to the indicated power.

21. $(2a)^4$ Answer: _____

22. $(5x)^2$ Answer: _____

23. $(3a)^3$ Answer: _____

24. $(4u)^2$ Answer: _____

25. $(6y)^2$ Answer: _____

26. $(2pq)^3$ Answer: _____

27. $(-2ab)^4$ Answer: _____

28. $(-4xy)^3$ Answer: _____

29. $(10cd)^5$ Answer: _____

30. $(-10yz)^6$ Answer: _____

31. $\left(\dfrac{3}{x}\right)^2$ Answer: _____

32. $\left(\dfrac{a}{5}\right)^2$ Answer: _____

33. $\left(\dfrac{a}{10b}\right)^2$ Answer: _____

34. $\left(\dfrac{x}{3}\right)^3$ Answer: _____

35. $(2a^2)^2$ Answer: _____

Answers to Mini-Practice:

(a) x^3 (b) a^6 (c) x^3 (d) x^2y^2 (e) $16a^4b^6$ (f) $\dfrac{8}{x^3}$

36. $(ab^2)^5$ *Answer:* _____   **37.** $(x^2y^3)^2$ *Answer:* _____   **38.** $(4x^2y^3)^3$ *Answer:* _____

39. $(10ab^2c^3)^4$ *Answer:* _____   **40.** $(x^4yz^3)^5$ *Answer:* _____   **41.** $\left(\dfrac{3b}{a^2}\right)^2$ *Answer:* _____

42. $\left(\dfrac{10x^2}{3y^3}\right)^3$ *Answer:* _____

5.3 NUMERICAL COEFFICIENTS

In the term $7xyz$, which is the constant factor? _____

The constant factor, in this case 7, is called the **numerical coefficient** or simply the **coefficient**.

What is the coefficient of the term $6x$? _____

What is the coefficient of the term x? _____

When there is a constant multiple, such as 6, we call that constant the coefficient. In the case of "x," there is a 1 understood before the variable and that is the coefficient.

Example 1   What is the numerical coefficient of

$$5x^2y$$

Solution.   The constant factor is 5. Therefore, the (numerical) coefficient is 5.   ∎

Example 2   What is the numerical coefficient of

$$-xy$$

Solution.   There is a -1 understood in front of xy. Consequently, the coefficient is -1.   ∎

MINI-PRACTICE   Determine the numerical coefficient of the following terms.
(Use this space for your work.)

    (a) $6xy$ (b) $3z$ (c) x

Answers: (a) _____   (b) _____   (c) _____

Answers to Mini-Practice:
(a) 6   (b) 3   (c) 1

IN YOUR OWN WORDS   explain the meaning of *numerical coefficient.* _____

_____

_____

Answer:

We call the numerical factor of a term the (numerical) coefficient.

Practice Exercises

In Exercises 1–10, what is the numerical coefficient of each of the following terms?

1. $13x$ Answer: _____
2. $5xy$ Answer: _____
3. 2 Answer: _____

4. xy Answer: _____
5. x^2y Answer: _____
6. $-xy$ Answer: _____

7. $-xy^2$ Answer: _____
8. $\frac{1}{2}x^2$ Answer: _____
9. $\frac{-2}{3}x^2y^3$ Answer: _____

10. $0.04abc$ Answer: _____

5.4 ADDITION AND SUBTRACTION OF TERMS

Let the letter i represent inches.

Write in symbols: "square inches" _____

Write in symbols: "cubic inches" _____

What do we measure in inches? _____

What do we measure in square inches? _____

What do we measure in cubic inches? _____

$$\underbrace{i}_{\substack{\text{measures} \\ \text{length}}} \qquad \underbrace{i^2}_{\substack{\text{measures} \\ \text{area}}} \qquad \underbrace{i^3}_{\substack{\text{measures} \\ \text{volume}}}$$

What is the base in all three cases? _____

The base is the same in all three cases. But the exponents give the message that each measuring unit represents a different kind of measurement.

Complete the following table by giving the proper measurement unit:

Unit of Measurement	Length	Area	Volume
yards (y)	y	y^2	y^3
feet (f)			
miles (mi)			
meters (m)			
centimeters (cm)			
kilometers (km)			

Perform the following addition and subtraction problems.

(a)   7 square yards   Let y represent yards: $7\,y^2$

        + 3 square yards $+3\,y^2$

                                      Answer:

(b)   7 square meters   Let m represent meters: $7\,m^2$

        − 3 square meters $-3\,m^2$

                                        Answer:

(c)   7 cubic meters   Let m represent meters: $7\,m^3$

        + 3 cubic meters $+3\,m^3$

                                        Answer:

(d)   7 cubic inches   Let i represent inches: $7\,i^3$

        − 3 cubic inches $-3\,i^3$

                                        Answer:

Answers: (a) $10\,y^2$ (b) $4\,m^2$ (c) $10\,m^3$ (d) $4i^3$

In addition and subtraction of terms both base and exponents stay the same. The numerical coefficients are added or subtracted as required.

We cannot further simplify terms of different kinds. Thus,

$$7x + 3y$$

cannot be simplified.

We can only simplify when adding or subtracting terms of the same kind. Thus,

$$4a + 5a = (4 + 5)a = 9a$$

We call terms that are of the same kind **like terms**. Like terms have the same bases; each base is raised to the same exponent. For example,

$$4ab^2 \quad \text{and} \quad 3ab^2 \quad \text{are } like \text{ terms}$$
$$7ab^2 \quad \text{and} \quad 5ab \quad \text{are } unlike \text{ terms}$$

Although the bases of $7ab^2$ and $5ab$ are the same, b is raised to the second power in $7ab^2$ and to the first power in $5ab$.

Example 1   Subtract.

$$13a - 18a$$

Solution. We have like terms in which the base is a and the exponent is 1. Subtract the numerical coefficients:

$$13 - 18 = -5$$

Therefore,

$$13a - 18a = -5a \qquad \blacksquare$$

Example 2   Add.

$$4x^2y + 6x^2y$$

Solution. $4x^2y$ and $6x^2y$ are like terms.

$$4 + 6 = 10$$

Therefore,

$$4x^2y + 6x^2y = 10x^2y$$

The numerical coefficients are added. The exponents do not change. $\blacksquare$

MINI-PRACTICE
(Use this space for your work.)

Perform the following additions and subtractions:

(a) $6a + 4a$ (b) $6a - 4a$ (c) $8x - 2x$ (d) $8x + 2x$ (e) $7a + 5a$

Answers: (a) _____   (b) _____   (c) _____   (d) _____   (e) _____

IN YOUR OWN WORDS

explain how we combine like terms: _____

_____

_____

RULE:

> **To combine like terms, add or subtract the numerical coefficients and multiply the resulting number by the common variable part.**

Answers to Mini-Practice:

(a) $10a$ (b) $2a$ (c) $6x$ (d) $10x$ (e) $12a$

WATCH OUT!   *Like Terms*

Explain why the following simplification is *wrong*.

$$
\begin{array}{r}
8 \text{ feet} \\
+\ 3 \text{ pounds} \\
\hline
\end{array}
\qquad \longrightarrow \qquad
\begin{array}{r}
8\,f \\
+\ 3\,p \\
\hline
11\,fp
\end{array}
$$

_____

_____

Answer:

Only like terms can be simplified when adding or subtracting. $8f + 3p$ cannot be simplified further.

Practice Exercises

In Exercises 1–40, add or subtract.

1. $3xy - 8xy$ Answer: _____
2. $2xy^3 - 13xy^3$ Answer: _____

3. $12x + 3x - 7x + 10x$ Answer: _____
4. $6p^2q - 4p^2q$ Answer: _____

5. $18ab - 25ab$ Answer: _____
6. $-7x^3y + 3x^3y - 8x^3y$ Answer: _____

7. $27st - 33ts$ Answer: _____
8. $p - p - 4p - 6p$ Answer: _____

9. $-92y + y + 5y - 2y$ Answer: _____
10. $7a - 43a - 5a + 2a$ Answer: _____

11. $P + 3P$ Answer: _____
12. $4M - 2M$ Answer: _____

13. $4M + 2M$ Answer: _____
14. $3y^2 + 4y^2$ Answer: _____

15. $4y^2 - 3y^2$ Answer: _____
16. $6t^3 - 5t^3$ Answer: _____

17. $13a^2b^3 + 3a^2b^3$ Answer: _____
18. $4a + 5a - 7a$ Answer: _____

19. $8xy^2 - 15xy^2$ Answer: _____
20. $2x^2 - 5x^2 + 7x^2$ Answer: _____

21. $5pq - 13pq + 8pq$ Answer: _____
22. $-8st + 3st - 13st$ Answer: _____

23. $17st + 9st - 35st$ Answer: _____
24. $-14a^2b + 18ba^2$ Answer: _____

25. $8.2xy - 9.7xy$ Answer: _____
26. $-5.7xy^3 + 3.2xy^3$ Answer: _____

27. $6.4a^2b^3 - 3.5a^2b^3$ Answer: _____
28. $9.3t^3 - 11.6t^3$ Answer: _____

29. $7.5xy^2 - 10.5xy^2$ Answer: _____
30. $\dfrac{3}{2}a^3 - \dfrac{7}{2}a^3$ Answer: _____

31. $\dfrac{5}{8}xy - \dfrac{3}{8}xy$ Answer: _____
32. $\dfrac{7}{2}y^4 - \dfrac{9}{4}y^4$ Answer: _____

33. $\dfrac{1}{2}a^3c^2 - \dfrac{3}{5}a^3c^2$ Answer: _____
34. $-\dfrac{7}{6}x^2y + \dfrac{4}{3}x^2y$ Answer: _____

35. $a + 2a + 3a + 4a$ Answer: _____
36. $10x + x + 2x + 5x$ Answer: _____

37. $3y - 2y + 7y - 4y$ Answer: _____
38. $12z - z + 3z - 2z$ Answer: _____

39. $5xy - 6xy + 7xy - 8xy + 9xy$ Answer: _____

40. $x^2y + 3x^2y - 2x^2y + 12x^2y - 10x^2y$ Answer: _____

5.5 Polynomials

Consider a *nonzero* sum (or difference) of terms in which like terms have already been combined. If this expression has only one term, it is called a **monomial**. –2 and $3xy$ are each monomials. We call an algebraic expression that is the sum (or difference) of two unlike terms a **binomial**. $a + 4b$ and $6x - y$ are each binomials. We call an algebraic expression that is the sum of three terms, each unlike the other two, a **trinomial**. $a - 2b + 5c$ is a trinomial. In general, we call expressions that are the sums of one or more terms **polynomials**.

These prefixes are often used in English. *Mono-* indicates one, *bi-* indicates two, *tri-* indicates three, and *poly-* indicates many. You probably know many words with these prefixes.

Write down one with each prefix: _____

_____

You might have written *monogram, bicycle, tricycle* and *polychromatic*, for example.

Classify the following algebraic expressions by the number of terms they contain. The first one is done for you.

(a) $3xy$ <u>monomial</u>_____

(b) $8x - 2y + 1$ _____

(c) $a + b$ _____

(d) $5a - 3b + 2c - 36 + d$ _____

Answers: (b) trinomial, (c) binomial, (d) polynomial. Note that (a), (b), and (c) are also polynomials.

The expression $2x + 3y + 5$ contains 3 terms and is therefore a trinomial.

Why is $3x(2y)(5z)$ a monomial? _____

_____

Why is $\dfrac{2xy}{3}$ a monomial? _____

_____

Remember that within a single term you can multiply by a number or variable, and you can divide by a number. However, when adding or subtracting *unlike* terms, several different terms result.

Write your own example of each of the following:

(a) Monomial: _____

(b) Binomial: _____

(c) Trinomial: _____

(d) A polynomial with six terms, each unlike the others:

_____

Example 1   How many unlike terms does the following expression contain?

$$3x^2y + 5xy^3$$

Solution.   The + sign separates two unlike terms:

$$3x^2y \quad \text{and} \quad 5xy^3$$ ■

Example 2   Classify the following polynomial as a monomial, binomial, trinomial, or as *none* of these:

$$2x + 3xy - 5x^3y$$

Solution.   The expression contains three terms,

$$2x, \quad 3xy, \quad \text{and} \quad 5x^3y$$

each unlike the others, and is therefore a *trinomial*.   ■

Be careful! The polynomial $5x + 3x$ contains 2 *like* terms, which can be added to yield $8x$; it is therefore a monomial.

MINI-PRACTICE
(Use this space for your work.)

How many unlike terms do the following expressions contain?

    (a) $x - 2y$ (b) $3x + 4y - 5$ (c) $3xy^2 + 2x^2y$

Answers: (a) _____   (b) _____   (c) _____

IN YOUR OWN WORDS   explain the relationship between terms and polynomials: _____

_____

_____

Answer:

    Terms can be number symbols or variable symbols, posssibly combined by multiplication or by division by a number.

    Polynomials are terms or sums of terms.

_____

Answers to Mini-Practice:

(a) 2 (binomial)   (b) 3 (trinomial)   (c) 2 (binomial)

Practice Exercises

Classify the following polynomials as monomials, binomials, trinomials, or as *none* of these.

1. $\dfrac{2xy}{3}$ *Answer:* _____

2. $3x + (2y)(5z)$ *Answer:* _____

3. $a^2 - b^2 + 4ab$ *Answer:* _____

4. $3x^3 + 4x^2 - 5x + 10$ *Answer:* _____

5. 10 *Answer:* _____

6. $a + ab + a^2b + a^2b^2$ *Answer:* _____

7. $3x + 2x$ *Answer:* _____

8. $3x - 2x$ *Answer:* _____

9. $5a + b - a$ *Answer:* _____

10. $a^2 + a + 2a^2 + 5a$ *Answer:* _____

5.6 MULTIPLICATION

Simplify: $(3x^2)(2x^3)$ _____

What did you do to get your answer? _____

_____

The correct answer is $6x^5$. If you did not get this answer, do the following:

$$(3x^2)(2x^3) = 3 \cdot 2 \cdot x^2 \cdot x^3 = 6x^{2+3} = 6x^5$$

Example 1   Multiply: $(4xy)(6x^3y)$

Solution.

Multiply the coefficients: $4 \times 6 = 24$
Multiply $x \cdot x^3 = x^{1+3} = x^4$ (same base; we add exponents)
Multiply $y \cdot y = y^{1+1} = y^2$

$$(4xy)(6x^3y) = 24x^4y^2$$ ∎

Example 2   Multiply: $(-2a^3b)(4b^6)(-5a^2b^3)$

Solution.

Multiply the coefficients: $(-2)(4)(-5) = 40$
Multiply $a^3a^2 = a^5$
Multiply $b(b^6)b^3 = b^{10}$

$$(-2a^3b)(4b^6)(-5a^2b^3) = 40a^5b^{10}$$ ∎

MINI-PRACTICE   Multiply.
(Use this space for your work.)

(a) $(3a^2)(4a)$ *Answer:* _____

(b) $(4y^3)(2y^4)$ *Answer:* _____

(c) $(2x^4)(3x)$ *Answer:* _____

(d) $(3ab)(2a^2b)(4ab^3)$ *Answer:* _____

_____

Answers to Mini-Practice:

(a) $12a^3$ (b) $8y^7$ (c) $6x^5$ (d) $24a^4b^5$

IN YOUR OWN WORDS   explain how to multiply terms than contain both constants and variables: _____

_____

_____

Answer:

To multiply terms, we first multiply the numerical coefficients and then add the exponents of the variables that are the same.

Practice Exercises

Multiply.

1. $(3p^2)(p^4)$ *Answer:* _____

2. $(-2y)(y^2)$ *Answer:* _____

3. $(5r^2)(3r^2)$ *Answer:* _____

4. $(5ab)(2a^2b^2)$ *Answer:* _____

5. $(2x^2y^3)(3x^3y^4)$ *Answer:* _____

6. $(3p)(4p^2)(p^2r^2)$ *Answer:* _____

7. $(xy)(-x^3)(x^4y^2)$ *Answer:* _____

8. $3abc(4a^2b)(2b^3c^2)$ *Answer:* _____

9. $-4xy(x^3y^6)(x^2y)(6y^5)$ *Answer:* _____

10. $5x^0y^3(4xy)(7x^4y^7)$ *Answer:* _____

11. $\frac{5}{9}x^2y\left(\frac{-9}{10}y\right)$ *Answer:* _____

12. $-4p^2q\left(\frac{7}{8}pq\right)\left(\frac{p^2q}{21}\right)$ *Answer:* _____

13. $\frac{5}{8}cd^2(-3cd)$ *Answer:* _____

14. $-xy\left(\frac{9}{7}x\right)\left(\frac{7}{9}y^3\right)$ *Answer:* _____

5.7 THE DISTRIBUTIVE LAW

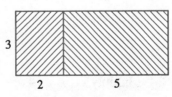

The picture shows three rectangles—2 smaller ones that make up the largest one. Their dimensions are:

$$3 \times 2 = \underline{\quad}, \quad 3 \times \underline{\quad} = \underline{\quad}, \quad \text{and} \quad 3 \times \underline{\quad} = \underline{\quad}$$

The sum of the areas of the two smaller rectangles equals the area of the largest rectangle. In other words,

$$3 \times 2 + 3 \times 5 = 3 \times (2 + 5)$$

or

$$3 \times 2 + 3 \times 5 = 3 \times 7$$

Now use the letters *a*, *b*, and *c* in the figure on page 140 to express the fact that the sum of the areas of the two smaller rectangles equals the area of the largest rectangle.

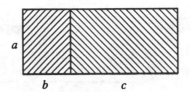

$$\underline{\hspace{2cm}} + \underline{\hspace{2cm}} = \underline{\hspace{2cm}}$$

In algebra we avoid using $\times$ for multiplication (because $\times$ can be confused with x), so we write:

$$a \cdot b + a \cdot c = a(b + c)$$

or in reverse order:

$$a(b + c) = a \cdot b + a \cdot c$$

In mathematics this rule or property is called the **distributive** law: we *distribute* multiplication over *addition*.

Evaluate $5(2 + 3)$ by using the order of operations: $\underline{\hspace{2cm}}$

Multiply $5(2 + 3)$ by using the distributive law: $\underline{\hspace{2cm}}$

In the second method you should have gotten: $5(2) + 5(3) = \underline{\hspace{1cm}}$

No doubt you got the same answer, 25, whichever method of evaluating you used.

Multiply: $5(x + 4) = \underline{\hspace{2cm}}$

As you probably realized, in this example we cannot add x and 4 to get a single number. Therefore, to multiply $5(x + 4)$, we have to use the distributive law to get

$$5(x + 4) = 5x + 20$$

Example 1   Multiply $3x(x^2 + x^5)$.

Solution. $$3x(x^2 + x^5) = 3x^3 + 3x^6$$ ■

Example 2   Multiply $2x(3x^3 - 4x^2 + 5x + 7)$.

Solution. $$2x(3x^3 - 4x^2 + 5x + 7) = 6x^4 - 8x^3 + 10x^2 + 14x$$ ■

MINI-PRACTICE   Multiply: (a) $4(x + y)$ (b) $3(x^2 + y^2)$
(Use this space for your work.)   (c) $4x(2x^2 + 3x)$ (d) $6x^2(3x^3 - 4x^4)$

Answers: (a) $\underline{\hspace{3cm}}$ (b) $\underline{\hspace{3cm}}$

(c) $\underline{\hspace{3cm}}$ (d) $\underline{\hspace{3cm}}$

Answers to Mini-Practice:

(a) $4x + 4y$ (b) $3x^2 + 3y^2$ (c) $8x^3 + 12x^2$ (d) $18x^5 - 24x^6$

IN YOUR OWN WORDS   explain how we multiply a monomial by a binomial: _____

_____

_____

Answer:

To multiply a monomial by a binomial, we use the distributive law and "distribute multiplication over addition."

Practice Exercises

Multiply.

1. $4(x^2 + 3y)$ *Answer:* _____
2. $3x(2x + 3x^2)$ *Answer:* _____
3. $2x^2(3x^3 + 4x^5)$ *Answer:* _____
4. $5x^3(x^2 + 3x + 4)$ *Answer:* _____
5. $7x^{-2}(x^{-3} + x^{-4})$ *Answer:* _____
6. $-4(3 - 2x - 5x^2)$ *Answer:* _____
7. $-2(y^2 - 4y + 7)$ *Answer:* _____
8. $-6pr^3(3p - 4pr + 7r^2 - 2p^2r^2)$ *Answer:* _____
9. $-4x^3(3x + 2xy - 4y^2)$ *Answer:* _____
10. $\frac{3}{4}xy(-x + y)$ *Answer:* _____
11. $-2abc(a - 2b + 3c)$ *Answer:* _____
12. $x^2y(-4x + y - 3z)$ *Answer:* _____
13. $10pq(p - r)$ *Answer:* _____
14. $abc(x - 2y)$ *Answer:* _____
15. $4a^2bc^3(a + 4ab - 3b^2c)$ *Answer:* _____
16. $10x^2yz^4(x + 5xy - 10xyz^2)$ *Answer:* _____

5.8 Division

Perform division: $\dfrac{12x^5}{2x^2}$

First reduce: $\dfrac{12}{2}$ = _____   and then reduce: $\dfrac{x^5}{x^2}$ = _____

Therefore, $\dfrac{12x^5}{2x^2} = 6x^3$.

Example 1   Divide $\dfrac{15a^2b^5}{5ab^3}$.

Solution. $\dfrac{15}{5} = 3$; $\dfrac{a^2}{a} = a$; $\dfrac{b^5}{b^3} = b^2$

Therefore, $\dfrac{15a^2b^5}{5ab^3} = 3ab^2$ ∎

Example 2   Divide $\dfrac{-48x^5y^2}{12x^3y}$.

Solution. $\dfrac{-48}{12} = -4$; $\dfrac{x^5}{x^3} = x^2$; $\dfrac{y^2}{y} = y$

Thus, $\dfrac{-48x^5y^2}{12x^3y} = -4x^2y$ ■

MINI-PRACTICE   Divide:
(Use this space for your work.)

(a) $\dfrac{15a^4b^3}{5a^2b}$ (b) $\dfrac{14x^6y^3}{7x^6y}$ (c) $\dfrac{12u^5v^3}{2uv^2}$

Answers: (a) _____   (b) _____   (c) _____

Negative Exponents

Simplify $\dfrac{5x^7y^5}{30x^2y^8}$ *Answer:* _____

You might have given your answer as

$$\dfrac{x^5y^{-3}}{6}$$

If you did, you are correct. It is common practice, however, not to leave the answer with a negative exponent.

How can we achieve this? _____

To get rid of a negative exponent, we invert:

$$y^{-3} = \dfrac{1}{y^3}$$

You now have a positive exponent in the denominator, rather than a negative exponent in the numerator. Therefore, the answer to

$$\dfrac{5x^7y^5}{30x^2y^8} \quad \text{simplifies to} \quad \dfrac{x^5}{6y^3}$$

Answers to Mini-Practice:

(a) $3a^2b^2$ (b) $2y^2$ (c) $6u^4v$

MINI-PRACTICE
(Use this space for your work.)

Rewrite using positive exponents only.

(a) $\dfrac{4x^{-5}y^{-1}}{4y^{-1}}$ (b) $\dfrac{a^7b^6}{a^9b^9}$ (c) $\dfrac{t^5u^{-5}v^2}{tv^3}$

Answers: (a) _____   (b) _____   (c) _____

IN YOUR OWN WORDS

explain the rule for dividing a single algebraic term by another algebraic term?

_____

_____

_____

Answer:

To divide one algebraic term by another, reduce the coefficients of the numerator and denominator, if possible, and subtract the exponents of the common variables. When negative exponents result, invert and change to positive exponents. (We assume that all factors have been multiplied out in both numerator and denominator.)

Practice Exercises

Divide:

1. $\dfrac{-20a^2}{-5a}$ *Answer:* _____

2. $\dfrac{22b^5}{-11b}$ *Answer:* _____

3. $\dfrac{-18r^2s^2}{-9r^2s}$ *Answer* _____

4. $\dfrac{12m^7}{4m^6}$ *Answer:* _____

5. $\dfrac{-5x^7y^5}{-30x^2y^3}$ *Answer:* _____

6. $\dfrac{7x^3y^3}{-14xy}$ *Answer:* _____

7. $\dfrac{3c^4d^3}{6c}$ *Answer:* _____

8. $\dfrac{-5p^3q}{-10pr}$ *Answer:* _____

9. $\dfrac{3n^5x^2y^3}{9nx^2y^3}$ *Answer:* _____

10. $\dfrac{4x^4y^2}{2x^5y^6}$ *Answer:* _____

11. $\dfrac{15a^4b^6c^{-2}}{20a^3b^8c}$ *Answer:* _____

12. $\dfrac{-12c^9de^{-2}}{18c^9d^4e^{-4}}$ *Answer:* _____

Answers to Mini-Practice:

(a) $\dfrac{1}{x^5}$ (b) $\dfrac{1}{a^2b^3}$ (c) $\dfrac{t^4}{u^5v}$

13. $\dfrac{30x^2}{6}$ *Answer:* _____

14. $\dfrac{24xy}{4}$ *Answer:* _____

15. $\dfrac{10x^2y^2}{5}$ *Answer* _____

16. $\dfrac{-8x^3}{2}$ *Answer:* _____

17. $\dfrac{27y}{-9}$ *Answer:* _____

18. $\dfrac{15a^2b^2c}{5}$ *Answer:* _____

19. $\dfrac{3x^2}{x}$ *Answer:* _____

20. $\dfrac{12x^4}{x^2}$ *Answer:* _____

21. $\dfrac{-36a^4b^2}{b}$ *Answer:* _____

22. $\dfrac{8x^2y^3}{y^2}$ *Answer:* _____

23. $\dfrac{-14ab^4}{b^3}$ *Answer:* _____

24. $\dfrac{4y^2}{2y}$ *Answer:* _____

25. $\dfrac{2x^2}{-2x}$ *Answer:* _____

26. $\dfrac{12x^4}{3x^2}$ *Answer:* _____

27. $\dfrac{-8y^2}{4y}$ *A nswer:* _____

28. $\dfrac{-16x}{4x}$ *Answer:* _____

29. $\dfrac{-16x}{-12y^2}$ *Answer:* _____

30. $\dfrac{27y^3}{-12y^2}$ *Answer:* _____

31. $\dfrac{5x^2}{-15x}$ *Answer:* _____

32. $\dfrac{a^4b^5}{a^3b}$ *Answer:* _____

5.9 DIVISION OF A POLYNOMIAL BY A MONOMIAL

Split $\dfrac{8+4}{2}$ into two separate fractions and then simplify:

_____ + _____ = _____

You should have:

$$\dfrac{8}{2} + \dfrac{4}{2} = 4 + 2 = 6$$

Split the fraction $\dfrac{8p+4q}{2}$ into two separate fractions:

_____ + _____ = _____

Why can we NOT simplify $4p + 2q$ when adding? _____

_____

Only like terms can be simplified when adding.

Example 1   Simplify $\dfrac{8p^2 + 4p^3}{2p}$.

Solution. $\dfrac{8p^2 + 4p^3}{2p} = \dfrac{8p^2}{2p} + \dfrac{4p^3}{2p} = 4p + 2p^2$ ∎

Example 2   Simplify $\dfrac{16x^4 + 12x^3 - 8x^2}{4x^2}$.

Solution.

$$\dfrac{16x^4 + 12x^3 - 8x^2}{4x^2} = \dfrac{16x^4}{4x^2} + \dfrac{12x^3}{4x^2} - \dfrac{8x^2}{4x^2} = 4x^2 + 3x - 2 \qquad ∎$$

Example 3   Simplify $\dfrac{25a^3b^5}{5a^2b^3} + \dfrac{6a^6b^2}{2a^5}$.

Solution. $\dfrac{25a^3b^5}{5a^2b^3} + \dfrac{6a^6b^2}{2a^5} = 5ab^2 + 3ab^2 = 8ab^2$ ∎

MINI-PRACTICE   Simplify.
(Use this space for your work.)

(a) $\dfrac{2x + 4}{2}$ (b) $\dfrac{5y^2 - 10y}{5y}$ (c) $\dfrac{6n^3 + 9n}{-3n}$ (d) $\dfrac{2x^3}{x^2} + \dfrac{3x^5}{x^4}$

Answers: (a) _____   (b) _____   (c) _____   (d) _____

IN YOUR OWN WORDS   explain how we divide a polynomial by a monomial. _____

_____

_____

Answer:

Each term of the polynomial is divided by the monomial.

WATCH OUT!   Consider $\dfrac{8p + 4q}{2}$

It is **wrong** to say:

"Two goes into eight four times, so the answer is $4p + 4q$."

Here we split the fraction into two parts:

$$\dfrac{8p}{2} + \dfrac{4q}{2} = 4p + 2q$$

_____

Answers to Mini-Practice:

(a) $x + 2$ (b) $y - 2$ (c) $-2n^2 - 3$ (d) $2x + 3x = 5x$

We could, instead, change $8p + 4q$ into $4(2p + q)$, reduce

$$\frac{4(2p + q)}{2} = 2(2p + q)$$

and again the answer would be $4p + 2q$.

Practice Exercises

Simplify.

1. $\dfrac{2x + 2}{2}$ Answer: _____

2. $\dfrac{5y + 5}{5}$ Answer: _____

3. $\dfrac{10a - 25}{5}$ Answer: _____

4. $\dfrac{16b - 40}{8}$ Answer: _____

5. $\dfrac{18y - 27}{-9}$ Answer: _____

6. $\dfrac{6a + 12}{3}$ Answer: _____

7. $\dfrac{-18t^2 - 6}{6}$ Answer: _____

8. $\dfrac{3a^2 + 2a}{a}$ Answer: _____

9. $\dfrac{6y^2 + 4y}{6}$ Answer: _____

10. $\dfrac{4b^3 - 3b}{b}$ Answer: _____

11. $\dfrac{12x^2 - 7x}{x}$ Answer: _____

12. $\dfrac{3x^2 - 6x}{3x}$ Answer: _____

13. $\dfrac{11z^2 + 22z}{11z}$ Answer: _____

14. $\dfrac{14k^2 - 12k}{2k}$ Answer: _____

15. $\dfrac{18 - 9z}{9}$ Answer: _____

16. $\dfrac{5x^2 - 10x}{-5x}$ Answer: _____

17. $\dfrac{x^4 - 3x^3}{x^2}$ Answer: _____

18. $\dfrac{a^3 - 5a^2}{a^2}$ Answer: _____

19. $\dfrac{8p^3 - 2p^2}{2p^2}$ Answer: _____

20. $\dfrac{27s + 18s^3}{9s}$ Answer: _____

21. $\dfrac{10a - 8a^2}{-2a}$ Answer: _____

22. $\dfrac{4x^4 - 6x^2}{-2x^2}$ Answer: _____

23. $\dfrac{9y^6 - 15y^3}{-3y^3}$ Answer: _____

24. $\dfrac{8y^2 - 6y + 2}{2}$ Answer: _____

25. $\dfrac{-3x^2 - 9x - 6}{-3}$ Answer: _____

26. $\dfrac{3m^2 + 6m + 9}{3m}$ Answer: _____

27. $\dfrac{16x^2 - 12x + 8}{-4}$ Answer: _____

28. $\dfrac{5x^2y^2 + 10xy}{5xy}$ Answer: _____

29. $\dfrac{8x^2y^2 - 24xy}{-8xy}$ Answer: _____

30. $\dfrac{50pq^2 + 30p^2q - 20p^2q^2}{10pq}$ Answer: _____

31. $\dfrac{27z^2 + 18z - 9}{-9}$ *Answer:* _____

32. $\dfrac{8x^5 - 32x^4 + 16x^3}{-8x^4}$ *Answer:* _____

33. $\dfrac{3x^2y + 6xy^2 - 9x^2y^2}{3xy}$ *Answer:* _____

34. $\dfrac{16a^2b - 20ab + 24ab^2}{4ab}$ *Answer:* _____

35. $\dfrac{22a^2b + 11ab + 33ab^2}{11ab}$ *Answer:* _____

36. $\dfrac{9x^2y + 6xy - 3xy^2}{3xy}$ *Answer:* _____

37. $\dfrac{5a^2b + 15ab - 30ab^2}{5ab}$ *Answer:* _____

38. $\dfrac{10y^5 + 15y^4 - 20y^3}{-5y^3}$ *Answer:* _____

39. $\dfrac{16a^2b^3c - 20a^3b^2c^4 + 24a^4bc^3}{4abc}$ *Answer:* _____

40. $\dfrac{5x^4y^3z^2 + 10x^3y^4z^2 - 15x^2y^3z^3}{5x^2y^2z^2}$ *Answer:* _____

41. $\dfrac{-6m^5n^5p^3 - 5m^3n^4p^2 - 4m^3n^4p^5}{-m^3n^4p}$ *Answer:* _____

42. $\dfrac{16a^2}{4ab} + \dfrac{20ab^2}{5ab}$ *Answer:* _____

43. $\dfrac{9y^6z^5}{3y^4z^2} + \dfrac{15y^2z^2}{5y^2z}$ *Answer:* _____

44. $\dfrac{5a^2b}{5a^2} + \dfrac{30ab^2}{3b^2}$ *Answer:* _____

45. $\dfrac{8x^2y^2}{8xy^2} - \dfrac{24xy^2}{6xy}$ *Answer:* _____

46. $\dfrac{6x^2y^3}{3xy} + \dfrac{20xy^5}{5y^3}$ *Answer:* _____

5.10 THE GREATEST COMMON FACTOR

What is the largest number that is a factor of 6 and 18? _____

Note that 6 is a factor of itself and of 18. Mathematicians call the number 6 the **greatest common factor** of 6 and 18. The expression "greatest common factor" is written in abbreviated form as *gcf*.

We can read *gcf* (6, 18) = 6 as: "The greatest common factor of 6 and 18 is 6."

Example 1   Find the greatest common factor of 12 and 45.

Solution.

Factor each of these numbers as far as possible. Begin anywhere.

$$12 = 4 \times 3 = 2 \times 2 \times 3 \quad \text{or} \quad 12 = 2 \times 6 = 2 \times 2 \times 3$$
$$45 = 9 \times 5 = 3 \times 3 \times 5 \quad \text{or} \quad 45 = 3 \times 15 = 3 \times 3 \times 5$$

The only positive common factor of 12 and 45, other than 1, is 3.

3 is the greatest common factor   ■

Example 2   Find the greatest common factor of 30 and 48.

Solution.

$$30 = 6 \times 5 = 2 \times 3 \times 5$$
$$48 = 8 \times 6 = 2 \times 2 \times 2 \times 2 \times 3 \quad \text{or} \quad 2^4 \times 3$$

6 is the largest number that goes into both 30 and 48.

6 is the greatest common factor   ■

MINI-PRACTICE   Find the *gcf* of each pair of numbers.
(Use this space for your work.)

(a) Find *gcf* (6, 24)   (b) Find *gcf* (16, 18)   (c) Find *gcf* (20, 24)

Answers: (a) _____   (b) _____   (c) _____

IN YOUR OWN WORDS   explain how to find the greatest common factor of two numbers: _____

_____

_____

Answer:

Find the largest number that divides evenly into both numbers, and you have the greatest common factor.

Practice Exercises

In Exercises 1–20, find the gcf of each pair of numbers.

1. 16 and 24   *Answer:* _____   2. 20 and 32   *Answer:* _____   3. 15 and 20   *Answer:* _____

4. 6 and 8   *Answer:* _____   5. 18 and 24   *Answer:* _____   6. 10 and 24   *Answer:* _____

7. 18 and 45   *Answer:* _____   8. 15 and 25   *Answer:* _____   9. 14 and 59   *Answer:* _____

10. 25 and 100   *Answer:* _____   11. 16 and 80   *Answer:* _____   12. 8 and 36   *Answer:* _____

13. 24 and 30   *Answer:* _____   14. 18 and 32   *Answer:* _____   15. 16 and 30   *Answer:* _____

16. 40 and 64   *Answer:* _____   17. 44 and 96   *Answer:* _____   18. 48 and 144   *Answer:* _____

19. 16 and 140   *Answer:* _____   20. 12 and 76   *Answer:* _____

Answers to Mini-Practice:
(a) 6   (b) 2   (c) 4

5.11 THE *gcf* OF ALGEBRAIC TERMS

Write $3x^2$ as the product of *three* factors: $3x^2 = $ _____

Write $3x^3$ as the product of *four* factors: $3x^3 = $ _____

Use your answers to complete this table:

Term	Factors
$3x^2$	
$3x^3$	

Encircle the factors that are common to both terms. You should have encircled 3, x, x.

$gcf\,(3x^2, 3x^3) = $ _____

The following table shows a summary of the procedure:

Term	Factors
$3x^2$	$3 \cdot x \cdot x$
$3x^3$	$3 \cdot x \cdot x \cdot x$

Your answer should be $3x^2$. Now, complete the following table and encircle the common factors:

Term	Factors
$9a^3b$	
$6a^2b$	

You should have encircled 3, a, a, and b.

The *gcf* of $9a^3b$ and $6a^2b$ is _____

Your answer should be $3a^2b$.

Complete the following table and encircle the common factors:

Term	Factors
$9a^3b^5$	
$6a^2b^3$	

$$gcf\,(9a^3b^5, 6a^2b^3) = \underline{\hspace{3cm}}$$

The answer is $3a^2b^3$.

Let's look again at these last examples. Can we write the *gcf* of two algebraic terms after looking at them?

(a) What is the *gcf* of 9 and 6? _____

(b) Of the expressions a^3 and a^2, which has the smaller exponent?

  _____

(c) Which one of the expressions b^5 and b^3 has the smaller exponent? _____

Now look back at the terms $9a^3b^5$ and $6a^2b^3$. The *gcf* is $3a^2b^3$.

Example 1   Find the greatest common factor of $5x^2y^3$ and $15xy^2$.

Solution.

The *gcf* of 5 and 15 is 5.
The *gcf* of x^2 and x is x.
The *gcf* of y^3 and y^2 is y^2.

   The *gcf* of $5x^2y^3$ and $15xy^2$ is $5xy^2$ ■

Example 2   Find the greatest common factor of $4a^3b^6$ and $8ab^4c$.

Solution.

The *gcf* of 4 and 8 is 4.
The *gcf* of a^3 and a is a.
The *gcf* of b^6 and b^4 is b^4.
Note that c is not a factor of the first term, so it is *not* a common factor.

   The *gcf* of $4a^3b^6$ and $8ab^4c$ is $4ab^4$ ■

MINI-PRACTICE   Find the *gcf* of each pair of terms.
(Use this space for your work.)

(a) $5a$ and $5b$ (b) $5a$ and $10\,ab$

(c) $2xy^2$ and $6x^2y$ (d) $12x^3y^2$ and $18xy^2$

Answers: (a) _____   (b) _____   (c) _____   (d) _____

IN YOUR OWN WORDS   explain how to find the *gcf* of two or more algebraic terms. _____

_____

_____

Answer:

To find the *gcf* of two or more algebraic terms, we first find the *gcf* of the numerical coefficients. Then we find the *gcf* of each variable expression by using the smallest exponent we have.

Practice Exercises

Find the *gcf* of the following terms.

1. $4xy^2$ and $6xy^3$ *Answer:* _____
2. $5xz^2$ and $7x^2z^3$ *Answer:* _____
3. $30a^3b^3$ and $20a^3b^2$ *Answer:* _____
4. $16x^2y$ and $24x^3y^4$ *Answer:* _____
5. $10\,pq$ and $12p^3$ *Answer:* _____
6. $25mn^4$ and $30m^4n$ *Answer:* _____
7. $5x$ and $20x^2$ *Answer:* _____
8. $25y^3$ and $50y^2$ *Answer:* _____
9. $15xy$ and $30xy^2$ *Answer:* _____
10. $2a^3b^2$ and $6a^2b^4$ *Answer:* _____
11. x^7 and x^3 *Answer:* _____
12. y^6 and y^{12} *Answer:* _____
13. x^2y^4 and xy^6 *Answer:* _____
14. $12x$ and $30x^2$ *Answer:* _____
15. $16a^3$ and $8a$ *Answer:* _____
16. $3x^4$ and $12x^2$ *Answer:* _____
17. x^3 and $12x^2$ *Answer:* _____
18. $14a^3$ and $49a^7$ *Answer:* _____
19. $8x^2y^3$ and $4x^3y$ *Answer:* _____
20. $9a^2b^4$ and $24a^4b^2$ *Answer:* _____
21. $-15a^4b^2$ and $9ab^6$ *Answer:* _____
22. $12a^2b^2$ and $16ab^5$ *Answer:* _____
23. $10a^3b^2c^4$ and $15ab^2c^3$ *Answer:* _____
24. $8x^3y^2z$ and $6x^4yz^5$ *Answer:* _____
25. $9ab^2c^3$ and $12ab^2c$ *Answer:* _____
26. $16x^2y^4z^6$ and $24xy^8z^2$ *Answer:* _____
27. ab^2c^3 and $5a^3b^2c$ *Answer:* _____
28. $16x^2$, $8x^4y^2$, and $12xy$ *Answer:* _____
29. $10a^3$, $20a^2b$, and $30ab^3$ *Answer:* _____
30. $25xy^2z$, $30x^2y^2$, and $40x^2y^2z$ *Answer:* _____

Answers to Mini-Practice:

(a) 5   (b) $5a$ (c) $2xy$ (d) $6xy^2$

5.12 THE *gcf* OF THE SUM OR DIFFERENCE OF UNLIKE TERMS

The greatest common factor of sums or differences of unlike terms is the *gcf* of the terms involved. For instance, to find the *gcf* of $4x^3y^2 + 6xy^3$, we first find $gcf(4x^3y^2, 6xy^3)$.

What is that? _____

Therefore, the *gcf* of $(4x^3y^2 + 6xy^3)$ is $2xy^2$.

Example 1   Find the *gcf* of $5x + 5y$.

Solution.

5 is the greatest common factor of the terms $5x$ and $5y$.
5 is the greatest common factor of the binomial $5x + 5y$, that is,

$$gcf(5x + 5y) = 5$$ ■

Example 2   Find the *gcf* of $9x^2y^3 - 15x^3y^2$.

Solution.

$3x^2y^2$ is the *gcf* of the terms $9x^2y^3$ and $15x^3y^2$.
$3x^2y^2$ is the *gcf* of the binomial $9x^2y^3 - 15x^3y^2$, that is,

$$gcf(9x^2y^3 - 15x^3y^2) = 3x^2y^2$$ ■

MINI-PRACTICE   Find the *gcf* of the following binomials:
(Use this space for your work.)

      (a) $16 + 24x$ (b) $xy - x^2y$ (c) $5ab + 20a^2$

Answers: (a) _____   (b) _____   (c) _____

IN YOUR OWN WORDS   explain how to find the greatest common factor of a binomial. _____

_____

_____

Answer:

To find the greatest common factor of a binomial, we find the greatest common factor of the terms of the binomial.

_____

Answers to Mini-Practice:

(a) 8   (b) xy (c) $5a$

Practice Exercises

In Exercises 1–10, find the *gcf* of the following binomials.

1. $5xz^2 + 7x^2z^3$ *Answer:* _____
2. $30a^3b^2 - 20a^2b$ *Answer:* _____
3. $16x^5y + 24x^2y^5$ *Answer:* _____
4. $8ab - 10ab^2$ *Answer:* _____
5. $10pq - 12p^3q$ *Answer:* _____
6. $25mn^4 + 30m^4n$ *Answer:* _____
7. $18x^5y^5 + 20x^5y^6$ *Answer:* _____
8. $12ab^3 + 18a^3b$ *Answer:* _____
9. $12x^3y + 9xy^3$ *Answer:* _____
10. $20x^2y^4z^5 + 40xy^6z^{10}$ *Answer:* _____

5.13 FACTORING COMPLETELY

Find the *gcf* of $16 + 24x$: _____

Divide $16 + 24x$ by 8: _____ + _____ (quotient)

Write $16 + 24x$ as the *gcf* multiplied by the quotient. _____

$$16 + 24x = 8(2 + 3x)$$

The polynomial $16 + 24x$ has been **factored completely**.

Complete the following table:

Expression	gcf	Quotient	Factored Expression
$16 + 24x$	8	$2 + 3x$	$8(2 + 3x)$
$6xy + 9x$			
$2a^2 + 5a$			
$7t^2 + 28t$			
$5x^4 - 15x^3$			

Example 1   Factor completely: $3x - 6$

Solution.   3 is the greatest common factor. Simplify the quotient of $3x - 6$ divided by 3.

$$\frac{3x-6}{3} = \frac{3x}{3} - \frac{6}{3} = x - 2$$

$$3x - 6 = 3(x - 2)$$ ■

Example 2   Factor completely: $12x^3 + 15x$

Solution. $3x$ is the greatest common factor.

$$\frac{12x^3}{3x} + \frac{15x}{3x} = 4x^2 + 5$$

$$12x^3 + 15x = 3x(4x^2 + 5) \qquad\blacksquare$$

The same method applies to polynomials with more than two terms.

Example 3   Factor completely: $20x^2y + 10xy^2 - 30x^2y^2$

Solution. $10xy$ is the greatest common factor.

$$\frac{20x^2y}{10xy} + \frac{10xy^2}{10xy} - \frac{30x^2y^2}{10xy} = 2x + y - 3xy$$

$$20x^2y + 10xy^2 - 30x^2y^2 = 10xy(2x + y - 3xy) \qquad\blacksquare$$

MINI-PRACTICE
(Use this space for your work.)

Find the *gcf* of each of the following polynomials, and then factor each polynomial.

(a) $4x^2 - 6xy^2$ (b) $9x^2 + 12x^3$

(c) $3x - 5z$ (d) $10ab + 20ac - 15a^2$

Answers: (a) _____   (b) _____

(c) _____   (d) _____

IN YOUR OWN WORDS

explain how to factor a polynomial. _____

_____

_____

Answer:

To factor a polynomial completely, we find the greatest common factor of the terms and then divide each term by the *gcf*.

WATCH OUT!   *Complete Factorization*

Use the distributive law to factor:

3 from $12x^2 + 24x^3$: _____

4 from $12x^2 + 24x^3$: _____

Answers to Mini-Practice:

(a) $2x(2x - 3y^2)$ (b) $3x^2(3 + 4x)$ (c) 1 is the *gcf*. $3x - 5z$ is already in factored form, and cannot be factored further.   (d) $5a(2b + 4c - 3a)$

$12x$ from $12x^2 + 24x^3$: _____

$12x^2$ from $12x^2 + 24x^3$: _____

Look at your answers and determine if any of the quotients you obtained could be factored further..

Which of the quotients could not be factored further? _____

_____

_____

Answer:

The first three quotients are factored, but only the last quotient is factored completely.

Practice Exercises

Find the *gcf* and factor.

1. $5a + 5b$ *Answer:*_____
2. $7b - 7c$ *Answer:*_____
3. $8x^2 + 4x$ *Answer:*_____
4. $10y - 25y^2$ *Answer:*_____
5. $16b - 8a^2$ *Answer:*_____
6. $12x - 12y$ *Answer:*_____
7. $8x + 12y$ *Answer:*_____
8. $12y^2 - 15y$ *Answer:*_____
9. $4x^2 - 8$ *Answer:*_____
10. $4a^2 + 5a$ *Answer:*_____
11. $3a^2 + 6a^5$ *Answer:*_____
12. $9x - 5x^2$ *Answer:*_____
13. $14y^2 - 15y$ *Answer:*_____
14. $6b^3 - 10b^2$ *Answer:*_____
15. $2x^4 + 4x^2$ *Answer:*_____
16. $30a - 6$ *Answer:*_____
17. $20b + 5$ *Answer:*_____
18. $16a - 24$ *Answer:*_____
19. $3y^4 - 9y$ *Answer:*_____
20. $10x^4 - 12x^2$ *Answer:*_____
21. $12a^5 - 32a^2$ *Answer:*_____
22. $8a^8 - 4a^5$ *Answer:*_____
23. $16y^4 + 8y^7$ *Answer:*_____
24. $10z^3 - 5x^2$ *Answer:*_____
25. $6b^2 + 7b$ *Answer:*_____
26. $6 + 3y$ *Answer*_____
27. $15t^3 - 5t$ *Answer:*_____
28. $9x^2y - 36xy$ *Answer:*_____
29. $15ab + 3a^2b^2$ *Answer:*_____
30. $12n^2 + 24n^3$ *Answer:*_____
31. $6a^2b - 3ab^2$ *Answer:*_____
32. $7rs^3 - 14r^2$ *Answer:*_____
33. $x^2y^2 + xy$ *Answer:*_____
34. $3x^2y^4 - 6xy$ *Answer:*_____
35. $12a^2b^5 + 9ab$ *Answer:*_____
36. $2a^5b - 3xy^3$ *Answer:*_____
37. $6a^2b^3 - 12b^2$ *Answer:*_____
38. $8x^2y^3 - 4x^2$ *Answer:*_____
39. $a^2b^2 + ab$ *Answer:*_____
40. $3p^2 + 3p - 3$ *Answer:*_____
41. $7a^2b + 14ab^2 - 2ab$ *Answer:* _____
42. $3x^4 + 3x^2 - 6x$ *Answer:* _____

43. $2x^5 - 4x^3 - 8x$ *Answer:* _____

44. $5y^3 - 10y + 15y^2$ *Answer:* _____

45. $15a^2b + 30a^2b^3 - 45ab$ *Answer:* _____

46. $4xy - 12x^2y^2 + 16x$ *Answer:* _____

SUMMARY

1. The rules for operations with expressions containing variables are the same as those for numbers.

2. Like terms can be added or subtracted and thereby simplified. The numerical coefficients are added/subtracted and the common variable part stays the same.

3. When multiplying terms, the numerical coefficients are multiplied and the exponents of the variables that are the same are added.

4. When multiplying a polynomial by a monomial, multiplication is distributed over addition and subtraction.

5. When dividing two terms, the numerical coefficients are divided and the exponents are subtracted for the variables that are the same. When dividing a polynomial by a monomial, divide each term of the polynomial by the monomial.

6. A negative exponent is changed to a positive exponent by inverting.

7. (a) The greatest common factor (*gcf*) of two or more numbers or unlike terms is the largest number or term that divides these numbers or terms.

(b) The *gcf* of two or more like variables is that variable with the smallest exponent that occurs.

(c) To factor a polynomial with more than one term, begin by writing it as a product of two or more factors, one of which is the *gcf* of the given expression.

Review Practice Exercises

In Exercises 1–37, perform the indicated operations.

1. $\dfrac{a^9}{a^5}$ *Answer:* _____   **2.** $x^4 \cdot x^5$ *Answer:* _____   **3.** $(2x)^3$ *Answer:* _____

4. $(y^2 \cdot z^3)^4$ *Answer:* _____   **5.** $\dfrac{x^5}{x^8}$ *Answer:* _____   **6.** n^0 *Answer:* _____

7. $x^4 \cdot x^2$ Answer _____   8. $\dfrac{x^{12}}{x^4}$ Answer: _____   9. $\dfrac{c^4}{c^4}$ Answer: _____

10. $\dfrac{y^2}{y}$ Answer: _____   11. $(3x)^2$ Answer: _____   12. $\dfrac{x^4}{x^6}$ Answer: _____

13. $(3x^2)^2$ Answer: _____   14. $(3x^4)^3$ Answer: _____   15. $(2a)^4$ Answer: _____

16. $(3x)^3$ Answer: _____   17. $(4x^3)^2$ Answer: _____   18. $(2ab^2)^3$ Answer: _____

19. $(3x^2y)^2$ Answer: _____   20. $(.2)^3(.2)^2$ Answer: _____   21. $(1.5x)^2$ Answer: _____

22. $(9.7)^0$ Answer: _____   23. $\left(\dfrac{3xy^3}{4}\right)^2$ Answer: _____   24. $\left(\dfrac{a}{b}\right)^5$ Answer: _____

25. $\left(\dfrac{3x^2}{2}\right)^2$ Answer: _____   26. $\left(\dfrac{4a^2}{b^3c^5}\right)^3$ Answer: _____

27. $(4a^2b)(-2ab^2)(6a^2b^3)$ Answer: _____   28. $(10xy)(-2xy^2z^3)(-5x^4z)$ Answer: _____

29. $(2ab)(b^2c)(-2ab^2c)(-abcd)$ Answer: _____   30. $\dfrac{6x^2y}{3xy}$ Answer: _____

31. $\dfrac{12a^4b^3c}{-4abc}$ Answer: _____   32. $\dfrac{28mn}{-14m^4n^3}$ Answer: _____

33. $\dfrac{20x^{-3}y}{10x^2y^4}$ Answer: _____   34. $\dfrac{12x + 4y}{4}$ Answer: _____

35. $\dfrac{8a^2 + 10ab}{2a}$ Answer: _____   36. $\dfrac{15x^3y^2 + 18x^2y^3 - 24x^3y^3}{3x^2y^2}$ Answer: _____

37. $\dfrac{10a^2bc - 20a^4b^2c + 30ac}{10ac}$ Answer: _____

In Exercises 38–56, perform the following addition and subtraction operations. If the expression cannot be simplified, explain why not.

38. $5x - 4x$ Answer: _____   39. $5x^2 - 4x$ Answer: _____   40. $5x^3 - 4x^3$ Answer: _____

41. $5x^4 - 4x^5$ Answer: _____   42. $10a + 10a^2$ Answer: _____   43. $3M + 3M^2$ Answer: _____

44. $7Y - 3Y$ Answer: _____   45. $8P + 3P^2$ Answer: _____   46. $6b - 3b$ Answer: _____

47. $3y^5 - 2y^5$ Answer: _____   48. $9M^3 + M$ Answer: _____   49. $2Q - Q^2$ Answer: _____

50. $2Q^2 + 9Q^2$ Answer: _____   51. $5.3x - 7.2x$ Answer: _____

52. $9.2x^3 - 10.7x^3$ Answer: _____   53. $7.3a^4b^5 - 9.3a^4b^5$ Answer: _____

54. $\dfrac{3}{4}xy^2 - \dfrac{7}{8}xy^2$ Answer: _____   55. $\dfrac{5}{2}cd^3 - \dfrac{3}{2}cd^3$ Answer: _____

56. $\dfrac{9}{7}pqr - \dfrac{8}{7}pqr$ Answer: _____

In Exercises 57–90, factor.

57. $4x + 4$ Answer: _____   58. $6y + 18$ Answer: _____

59. $9a^2 + 12a$ Answer: _____   60. $20x + 25$ Answer: _____

61. $12x - 18$ *Answer:* _____

62. $3x + 6$ *Answer:* _____

63. $3b^2 + 7b$ *Answer:* _____

64. $12y^3 + 15y$ *Answer:* _____

65. $4a^2b + 10ab^2$ *Answer:* _____

66. $9x^2 + 6x$ *Answer:* _____

67. $15pq^2 - 25p^2q$ *Answer:* _____

68. $6x^2 + 4x$ *Answer:* _____

69. $16a^2b - 20a$ *Answer:* _____

70. $14x^2y^2 + 21xy$ *Answer:* _____

71. $a^2 + 7a$ *Answer:* _____

72. $6a^2 - 6a$ *Answer:* _____

73. $3b^2 + 9$ *Answer:* _____

74. $4x^2 + 8x$ *Answer:* _____

75. $10a^2 - 25$ *Answer:* _____

76. $20x^3 + 30x$ *Answer:* _____

77. $9y^3 + 27y$ *Answer:* _____

78. $2x^2 + 4xy$ *Answer:* _____

79. $3a^2b^2 + 9ab$ *Answer:* _____

80. $12p^2q - 18pq^2$ *Answer:* _____

81. $12x^3y + 20xy^2$ *Answer:* _____

82. $8x^3y - 12x^2$ *Answer:* _____

83. $24x^3 + 16x^2 + 8x$ *Answer:* _____

84. $39x^4y^3 - 26x^3y^2 + 13x^2y$ *Answer:* _____

85. $15a^2b + 10ab^2 - 5ab$ *Answer:* _____

86. $14a^5 + 12a^4 - 8a^3 + 4a^2$ *Answer:* _____

87. $16x^2yz - 12xy^2z$ *Answer:* _____

88. $9x^2y^4 + 6x^2y^3 - 3xy^2 - 2y$ *Answer:* _____

89. $25a^4b^3c^2 - 15a^3b^2c + 10a^2bc^2$ *Answer:* _____

90. $16x^2y^4z^3 - 8x^4y^2z^3 + 12x^3y^3z^4$ *Answer:* _____

If you have questions about this chapter, write them here: _____

_____

_____

_____

_____

6 Translations Using Variables

We continue to compare English and mathematical sentences. Until now you have worked only with constants. Now you will also use variables as symbols for numbers. This will help you to develop skills in writing equations and in solving word problems later on.

6.1 WORD PROBLEMS

Read the following folk story:

A poor, hungry boy was walking along a bright road, when he ran into a kind fisherman. "I caught so many fish today;" said the fisherman, feeling sorry for the boy, "take some home." And he handed the boy nine fish.

The boy continued happily along the road, until he met an old friend, who also was hungry. "Here's some fish for you," he said. And he gave his friend three fish.

Farther down the road he met another friend. "Here," he said, "take half of my fish!"

The boy walked home in the sunshine with the fish he had left and gave them to his mother.

Let's figure out what's going on in the story:

How many fish did the first boy get from the fisherman? _____

159

Attending Regularly

1. **Each class teaches you something new.**
   • Try not to miss even one class.
   • Come to class every day, except in case of an emergency.

2. **What you miss when you are absent:**
   • Your instructor will emphasize the most important ideas.
   • A classmate may think of a question you didn't think of.
   • Your instructor's explanation will tie things together for you.

3. **If you miss a class:**
   • Have the phone numbers of two classmates. Call to get the homework.
   • Do the homework before the next class in order to catch up.
   • Know your instructor's office hours. Stop in and get help.
   • Go to the math lab for help.

What did he do with the fish? _____

   Translate this to mathematics: _____

What did he do with the fish he now had left? _____

   Translate this to mathematics: _____

Retrace your steps and write all mathematical operations in *one sequence without doing any calculations*:

_____

You probably came up with $\frac{9-3}{2}$ without too much difficulty. What you did was to state and solve a word problem in mathematical symbols. True, the word problem you just looked at is not typical. Most word problems do not look like folk tales. But with a little training, you can come to see them as stories.

   Let's look at another word problem: Half of the difference of a number minus three.

   How did you feel when you read this? _____

You may have written "angry," "depressed," "nervous." Maybe you felt challenged. Most people do not like word problems. There are no bright roads and no kind fishermen. But if you hang on — there's a story of sorts. There is a challenge, if you can only try to approach the problem without fear. As you saw from the folk tale, you have the skills to solve these problems. Go back to the problem: Half of the difference of a number and 3.

What if the number were nine? Translate the difference of nine minus three: _____

Without writing the answer, translate half of that difference:

_____

Did you get $\dfrac{9-3}{2}$? _____

Compare this result with the number of fish in the folk tale! They should be the same. You are looking at a problem you already solved!

Now, if we didn't have 9, but, instead, had "a number?" _____

Let n mean "number," so that we then have $\dfrac{n-3}{2}$.

There are a few useful hints for translating from English into algebra:

> **1. Calm down.**
>
> **2. Try to remember another similar statement you have translated earlier.**
>
> **3. Go slowly, step by step. Do not try to swallow the whole statement at once.**

Translations from English into Mathematics

Operation: Addition

Translate: The sum of six and four: _____

Now translate: The sum of *a number* and four: _____

What number do you think of as being "a number?" _____

Could a different number be correct? _____

The answer is "yes" since if one of the numbers is unknown, we are free to think of any number.

How can we represent an unknown number? _____

An unknown number is represented by a letter, such as n, x, or y. The symbol for "sum" is of course +. Other possible representations for "the sum of a number and four" are:

(a) $n+4$ (b) $y+4$ (c) $t+4$ (d) $p+4$

Translations from Mathematics into English

Now that we have translated some English sentences into mathematics, let us reverse the problem. In Chapter 4 you practiced translating mathematical expressions into English phrases. At present, we see what expressions involving unknowns stand for in English.

What does x mean in the expression $x+3$? _____

You probably realized that x stands for "a number." So, the expression $x + 3$ can be translated as "the sum of a number and three" or "three more than a number." If we used y instead of x for the unknown, $y + 3$ would be translated in exactly the same way.

Translate $x + 5$ into English. First use the word in (a); then use the word in (b).

(a) sum _____

(b) more _____

Operation: Subtraction

Translate into mathematics:

(a) The difference of a number and six (in this order): _____

(b) The difference of six and a number (in this order): _____

(c) Six less than a number: _____

(d) A number subtracted from six: _____

The answer to (a) and (c) is $x - 6$. The answer to (b) and (d) is $6 - x$.

Translate $x - 5$ into English using the word

(a) difference _____

(b) less _____

Remember, that the word *difference* tells us to translate in *direct* order and the words *less than* tell us to translate in *reverse* order. Therefore, $x - 5$ is "the difference of a number and five" or "five less than a number."

Operation: Multiplication

Translate into mathematics:

The product of a number and five: _____

The product of five and a number: _____

A number times five: _____

Five times a number: _____

The answer to all these problems can be written as $x(5)$ or $5x$. Since multiplication is commutative, the expressions are equal.

Translate $5 \cdot x$ into English: _____

You might have said "the product of five and a number" or "five times a number."

Now translate $\frac{1}{5} \cdot x$ into English: _____

Again, we can use the words "product" or "times" in the translation, but here we could also say

$$\frac{1}{5} \text{ "of" a number}$$

Remember that the dot for multiplication is usually not written out:

$$5x \quad \text{is the same as} \quad 5 \cdot x$$

and

$$\frac{1}{5}x \quad \text{is the same as} \quad \frac{1}{5} \cdot x$$

WATCH OUT!   *Twice a number versus the square of a number*

Translate x^2 into English: _____

You probably wrote "the square of a number" or "a number to the second power."

Translate "Twice a number" into mathematics: _____

Hopefully, you remembered that "twice a number" is the same as "two times a number" or $2x$.

Operation: Division

Translate into mathematics:

(a) A number divided by five: _____

(b) Five divided by a number: _____

The answer to (a) is $x \div 5$. The answer to (b) is $5 \div x$.

Translate $x \div 2$ into English: _____

You might have said "A number divided by 2," or "2 into a number."

Example 1   Translate into mathematics: Seven less than a number.

Solution.   Reverse order:

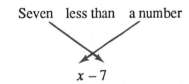

$$x - 7$$

∎

Example 2   Translate into English: $6 \div x$.

Solution.

Straight translation: The quotient of six divided by a number
                        Six divided by a number

Reverse translation: A number divided into six

∎

MINI-PRACTICE
(Use this space for your work.)

Translate into mathematics:

(a) A number subtracted from six   *Answer:* _____

(b) The quotient of seven divided by a number   *Answer:* _____

Translate into English:

(c) $4 - x$ *Answer:* _____

(d) $x \div 5$ *Answer:* _____

WATCH OUT!   *Multiplication and Division*

Take a whole apple. Cut it into two equal parts.

What do we call each part we have? _____

You probably wrote "half an apple."

When we divide an apple in two equal parts, each part is the same as one half of the apple. In symbols we write:

$$a \div 2 = \frac{1}{2}a$$

When we divide fractions, how do we carry out the division?

_____

You probably remembered that division is changed to multiplication when the second number is inverted. This is the same as with the apples:

$$a \div 2 = a \cdot \frac{1}{2}$$

Why is $a \cdot \frac{1}{2} = \frac{1}{2} \cdot a$? _____

The reason is that multiplication is commutative. In mathematics we say

$\frac{a}{2}$ is equal to $\frac{1}{2} \cdot a$ or $\frac{a}{2} = \frac{1}{2}a$

Practice Exercises

In Exercises 1–20, translate the following into mathematics.

1. The sum of four and a number   *Answer:* _____

2. A number increased by four   *Answer:* _____

3. Four more than a number   *Answer:* _____

4. A number added to four   *Answer:* _____

Answers to Mini-Practice:

(a) $6 - x$ (b) $7 + x$ (c) A number subtracted from four   (d) The quotient of a number divided by five

5. Four plus a number   *Answer:*_____

6. The difference of three and a number in this order   *Answer:*_____

7. Three less than a number   *Answer:*_____

8. The difference of a number and four in this order   *Answer:*_____

9. Take away four from a number.   *Answer:*_____

10. A number decreased by four   *Answer:*_____

11. The difference of four and a number in this order   *Answer:*_____

12. A number minus four   *Answer:*_____

13. Four decreased by a number   *Answer:*_____

14. Four less than a number   *Answer:*_____

15. Two times a number   *Answer:*_____

16. The product of two and a number   *Answer:*_____

17. A number divided by three   *Answer:*_____

18. Three divided by a number   *Answer:*_____

19. The quotient of a number divided by three   *Answer:*_____

20. The difference of three and a number in this order, divided by nine   *Answer:*_____

In Exercises 21–30, translate the following into English.

21. $7 - x$ *Answer:* _____

22. $8 \cdot x$ *Answer:* _____

23. $x + 7$ *Answer:* _____

24. $x - 15$ *Answer:* _____

25. $8 \div x$ *Answer:* _____

26. $a \cdot 19$ *Answer:* _____

27. $23 + m$ *Answer:* _____

28. $21 - n$ *Answer:* _____

29. $t - 33$ *Answer:* _____

30. $y \cdot 96$ *Answer:* _____

6.2 MORE THAN ONE OPERATION

Translate the algebraic expression $4 + 2x$ into English:

_____

You probably gave one of the following answers:

(a) The sum of four and two times a number

(b) Four increased by twice a number
(c) Twice a number added to four
(d) Four plus the product of two times a number

In this example, how many operations were involved? _____

There is both addition and multiplication.

List the different operations in the expression $y + \dfrac{5}{y}$:

_____

Translate the expression $y + \dfrac{5}{y}$ into English in two different ways:

_____

_____

Probably your answers can be matched with two of the following possible translations:

(a) The sum of a number and five divided by the number. $\Big($This can be misleading because it could also be the translation of $\dfrac{x+5}{x}$.$\Big)$

(b) A number increased by the quotient of 5 divided by the number

(c) The quotient of five and a number added to the number

In the following table, match each expression with its correct English translation by using arrows.

Expression	*English Translation*
(a) $4 - 6$	1. Three more than the product of two and five
(b) $6 - 4$	2. Eight is more than five.
(c) $2x + 3x$	3. Six subtracted from a number
(d) $\dfrac{3}{x}$	4. The sum of three and a number is eight.
(e) $3 + n = 8$	5. The sum of twice a number and three times the same number
(f) $\dfrac{x}{3} - 5$	6. Three divided by a number
(g) $5 - \dfrac{x}{3}$	7. Four less than a number
(h) $2 \cdot 5 + 3$	8. Eight more than five
(i) $x - 6$	9. One-third of a number is subtracted from five.
(j) $8 > 5$	10. Six less than four
(k) $5 + 8$	11. The difference of six and four in this order
(l) $x - 4$	12. Five subtracted from a number that is divided by three

The correct matchings are: (a)–10; (b)–11; (c)–5; (d)–6; (e)–4; (f)–12; (g)–9; (h)–1; (i)–3; (j)–2; (k)–8; (l)–7.

Example 1   Translate: The sum of three times a number and four

Solution.

$$\underset{3x}{\underset{\downarrow}{\text{three times a number}}} \quad \overset{\text{The sum of}}{\underset{+}{\underset{\downarrow}{}}} \quad \underset{4}{\underset{\downarrow}{\text{and four}}}$$

■

Example 2   Translate: Twelve decreased by the sum of a number and five

Solution.

$$\underset{12}{\underset{\downarrow}{\text{Twelve}}} \quad \underset{-}{\underset{\downarrow}{\text{decreased by}}} \quad \underset{(x+5)}{\underset{\downarrow}{\text{the sum of a number and five}}}$$

The sum, $(x + 5)$, must be in parentheses.   ■

MINI-PRACTICE

(Use this space for your work.)

Translate into mathematics:

(a) Ten subtracted from twice a number   *Answer:* _____

(b) Six divided by the sum of a number and ten   *Answer:* _____

(c) Three times a number that has been decreased by four   *Answer:* _____

Practice Exercises

Translate into mathematics:

1. Six more than twice a number   *Answer:* _____

2. A number increased by six times that number   *Answer:* _____

3. Five less than the product of a number and eight   *Answer:* _____

4. The product of a number and the sum of the same number and five   *Answer:* _____

5. Twelve less than one-third of a number   *Answer:* _____

6. The sum of five-sixths of a number and three-eighths of that same number *Answer:* _____

7. A number is decreased by five and then increased by 12.   *Answer:* _____

8. Five times the sum of a number and three   *Answer:* _____

9. The sum of five times a number and two   *Answer:* _____

Answers to Mini-Practice:

(a) $2x - 10$ (b) $6 \div (x + 10)$ (c) $3(x - 4)$

10. Sixteen decreased by the sum of twice a number and eleven   *Answer:* _____

11. Six times a number plus four times the same number   *Answer:* _____

12. Eight is decreased by the sum of six and a number.   *Answer:* _____

13. Twice a number is increased by three more than the given number.   *Answer:* _____

14. Fifteen more than twice a number   *Answer:* _____

15. Add three to a number and then divide the result by two.   *Answer:* _____

16. Three times a number is subtracted from twice that same number.   *Answer:* _____

17. Take a number and subtract three; double the result, and then divide by four.   *Answer:* _____

18. Take away six from the square of a number.   *Answer:* _____

19. The sum of three times a number and twice the square of that same number   *Answer:* _____

20. Four times the sum of three times a number and the cube of that same number   *Answer:* _____

21. A number added to the product of three and the number   *Answer:* _____

22. Four times the sum of a number and nine   *Answer:* _____

23. The sum of twelve times a number and twice that same number   *Answer:* _____

24. Four increased by the quotient of six divided by a number   *Answer:* _____

25. The sum of three times a number and four   *Answer:* _____

26. Seven subtracted from twice a number   *Answer:* _____

27. Seven less than the product of a number and ten   *Answer:* _____

28. The sum of one-half a number and six   *Answer:* _____

29. Twice a number is subtracted from the square of that same number.   *Answer:* _____

30. The quotient of a number divided by the square of that same number   *Answer:* _____

SUMMARY

When translating from English into mathematics:

- Letters, such as n, x, or y represent the unknown number.
- *Difference* tells us to translate in direct order; *less than* tells us to translate in reverse order.
- The dot for multiplication is usually not written out.
- *Divided by* tells us to translate in direct order; *divided into* tells us to translate in reverse order.
- Sometimes there is more than one operation involved.

Review Practice Exercises

In Exercises 1-20, translate the following into mathematics.

1. The sum of a number and ten   *Answer:* _____

2. Five more than a number   *Answer:* _____

3. Five less than a number   *Answer:* _____

4. Take away six from a number.   *Answer:* _____

5. A number decreased by twelve   *Answer:* _____

6. Twelve decreased by a number   *Answer:* _____

7. Five times a number   *Answer:* _____

8. A number divided by five   *Answer:* _____

9. Five divided by a number   *Answer:* _____

10. Eight more than three times a number   *Answer:* _____

11. Six less than the product of a number and five   *Answer:* _____

12. The square of a number is decreased by one.   *Answer:* _____

13. One more than twice a number   *Answer:* _____

14. Ten less than one-fourth of a number   *Answer:* _____

15. The sum of the square of a number and three   *Answer:* _____

16. Subtract five from a number and then divide the result by three.   *Answer:* _____

17. Take away ten from the product of five and a number.   *Answer:* _____

18. Three times a number divided by ten   *Answer:* _____

19. The sum of two-thirds of a number and twice the same number   *Answer:* _____

20. Seventeen increased by the quotient of eight divided by a number   *Answer:* _____

In Exercises 21–30, translate the following into English.

21. $10 \cdot x$ *Answer:* _____

22. $\dfrac{x}{4}$ *Answer:* _____

23. $\dfrac{4}{x}$ *Answer:* _____

24. $9 - x$ *Answer:* _____

25. $9 + x$ *Answer:* _____

26. $y(23)$ *Answer:* _____

27. $y + 14$ *Answer:* _____

28. $\dfrac{14}{y}$ *Answer:* _____

29. $\dfrac{y}{14}$ *Answer:* _____

30. $20z$ *Answer:* _____

If you have questions about this chapter, write them here: _____

_____

_____

_____

_____

_____

PART III
Algebraic Expressions

7 Evaluation

Remember to

* *USE PARENTHESES WHEN SUBSTITUTING FOR VARIABLES*
* *KEEP A LIST OF FORMULAS FOR REFERENCE*
* *PREPARE BEFORE EACH CLASS BY READING AHEAD*
* *BRING QUESTIONS TO CLASS*

In this chapter you will replace the variables in expressions with constants and then evaluate the expressions. You will learn a little about codes and also about the importance of formulas.

7.1 CODES

Look at the following:

$$\cdot\,\cdot\,\cdot \quad -\,-\,- \quad \cdot\,\cdot\,\cdot$$
$$\text{S} \qquad \text{O} \qquad \text{S}$$

What does it indicate? _____

No matter where you come from, you probably said that SOS indicates a call for HELP!

The series of dots and dashes is Morse Code for SOS, which is popularly, though erroneously, thought to mean "Save Our Ships." But it is a kind of shorthand notation to call for help. The International Morse Code was developed around 1840 by the American artist and inventor, Samuel Morse. He translated all letters and numbers into series of dots and dashes, which could be transmitted through wires over long distances.

Preparing for Class

1. **Before class:**
   * Look over the preceding lesson. This will help you follow the present lesson.
   * Write down a list of questions on your homework. Ask them in class.
   * Come early to class. It's harder to follow class if you miss the first few minutes.

2. **During class:**
   * Ask as many questions as you need to about what you don't understand.
   * Don't wait for someone else to ask your questions.
   * Answer the questions you can.
   * Don't be afraid to take part in discussion.

3. **After class:**
   * If you still don't understand something, ask your instructor right after class.
   * If there is not enough time, see your instructor during office hours.
   * Attend the math lab.

You are probably familiar with certain codes. Below are some codes; check those that you have seen before:

007	HAL
911	212
F	K

Each of the codes above stands for something different. Some codes stand for certain telephone areas: 212, for example, indicates Manhattan and the Bronx. 911 is the telephone code for emergency.

F is a code for a failing grade, and K stands for 1000. Adventure-movie fans know that 007 is a fictional secret agent's code name.

To get the feel of a code, let us assign numbers to the alphabet as follows: a = 1, b = 2, c = 3, etc. Complete the following table by assigning to consecutive letters of the alphabet consecutive positive integers.

Letter:	a	b	c	d	e	f	g	h	i	j	k	l	m
Number:	1	2	3	4	__	__	__	__	__	__	__	__	__

Letter:	n	o	p	q	r	s	t	u	v	w	x	y	z
Number:	__	__	__	__	__	__	__	__	__	__	__	__	__

Now, decode the following message:

1 12 7 5 2 18 1   9 19   6 21 14

_____

Let's hope you will agree with the message: "Algebra is fun!"

Variables   Evaluate the following expressions:

$2 + 3 =$ _____ $x + y =$ _____

No doubt you were able to find the value of

$2 + 3$ as 5

But what did you write as the value of $x + y$? _____

It just can't be done. We cannot evaluate an expression in algebra until we give a specific value to each letter. When we do this, we are doing a form of coding.

Replace x with any number you wish: $x =$ _____

Replace y with any number you wish: $y =$ _____

Now, evaluate $x + y$: _____ + _____ = _____

In mathematics, when we replace a letter by a number, we say we **substitute** a number for a variable or **assign** a number to a variable.

In the expression, $3x^2 + x - 2$, how many different variables do we have? _____

We have only one variable, and that one is x. A letter always stands for the same number in an expression.

Example 1   How many different variables are there in the term xy?

Solution.   In the term

$$xy$$

there are two different letters, x and y. Therefore, xy consists of *two* variables. ∎

Example 2   How many different variables are there in the polynomial

$$x^3 - x^2 + x + 4$$

Solution.   In the polynomial
$$x^3 - x^2 + x + 4$$

there is only one variable, namely x. ∎

MINI-PRACTICE
(Use this space for your work.)

How many different variables are there in the following polynomials:

(a) $2x^2 + 3x$ (b) $3x - 5y - 1$ (c) $xy - x + y$

Answers: (a) _____   (b) _____   (c) _____

IN YOUR OWN WORDS

explain how we know how many variables there are in a polynomial: _____

_____

_____

Answer:

Count the number of letters involved. Any letter always stands for the same value in an expression. Different letters could be assigned the same value, but the same letter *must* be assigned the same value. We count the variables by counting the number of *different* letters.

Practice Exercises

How many different variables do we have in each of the following expressions?

1. $y^3 + y^2$ *Answer:* _____   2. $4xy - y$ *Answer:* _____

3. $3x^2 + 2x - 5$ *Answer:* _____   4. $5a^5b^3$ *Answer:* _____

5. $2a^2bc - 6ab^2c^3 + 4a^3b^3c^4$ *Answer:* _____   6. $13s - 9t$ *Answer:* _____

7. $39pq + 2q$ *Answer:* _____   8. $x^3 + 2x^2 + x - 2$ *Answer:* _____

9. $8a^2b^8c^{12}$ *Answer:* _____   10. $2mn^7 + 5m^3n^2 - 3m^3n^3$ *Answer:* _____

7.2 EVALUATING ALGEBRAIC EXPRESSIONS

When we are asked to evaluate algebraic expressions, we first need some information. To help us understand what information is needed, it is useful to translate the expression into English. Translate the following expression into English:

$2x + 3y + 1$: _____

_____

You probably wrote something like: "Twice a number plus three times another number plus one." Since $2x$ tells us that the number 2 is multiplied by x, and $3y$ tells us that the number three is multiplied by y, it is understood that $2x + 3y + 1$ means $2(x) + 3(y) + 1$.

———————————————

Answers to Mini-Practice:

(a) 1   (b) 2   (c) 2

Write "*" where you get a message to multiply.
Write "/" where you get a message to divide.

1. −8*ab* _____   **2.** 5 + 4*a* _____

3. *a* ÷ *b* _____   **4.** 3*a* ÷ *b* _____

Answers: **1.** −8 * *a* * *b*   **2.** 5 + 4 * *a*   **3.** *a*/*b*   **4.** 3 * *a*/*b*

Now replace *a* with 6 and *b* with 2 and evaluate each of the preceding answers.

1. _____

2. _____

3. _____

4. _____

Answers: **1.** −96   **2.** 29   **3.** 3   **4.** 9

Go back to the code we used on page 174. Use the same numbers and find the values of the following expressions:

Expression	Expression Written with Number Replacement	Value
a + *b*		
ab		
p + *q* − *k*		
like		
l + *i* + *k* + *e*		

Did you get the same value for *l* + *i* + *k* + *e* as for *like*? _____

If not, why not? _____

What is the operation involved in the expression *l* + *i* + *k* + *e*?

_____

What is the operation involved in the expression *like*? _____

Recall that if there is no visible operation symbol between two letters, it is understood that there is always an invisible multiplication sign.

Example 1   Evaluate 3*x* when $x = \frac{1}{6}$.

Solution. $3\left(\frac{1}{6}\right) = \frac{3}{1} \cdot \frac{1}{6} = \frac{1}{2}$ ∎

Example 2   Evaluate $2a^2 + 3b$ when $a = 3$ and $b = -2$.

Solution.

$$2a^2 = 2(3)^2 = 2(9) = 18$$
$$3b = 3(-2) = -6$$
$$2a^2 + 3b = 18 + (-6) = 12$$ ∎

MINI-PRACTICE
(Use this space for your work.)

Suppose that $p = 3$, $q = 7$, and $r = 12$. Substitute the value of p, q, and r to determine the value of the following polynomials.

(a) $p + q$ (b) $2p + q$ (c) $3p + q - r$ (d) pq (e) pqr

Answers: (a) _____   (b) _____   (c) _____   (d) _____   (e) _____

IN YOUR OWN WORDS   explain how we evaluate a polynomial for given values: _____

_____

_____

Answer:

Each variable is replaced with a given number that is put in parentheses. The operations are then performed according to the rules for the order of operations.

WATCH OUT!   It is often important to use parentheses around variables when we substitute numbers for them. For example, if

$$y = \frac{1}{2}$$

then

$$3y = 3\left(\frac{1}{2}\right) = \frac{3}{1} \cdot \frac{1}{2} = \frac{3}{2} = 1\frac{1}{2}$$

If we do not use parentheses when we substitute $\frac{1}{2}$ for y, we have

$$3y = 3\frac{1}{2}   \text{(a mixed number)}$$

which actually means $3 + \frac{1}{2}$. That is, in this case we are adding instead of multiplying. In other words,

$$3\left(\frac{1}{2}\right) = \frac{3}{1} \cdot \frac{1}{2} = \frac{3}{2} = 1\frac{1}{2} = 1.5$$

whereas

$$3\frac{1}{2} = 3 + \frac{1}{2} = 3.5$$

Answers to Mini-Practice:

(a) 10   (b) 13   (c) 4   (d) 21   (e) 252

Practice Exercises

In Exercises 1–8, go back to the code on page 174 and evaluate the following polynomials for those values.

1. $a + a$ Answer:_____

2. sss Answer:_____

3. mnm Answer:_____

4. $2r + 5 + r$ Answer:_____

5. $3p - 2q$ Answer:_____

6. $8m - 2n^2$ Answer:_____

7. $(5s + 9t)^2$ Answer:_____

8. $4f + 13g - 9h$ Answer:_____

9. Evaluate the following polynomial when $x = 8$.

$$2x + 5x^2$$ Answer:_____

10. Evaluate the following polynomial when $x = 5$.

$$3(x + 1)$$ Answer:_____

11. Evaluate the following polynomial when $x = -8$.

$$-9x + 2x^2$$ Answer:_____

12. Evaluate the following polynomial when $x = -5$.

$$-14(x + 3)$$ Answer:_____

13. Evaluate the following polynomial when $x = -4$ and $y = 3$.

$$7x + 2(x + y)$$ Answer:_____

14. Evaluate the following polynomial when $x = 7$ and $y = -12$.

$$x(x + y) + 36$$ Answer:_____

15. Evaluate the following polynomial when $x = 3$ and $y = 5$.

$$3x + 2y - x(y - 2)$$ Answer:_____

16. Evaluate the following polynomial when $x = -6$ and $y = -11$.

$$3x(x^2 - y^2) - xy$$ Answer:_____

17. Evaluate the following polynomial when $a = 8$, $b = -2$, and $c = 5$.

$$a(b + c) - c^2$$ Answer:_____

18. Evaluate the following polynomial when $a = -9$, $b = -1$, and $c = 13$.

$$a^2 + b + c^2$$ Answer:_____

19. Evaluate the following polynomial when $a = -17$, $b = 0$, and $c = 25$.

$$b(3a^2 + 8b^5 + 2c)$$ Answer:_____

20. Evaluate the following polynomial when $a = -1$, $b = -2$, and $c = -3$.

$$(7a + b^2) - c^3$$ Answer:_____

In Exercises 21–24, evaluate the following polynomials when $a = 1$, $b = 0$, $c = 2$, $x = 3$, and $y = 4$.

21. $a + b + c$ Answer:_____

22. $c - b + y$ Answer:_____

23. $3a - by + x$ Answer:_____   **24.** $a^2 + c(y + cx)^2$ Answer:_____

In Exercises 25–28, evaluate the following polynomials when a = 1.5, b = 0, c = 0.5, x = 0.87, and y = 1.35.

25. $a + b + c$ Answer:_____   **26.** $c - b + y$ Answer:_____

27. $3a - by + x$ Answer:_____   **28.** $(a + y)^2 + (bx)^2 - c$ Answer:_____

In Exercises 29–32, assign your own values to the different variables. Evaluate each polynomial for the values you have chosen.

$$a = \_\_\_\_\_, \quad b = \_\_\_\_\_, \quad c = \_\_\_\_\_, \quad x = \_\_\_\_\_, \quad y = \_\_\_\_\_$$

29. $a + b + c$ Answer:_____   **30.** $c - b + y$ Answer:_____

31. $3a - by + x$ Answer:_____   **32.** $(a + c)^2 - b^3$ Answer:_____

In Exercises 33–38, evaluate 2x + 3y + 1 for the following values of x and y.

33. $x = 4, \quad y = 3$ Answer:_____   **34.** $x = 0, \quad y = 1$ Answer:_____

35. $x = 0, \quad y = -1$ Answer:_____   **36.** $x = -2, \quad y = 3$ Answer:_____

37. $x = -1, \quad y = 2$ Answer:_____   **38.** $x = -2, \quad y = -2$ Answer:_____

In Exercises 39–42, evaluate the following polynomials when a = –1, b = –2, c = –3, x = 0, and y = –4.

39. $a + b + c$ Answer:_____   **40.** $c - b + y$ Answer:_____

41. $3a - by + x$ Answer:_____   **42.** $(a + y)^2 + (bx)^2 - c$ Answer:_____

7.3 FORMULAS FROM GEOMETRY

We use formulas in many areas of our lives. Some say there is a formula for a happy marriage. You might have used a formula for a baby's bottle, or a formula in a chemistry lab.

Commonly used expressions or rules for obtaining certain solutions are called **formulas**. When we bake, we often use recipes. A recipe is a formula. Pharmacists use formulas to fill prescriptions.

Formulas in science are often developed experimentally. Sometimes we can understand how the formula was developed. However, in more advanced work, we may not be able to understand how each formula we come across was developed. Nevertheless, we must know how to *use* these formulas. Formulas are used a lot in business and science. When you use formulas, you *apply* your knowledge of *mathematics*. If you think about it, you have used many formulas at work or at school.

Name one! _____

You might have said "the area of a rectangle is length times width," for example. If you worked in a bank, you might be familiar with formulas used for calculating interest. Many people use formulas for different kinds of percent problems.

Consider a formula as a code. You have to replace the variables of the formula with numbers, and then evaluate the numerical expression you get. We first consider some important formulas from geometry.

Area of Rectangle   Draw a rectangle that is 1 inch by 2 inches:

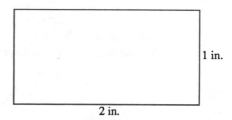

What is the area of the rectangle? _____

Your rectangle is (1×2) square inches.

If the length were l and the width w, what would the area be?

_____

$$A = l \cdot w$$

would be the area of the rectangle.

Example 1   Find the area of a rectangle when $l = 5$ cm. and $w = 2$ cm.

Solution. $A = l \cdot w$ becomes $A = 5 \cdot 2 = 10$

The area is 10 cm^2.   ■

Area of Triangle   If, in the drawing you made of the rectangle, the length had been b and the width h, what would the area have been? _____

In the rectangle, draw a diagonal (a line connecting two opposite corners) so that the rectangle is divided into two parts.

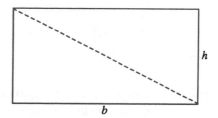

What is each part called? _____

What is the area of each triangle? _____

$$A = \frac{b \cdot h}{2}$$

is the formula for the area of a triangle, where b represents the base and h represents the height.

Example 2   Find the area of a triangle with base 10 inches and height 2 inches.

Solution. $A = \dfrac{b \cdot h}{2}$ becomes $A = \dfrac{10 \cdot 2}{2} = 10$

The area is 10 square inches.   ■

Area of Trapezoid   A four-sided figure with exactly two parallel sides is called a **trapezoid**. The following figures are examples of trapezoids.

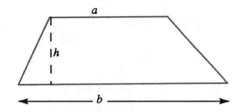
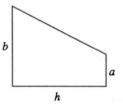

The two parallel sides are called **bases** and can be represented by a and b. The area of the trapezoid is

$$A = \frac{h(a+b)}{2}$$

where a represents the smaller base, b the larger base, and h the height. (**Height** is defined as the distance between the parallel sides.) As an example, let's find the area of the following trapezoid. In this figure cm. stands for centimeters.

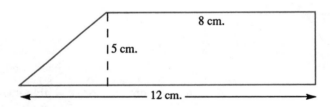

What is the value of a? _____

What is the value of b? _____

What is the value of h? _____

Since 8 cm. is the length of the shorter base, we have $a = 8$ and $b = 12$. The height is 5 cm., so that $h = 5$. Substitute these values in the formula:

$$A = \frac{\underline{\quad}(\underline{\quad} + \underline{\quad})}{2}$$

$$A = \frac{h(a+b)}{2} = \frac{5(8+12)}{2} = \frac{5(20)}{2} = \frac{100}{2} = 50$$

The area is 50 cm.^{2}

Example 3   Find the area of a trapezoid with base $a = 2$ in., base $b = 4$ in., and the height $h = 3$ in.

Solution.   The area of a trapezoid is given by

$$A = \frac{h(a+b)}{2}$$

Here

$$\frac{h(a+b)}{2} = \frac{3(2+4)}{2} = \frac{3 \cdot 6}{2} = \frac{18}{2} = 9$$

The area is 9 square inches.   ■

MINI-PRACTICE
(Use this space for your work.)

(a) The length of a rectangle is 7 in. and the width is 4 in. Find the area.

    Answer: _____

(b) The base of a triangle is 10 cm. and the height is 6 cm. Find the area.

    Answer: _____

(c) The bases of a trapezoid are 3 inches and 7 inches, and the height is 5 inches. Find the area.

    Answer: _____

Perimeter   The **perimeter** is the total length around a geometric figure. Go back to your rectangle and measure the perimeter.

$$P = 2l + 2w$$

is the formula used to find the perimeter of (total length around) a rectangle. $2l + 2w$ means twice the length plus twice the width.

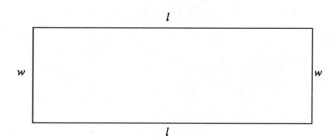

If the length is 3 inches and the width 1 inch, what is the perimeter?

_____

$$P = 2(3) + 2(1) = 6 + 2 = 8$$

The perimeter is 8 inches.

_____

Answers to Mini-Practice:

(a) 28 square inches   (b) 30 cm.^{2} (c) 25 square inches

Example 4   Find the perimeter of a rectangle with length 5 inches and width 3 inches.

> *Solution.* $P = 2l + 2w = 2(5) + 2(3) = 16$
>
> The perimeter is 16 inches.   ■

A **square** is a rectangle with all four sides equal.

Example 5   Find the perimeter of a square with a side of 4 cm.

> *Solution.* $P = 2(4) + 2(4) = 8 + 8 = 16$ or $P = 4(4) = 16$
>
> The perimeter is 16 inches.   ■

MINI-PRACTICE
(Use this space for your work.)

Find the perimeter of each figure.

(a) A rectangle with length 6 inches and width 4 inches   *Answer:* _____

(b) A triangle with sides 3, 4, and 5 inches   *Answer:* _____

(c) A trapezoid with bases 10 and 6 inches and the other sides 4 inches each

   Answer: _____

Practice Exercises

In Exercises 1–8, find the areas of rectangles having the indicated measurements:

1. Length = 5 inches and width = 3 inches   *Answer:* _____

2. Length = 2.3 inches and width = 1.4 inches   *Answer:* _____

3. Length = 24 cm. and width = 12 cm.   *Answer:* _____

4. Length = 10.3 cm. and width = 5.4 cm.   *Answer:* _____

5. Length = $\frac{1}{2}$ inch and width = $\frac{3}{4}$ inch   *Answer:* _____

6. Length = $\frac{3}{2}$ cm. and width = $\frac{5}{2}$ cm.   *Answer:* _____

7. Length = 0.02 km. and width = 2.3 km.   *Answer:* _____

8. Length = $\frac{7}{8}$ feet and width = $\frac{8}{7}$ feet   *Answer:* _____

In Exercises 9–16, find the areas of the following triangles.

9. Base = 3 inches and height = 5 inches   *Answer:* _____

10. Base = 2 cm. and height = 4 cm.   *Answer:* _____

Answers to Mini-Practice:

(a) 20 inches   (b) 12 inches   (c) 24 inches

11. Base = 2.5 inches and height = 3.8 inches   *Answer:* _____

12. Base = 4.2 inches and height = 5.3 inches   *Answer:* _____

13. Base = $\frac{1}{2}$ inch and height = $\frac{3}{4}$ inch   *Answer:* _____

14. Base = $\frac{3}{2}$ cm. and height = $\frac{5}{2}$ cm.   *Answer:* _____

15. Base = 3.02 feet and height = 4.8 feet   *Answer:* _____

16. Base = $\frac{3}{14}$ km. and height = $\frac{42}{9}$ km.   *Answer:* _____

In Exercises 17–22, find the areas of the following trapezoids.

17. a = 4.2 inches, b = 5.5 inches, h = 3.1 inches   *Answer:* _____

18. a = 2.1 inches, b = 3.6 inches, h = 1.3 inches   *Answer:* _____

19. a = 1.7 cm., b = 2.3 cm., h = 2.9 cm.   *Answer:* _____

20. a = $\frac{1}{2}$ inch, b = $\frac{3}{4}$ inch, h = $\frac{5}{2}$ inches   *Answer:* _____

21. a = $\frac{3}{2}$ cm., b = $\frac{1}{2}$ cm., h = $\frac{3}{5}$ cm.   *Answer:* _____

22. a = 2.03 feet, b = 8.1 feet, h = 3.15 feet   *Answer:* _____

In Exercises 23–28, find the perimeter of the rectangle with the given dimensions.

23. Length = 2.3 inches and width = 1.4 inches   *Answer:* _____

24. Length = 24 cm. and width = 12 cm.   *Answer:* _____

25. Length = 10.3 cm. and width = 5.4 cm.   *Answer:* _____

26. Length = $\frac{1}{2}$ inch and width = $\frac{3}{4}$ inch   *Answer:* _____

27. Length = $\frac{4}{5}$ cm. and width = $\frac{3}{10}$ cm.   *Answer:* _____

28. Length = 17.02 km. and width = 7.9 km.   *Answer:* _____

In Exercises 29–34, find the perimeter of the triangle with the given sides.

29. 4.5 inches, 6.8 inches, and 7.2 inches   *Answer:* _____

30. 15.4 cm., 23.8 cm., and 30 cm.   *Answer:* _____

31. $\frac{1}{2}$ inch, $\frac{3}{4}$ inch, and $\frac{5}{4}$ inches   *Answer:* _____

32. $\frac{4}{5}$ cm., $\frac{3}{10}$ cm., and $\frac{6}{5}$ cm.   *Answer:* _____

33. 28.2 km., 48.05 km., and 14 km.   *Answer:* _____

34. $\frac{3}{8}$ cm., $\frac{5}{16}$ cm., and $\frac{1}{4}$ cm.   *Answer:* _____

7.4 FORMULAS FROM ACCOUNTING

Assets   In accounting, people work with assets, liabilities, and owner's equity. **Assets** are economic resources owned by a business. Cash, money owed to the company, land, buildings, and equipment are examples of assets. **Liabilities** are debts, that is, they are things the company owes to another company or person. Salary owed to employees, taxes owed, and payment owed to a supplier are examples of liabilities.

By **owner's equity** we mean what would be left over to the owner of the business if all liabilities were paid. For example, if the company owns a building worth $40,000, has $50,000 in the bank, and expects to be paid $20,000 from customers, the assets are $110,000. Furthermore, if the company is supposed to pay out $30,000, the liabilities are $30,000 and the owner's equity is $80,000.

The connection between assets, liabilities, and owner's equity can be written as a formula.

$$\text{Assets} = \text{Liabilities} + \text{Owner's Equity}$$

In symbols we can write

$$A = L + O$$

Example 1   Find the assets of a business, given a liability of $5000 and an owner's equity of $800.

Solution. $A = L + O$

Liabilities L	Owner's Equity O	Assets A
$5000	$800	$5000 + $800 = $5800

The assets amount to $5800.   ∎

Net Income $$\text{Net Income} = \text{Revenue} - \text{Cost}$$

$$N = R - C$$

Revenue is all the money coming into the company. **Cost** represents all the expenses a company has to pay out to run the business. **Net income** is what's left over from the revenue when the expenses are paid.

If you, as a small-scale operator, buy candy for 10¢ apiece and sell it for 12¢ apiece, and have no other expenses, your revenue would be 12¢, expenses 10¢, and net income 2¢. According to the formula,

$$N = 12 - 10 = 2$$

What happens if cost is higher than revenue? For example, let revenue (R) be $5000 and cost (C) be $7000. Then

$$N = 5000 - 7000 = -2000$$

that is, more money was spent than recovered. A negative net income implies that the company is in debt.

Example 2   Find net income, given a revenue of $4000 and a cost of $97.58.

Solution. $N = R - C$

Revenue *R*	Cost *C*	Net Income *N*
$4000	$97.58	$4000 – $97.58 = $3902.42

There is a net income of $3902.42.   ■

Net Pay   **Net pay** represents your take-home pay, **gross salary** represents the amount of money you earn for your work, and **deductions** represent all the money taken from your earnings to pay taxes, social security, medical benefits, and so on.

Net Pay = Gross Pay – Deductions

$$N = G - D$$

As an example, let's find the net pay when the gross salary is $25,000 and the deductions amount to $5000.

$$N = 25,000 - 5000 = 20,000$$

The net pay is $20,000.

Example 3   Find net pay, given a gross salary of $575.85 and deductions amounting to $125.80.

Solution. $N = G - D$

Gross Salary *G*	Deductions *D*	Net Pay *N*
$575.85	$125.80	$575.85 – $125.80 = $450.05

Net pay is $450.05.   ■

Annual Depreciation   **Depreciation** is the decrease in value of an item or a property. **Residual value** is how much the item is worth after the decrease in value. A depreciation formula used in accounting is

$$\text{Annual Depreciation} = \frac{\text{Cost} - \text{Residual Value}}{\text{Estimated Life (in years)}}$$

In letters we would write

$$d = \frac{c-r}{e}$$

where d stands for annual depreciation, c for cost, r for residual value, and e for estimated life.

Example 4   Find the annual depreciation for the following values of cost, residual value, and estimated life: cost = $150, residual value = $35.88, estimated life = 2 years.

Solution. $d = \frac{c-r}{e}$

Cost c	Residual Value r	Estimated Life e	Annual Depreciation d
150	35.88	2	$\frac{150-35.88}{2} = 57.06$

The annual depreciation is $57.06.   ■

MINI-PRACTICE
(Use this space for your work.)

1. Find the assets, A, when L = $5000 and O = $4000.   *Answer:* _____

2. Find the net income when R = $10,000 and C = $6000.   *Answer:* _____

3. Find the net pay when the gross pay is $3000 and the deductions amount to $1500.   *Answer:* _____

4. Find the annual depreciation, d, when

   (a) c = $200, r = $100, and e = 2 years   *Answer:* _____

   (b) c = $200, r = $100, and e = $1\frac{1}{2}$ years   *Answer:* _____

Practice Exercises

In Exercises 1–6, find the assets, A, for the given values of liabilities, L, and owner's equity, O.

1. L = $13,000 and O = $350   *Answer:* _____

2. L = $17,000 and O = $1510   *Answer:* _____

Answers to Mini-Practice

1. $9000   **2.** $4000   **3.** $1500   **4.** (a) $50   (b) $66.67

 3. $L = \$52,435.23$ and $O = \$10,239$ Answer: _____

 4. $L = \$36,428$ and $O = \$15,421$ Answer: _____

 5. $L = \$72,571.28$ and $O = \$31,897.56$ Answer: _____

 6. $L = \$1,435,439$ and $O = \$645,779$ Answer: _____

In Exercises 7–14, find the net income, N, for the given values of revenue, R, and cost, C.

 7. $R = \$18,300$ and $C = \$3580.25$ Answer: _____

 8. $R = \$25.40$ and $C = \$5.90$ Answer: _____

 9. $R = \$1498.36$ and $C = \$7.39$ Answer: _____

10. $R = \$2000$ and $C = \$938$ Answer: _____

11. $R = \$3500$ and $C = \$5428.60$ Answer: _____

12. $R = \$480$ and $C = \$795.35$ Answer: _____

13. $R = \$380$ and $C = \$395.38$ Answer: _____

14. $R = \$56,426.31$ and $C = \$63,721.79$ Answer: _____

In Exercises 15–22, find the net pay, N, for the given values of gross salary, G, and deductions, D.

15. $G = \$482.75$ and $D = \$67.36$ Answer: _____

16. $G = \$149.50$ and $D = \$7.38$ Answer: _____

17. $G = \$258$ and $D = \$37.80$ Answer: _____

18. $G = \$1528.99$ and $D = \$351.62$ Answer: _____

19. $G = \$2321.48$ and $D = \$568.33$ Answer: _____

20. $G = \$5471$ and $D = \$321.42$ Answer: _____

21. $G = \$4222.03$ and $D = \$463$ Answer: _____

22. $G = \$938.21$ and $D = \$23$ Answer: _____

In Exercises 23–30, find the annual depreciation, d, for the given values of cost, c, residual value, r, and estimated life, e.

23. $c = \$785.95$, $r = \$110.74$, and $e = 3$ years   Answer: _____

24. $c = \$493$, $r = \$78.90$, and $e = 2$ years   Answer: _____

25. $c = \$1396.50$, $r = \$235.75$, and $e = 2\frac{1}{2}$ years   Answer: _____

26. $c = \$7984.33$, $r = \$537.83$, and $e = 5$ years   Answer: _____

27. $c = \$2943.57$, $r = \$195.37$, and $e = 4$ years   Answer: _____

28. $c = \$1548$, $r = \$1236$, and $e = 2$ years   Answer: _____

29. $c = \$142.02$, $r = \$86.30$, and $e = 4$ years   Answer: _____

30. $c = \$23,548.20$, $r = \$18,321.36$, and $e = 4.5$ years Answer: _____

7.5 FORMULAS FROM SCIENCE

Temperature   The **Fahrenheit** standard for temperature was invented by a German instrument-maker named Gabriel Daniel Fahrenheit. Sometimes temperature is measured in degrees **Celsius** instead of in degrees Fahrenheit. This standard for measuring temperature was invented by the Swedish scientist Anders Celsius, and is called the **Celsius scale.**

The temperature at which water freezes is 0° on the Celsius scale and 32° on the Fahrenheit scale. The temperature at which water boils is 100° on the Celsius scale and 212° in Fahrenheit.

°**C**	°**F**
100°	212°
0°	32°

By comparing the two scales in the table, what do you think 50° corresponds to in Fahrenheit degrees? _____

To convert from degrees F to degress C, we use the formula

$$C = \frac{5}{9}(F - 32)$$

To convert degrees Celsius into degrees Fahrenheit, we use the formula

$$F = \frac{9}{5}C + 32$$

Example 1   Convert 10°F to degress Celsius.

Solution. $C = \frac{5}{9}(F - 32)$

Let $F = 10$. Then

$$C = \frac{5}{9}(10 - 32) = \frac{5}{9}(-22) = \frac{-110}{9} = -12.22 = -12 \text{ (approximately)}$$

Thus, 10°F is approximately $-12°C$.   ∎

Example 2   Convert $-40°C$ to degrees Fahrenheit.

Solution. $F = \frac{9}{5}C + 32$

Let $C = -40$. Then

$$F = \frac{9}{5}(-40) + 32 = \frac{-360}{5} + 32 = -72 + 32 = -40$$

Thus, $-40°C = -40°F$.   ∎

MINI-PRACTICE
(Use this space for your work.)

(a) Find the temperature in degrees Celsius when $F = 32$.

Answer: _____

(b) Find the temperature in degrees Fahrenheit when $C = -10$.

Answer: _____

Mass, Volume, Density

In science one often has to use a formula that gives the relationship between mass, volume, and density. **Mass** can be regarded as the same as weight; **volume** is the space an object takes up; and **density**, which measures the compactness of an object, is mass/volume. For example, one pound of copper takes up less volume than one pound of feathers because copper has a higher density.

The formula for the relationship is given by

$$\text{Mass} = \text{Volume} \cdot \text{Density} \quad \text{or} \quad M = V \cdot D$$

Example 3

Find the mass when the volume is 38.6 milliliters (ml.) and the density is 1.75 grams per milliliter (g./ml.).

Solution.

$$M = V \cdot D$$

$$V = 38.6 \text{ ml.} \qquad D = 1.75 \text{ g./ml.}$$

$$M = V \cdot D = 38.6(1.75) = 67.55$$

The mass is 67.55 grams. ■

Rate, Distance, Time

If you drive 200 miles in four hours, the rate at which you travel is given by

$$\frac{200}{4} = 50 \text{ (miles per hour)}$$

The formula we used here is

$$\text{Rate} \cdot \text{Time} = \text{Distance} \quad \text{or} \quad \text{Rate} = \frac{\text{Distance}}{\text{Time}}$$

In symbols,

$$R \cdot T = D \quad \text{or} \quad R = \frac{D}{T}$$

Example 4

Find the rate when the distance traveled is 90 miles and the time traveling is 2 hours.

Solution.

$$R = \frac{D}{T}$$

$D = 90$ miles, $T = 2$ hours

$$R = \frac{D}{T} = \frac{90}{2} = 45$$

The rate is 45 miles per hour. ■

Answers to Mini-Practice:

(a) 0°   (b) 14°

MINI-PRACTICE   (a) Find the mass when $V = 50$ ml. and $D = 0.5$ g./ml.   *Answer:* _____

(Use this space for your work.)

(b) Find the rate when $D = 600$ mi. and $T = 4$ hr.   *Answer:* _____

Practice Exercises

In Exercises 1–10, find the temperature in degrees Celsius that corresponds to the temperature given in degrees Fahrenheit.

1. 95°F   *Answer:* _____   2. 59°F   *Answer:* _____   3. 140°F   *Answer:* _____

4. 5°F   *Answer:* _____   5. 41°F   *Answer:* _____   6. 23°F   *Answer:* _____

7. –4°F   *Answer:* _____   8. 230°F   *Answer:* _____   9. 131°F   *Answer:* _____

10. 104°F   *Answer:* _____

In Exercises 11–20, find the temperature in degrees Fahrenheit that corresponds to the temperature given in degrees Celsius. Express your answer to the nearest tenth of a degree.

11. 15°C   *Answer:* _____   12. 50°C   *Answer:* _____   13. 100°C   *Answer:* _____

14. –20°C   *Answer:* _____   15. –10°C   *Answer:* _____   16. 20°C   *Answer:* _____

17. 25°C   *Answer:* _____   18. 37°C   *Answer:* _____   19. 35°C   *Answer:* _____

20. –75°C   *Answer:* _____

In Exercises 21–28, find the mass, M, for the given volume, V, and density, D.

21. $V = 13$ cm.^{3} and $D = 16$ g./cm.^{3} *Answer:* _____

22. $V = 8$ m.^{3} and $D = 5$ kg./cm.^{3} *Answer:* _____

23. $V = 4$ ml. and $D = 3$ g./ml.   *Answer:* _____

24. $V = 96$ cm.^{3} and $D = 5$ oz./cm.^{3} *Answer:* _____

25. $V = 121$ ml. and $D = 17$ oz./ml.   *Answer:* _____

26. $V = 213$ cm.^{3} and $D = 11$ oz./cm.^{3} *Answer:* _____

27. $V = 23.5$ ml. and $D = 26$ g./ml.   *Answer:* _____

28. $V = 17.4$ ml. and $D = 25$ g./ml.   *Answer:* _____

In Exercises 29–36, find the rate, R, for the given distance, D, and time, T.

29. $D = 75$ miles and $T = 1\frac{1}{2}$ hours   *Answer:* _____

30. $D = 100$ miles and $T = 2$ hours   *Answer:* _____

31. $D = 355$ miles and $T = 5$ hours   *Answer:* _____

32. $D = 2586$ miles and $T = 3$ hours   *Answer:* _____

33. $D = 4832$ km. and $T = 2$ hours   *Answer:* _____

34. $D = 56$ m. and $T = 8$ seconds   *Answer:* _____

35. $D = 428$ m. and $T = 4$ minutes   *Answer:* _____

36. $D = 3055$ km. and $T = 13$ hours   *Answer:* _____

Answers to Mini-Practice:

(a) 25 g.   (b) 150 m.p.h.

Summary

Evaluation. When we substitute to evaluate expressions,

1. we substitute given values for all variables;
2. we follow the order of operations.
    (a) Apply all parentheses.
    (b) Apply all exponents.
    (c) Perform multiplication/division from left to right.
    (d) Perform addition/subtraction in any order.

Formulas. The following is a summary of all the formulas that have been used in this section:

Area

Rectangle:	$A = l \cdot w$
Square:	$A = s^2$ (s = side)
Triangle:	$A = \dfrac{b \cdot h}{2}$
Trapezoid:	$A = \dfrac{h(a+b)}{2}$

Perimeter

Rectangle:	$P = 2l + 2w$
Triangle:	$P = a + b + c$
Square:	$P = 4s$

Accounting

Annual Depreciation:	$d = \dfrac{c - r}{e}$
Assets:	$A = L + O$
Net income:	$N = R - C$
Net pay:	$N = G - D$

Science

Temperature Conversion

Fahrenheit to Celsius:	$C = \dfrac{5}{9}(F - 32)$
Celsius to Fahrenheit:	$F = \dfrac{9}{5}C + 32$
Mass:	$M = V \cdot D$
Distance:	$D = R \cdot T$

Review Practice Exercises

In Exercises 1–30, find the value of each polynomial for the given values of the variables.

1. $4x$ when $x = -3$ *Answer:* _____

2. $y + 5$ when $y = -2$ *Answer:* _____

3. $b^2 + 1$ when $b = 5$ *Answer:* _____

4. 4^a when $a = 2$ *Answer:* _____

5. $2a^2 - 3a$ when $a = 5$ *Answer:* _____

6. $p^3 - 2p$ when $p = 10$ *Answer:* _____

7. $2a + 5b$ when $a = -3, b = 4$ *Answer:* _____

8. $2c - 4d$ when $c = 9, d = -4$ *Answer:* _____

9. $a^2 + 2b^2$ when $a = 5, b = 3$ *Answer:* _____

10. $5x^2 - 2x - 3$ when $x = 3$ *Answer:* _____

11. $\dfrac{6y + 2y^2}{y - 4}$ when $y = 5$ *Answer:* _____

12. $\dfrac{3y^2 - 2y - 8}{3y + 4}$ when $y = 3$ *Answer:* _____

13. $4(5x) - 3(5 - x)$ when $x = 4$ *Answer:* _____

14. $7(4 + 12g)$ when $g = -2$ *Answer:* _____

15. $2 + \dfrac{5x}{2}$ when $x = 6$ *Answer:* _____

16. $7y + \dfrac{y^2 - 3}{y - 3}$ when $y = 5$ *Answer:* _____

17. $(a + b)^2$ when $a = 2$ and $b = -7$ *Answer:* _____

18. $\dfrac{x^2 - 2xy + y^2}{x - y}$ when $x = 7$ and $y = 2$ *Answer:* _____

19. $(3x^2 - 5y^2 - 1)xy$ when $x = 3$ and $y = 2$ *Answer:* _____

20. $y^2 + \dfrac{x^3 + y^3}{x + y}$ when $x = 5$ and $y = 4$ *Answer:* _____

21. $\dfrac{x^2 - 3y^2 + 24}{2xy}$ when $x = 6$ and $y = 4$ *Answer:* _____

22. $\dfrac{2x^2 - 3xy - 10y^2}{x - 5y}$ when $x = 20$ and $y = 3$ *Answer:* _____

23. $(x + y)(x^2 - xy + y^2)$ when $x = 3$ and $y = 2$ *Answer:* _____

24. $(a^2 + b^2)(a^2 - b^2)$ when $a = 4$ and $b = 3$ *Answer:* _____

25. $(3a^2 - 10b)(a + 5b)$ when $a = 10$ and $b = 8$ *Answer:* _____

26. $3a^3 + 2(b^2 - 3c)$ when $a = -1, b = 4$, and $c = 4$ *Answer:* _____

27. $6m + 2n^2 - 8p$ when $m = 3, n = -5$, and $p = 7$ *Answer:* _____

28. $7s(4t^2 - 8u) + stu$ when $s = -2, t = 1$, and $u = 2$ *Answer:* _____

29. $3x^2 + 4y - (3v + w^2)$ when $x = 5, y = -7, v = -17$, and $w = 1$ *Answer:* _____

30. $(5x - y^2) + 2(v + w) + xyvw$ when $x = 1, y = -1, v = 2$, and $w = -2$ *Answer:* _____

In Exercises 31–35, evaluate each polynomial when $a = 2, b = 0.5, c = 1$, and $d = 2.1$.

31. $b + d - c$ *Answer:* _____

32. $ab + bc + cd$ *Answer:* _____

33. $abcd$ *Answer:* _____

34. $ab + ac + ad + bc + bd$ *Answer:* _____

35. $(a + b)^2 - (b + c)^2 + (d - b)^2$ *Answer:* _____

In Exercises 36–40, use the appropriate formula. It is not necessary to memorize the formulas. Simply refer to the list on page 193.

36. Find the assets if

(a) liabilities = \$358.75 and owner's equity = \$500

Formula: _____   *Answer:* _____

(b) liabilities = \$198.36 and owner's equity = \$487.21   *Answer:* _____

37. Find net income if

   (a) revenue = $1438.50 and cost = $938.51

      Formula: _____   *Answer:* _____

   (b) revenue = $2,350 and cost = $289.34   *Answer:* _____

38. Convert degree Fahrenheit to degrees Celsius or vice versa.

   (a) 104°F   Formula: _____   *Answer:* _____

   (b) 77°F   *Answer:* _____

   (c) 55°C   Formula: _____   *Answer:* _____

   (d) 15°C   *Answer:* _____

39. Find the mass when

   (a) volume = $12\frac{1}{2}$ ml. and density = 6 g./ml.

      Formula: _____   *Answer:* _____

   (b) volume = 10 ml. and density = 3.8 g./ml.   *Answer:* _____

40. Find the rate when

   (a) distance = 42 miles and time = 2 hours

      Formula: _____   *Answer:* _____

   (b) distance = 208 cm. and time = 4 seconds   *Answer:* _____

   If you have questions about this chapter, write them here: _____

_____

_____

_____

_____

_____

8 Simplification

Remember to

* ✳ *Be careful with parentheses*
* ✳ *Write neatly to avoid mistakes*
* ✳ *Review translations of math operations*
* ✳ *Go step by step*

We will now simplify expressions involving several operations and/or grouping symbols. To simplify means to make something look easier than before. You have already simplified algebraic expressions in Chapter 5 when you combined like terms and when you multiplied or divided using variables.

8.1 Multiplication Combined with Addition/Subtraction

Multiply: $5(x + 2) = $ _____

Multiply: $3x(x - 4) = $ _____

Multiply: $5x(x^2 - 2x + 5) = $ _____

When multiplying a monomial by a polynomial with several terms, we always use the distributive law. We *distribute multiplication* over *addition and/or subtraction*.

Apply the distributive law to the expressions:

$4(x + 3y)$ _____   and $5(2x + 4y)$ _____

Then apply the distributive law twice in the expression

$4(x + 3y) + 5(2x + 4y)$ _____

Your answer should look like this: $4x + 12y + 10x + 20y$

Combine like terms: _____

The final answer is $14x + 32y$.

Next, simplify $3x - 2(x - 6)$ by distributing -2 over $x - 6$:

_____

Combine like terms: _____

Your answer should be $x + 12$.

Example 1   Simplify $5 - (x - 2)$.

Solution.   Distribute the first $-$:
$$5 - x + 2$$
Simplify:
$$7 - x$$
$$5 - (x - 2) = 7 - x$$ ■

Example 2   Simplify $2(a - b) + 3(2a + 4b)$.

Solution.   Distribute both 2 and 3:
$$2a - 2b + 6a + 12b$$
Combine like terms:
$$8a + 10b$$
$$2(a - b) + 3(2a + 4b) = 8a + 10b$$ ■

MINI-PRACTICE   Simplify:
(Use this space for your work.)

(a) $8(a - 2b) + 5a$ (b) $5(x + 3y) + 13(2x + y)$

(c) $7(2x + 2) - 3(4x - 2)$

Answers: (a) _____   (b) _____   (c) _____

IN YOUR OWN WORDS   explain how to simplify an expression: _____

_____

_____

Answer:

We simplify an expression by using the distributive law to remove all parentheses. Then we combine all like terms.

Answers to Mini-Practice:

(a) $13a - 16b$ (b) $31x + 28y$ (c) $2x + 20$

Office Hours

1. **These are hours when your instructor is available for questions.**
   - Find out when these office hours are held.
   - They are usually the same hours every week.
   - Find out where your instructor's office is.

2. **Use these office hours.**
   - Drop in as often as you like.
   - You will get individual attention.
   - Show your instructor that you care about your work.

Practice Exercises

In Exercises 1–26, simplify.

1. $7(a + 8b) + 8a$ Answer: _____
2. $12(a - 2b) - 14a$ Answer: _____

3. $x - 8(y - x)$ Answer: _____
4. $9(a + 2b) + 4(3a + 2b)$ Answer: _____

5. $5(a - b) + 2(4a - 2b)$ Answer: _____
6. $16b^2 + 5(a - 4b^2)$ Answer: _____

7. $13a^2 - 2(4a^2 - b^2)$ Answer: _____
8. $4(x - y) - 5(x - 3y)$ Answer: _____

9. $-2(xy - x) - 7(xy - x)$ Answer: _____
10. $3(2x + 1) - 2(x - 2)$ Answer: _____

11. $(y^5 + 3y^3 + y) - (y^5 - 2y^3 - y)$ Answer: _____
12. $3(x + 2y) + 2(3x + 2y)$ Answer: _____

13. $(x - 2y)2 + (x + 3y)$ Answer: _____
14. $3(2a + b) + 8(3a - 2b)$ Answer: _____

15. $3x(y - 3) - y(x + 5)$ Answer: _____
16. $x^2 + xy - 1 - 2x(x - 6y)$ Answer: _____

17. $2x^3 - 4x^2(x^2 + x - 3) - x^4$ Answer: _____

18. $(x^3 + x^2 - 5x) - 2x(x^2 - x + 2)$ Answer: _____

19. $y^4 - y(y^3 - y^2) + y^3 - y$ Answer: _____

20. $y^6 + 3y^5 - 2y^4 - y^2(y^4 - 4y^3 + 7y)$ Answer: _____

21. $-3b^2 - b(2b^3 + 5b) + 4b^2(-3 - b^2)$ Answer: _____

22. $a(5a - 2b + 4c) - 2b(3a - 5b + c) + c(a - b - c)$ Answer: _____

23. $r^2(2s + 3t) - 3s(r^2 - 5t) + 8t(r^2 - 6s) + rs$ Answer: _____

24. $c^2d^2(cd - 6c + 9d) - 3c^3d^3 + 3cd(c^2d^2 - 4c^2d + d^2)$ Answer: _____

25. $fgh(f^2gh^2 - 9fg^2h^2) - 7g^3h(f^2h^2 - 3f^3h^2)$ Answer: _____

26. $x^2y(x^2y) - 5ab(ab) + x^4y^2 - a^2b^2 + 3xy(x^3y - 2ab)$ Answer _____

8.2 TRANSLATING ADDITION AND SUBTRACTION OF POLYNOMIALS

Translate into mathematics:

(a) Subtract 3 from 5. _____

(b) Subtract $(2 + x)$ from $(5 - x)$. _____

(c) Subtract $x^2 + 2$ from $6x^2 - x$. _____

As you probably remember, when we subtract *from*, we must reverse the order in subtraction; thus,

(a) becomes $5 - 3$ and   (b) becomes $(5 - x) - (2 + x)$

In (c) we take the quantity $x^2 + 2$ from something. There are no parentheses in $x^2 + 2$, but when we subtract the expression, we *must* show with parentheses that $(x^2 + 2)$ belongs together. (The minus sign applies to both x^2 and 2.)

$$(6x^2 - x) - (x^2 + 2)$$

Add $3x^2 + x$ to $x^2 + 4$: _____

Do we need parentheses here? _____

In addition (a commutative operation) we can switch the order and we do not need parentheses:

$$(3x^2 + x) + (x^2 + 4) = 3x^2 + x + x^2 + 4 = 4x^2 + x + 4$$
$$(x^2 + 4) + (3x^2 + x) = x^2 + 4 + 3x^2 + x = 4x^2 + x + 4$$

Example 1   Add $x^2 - 6$ to $2x^2 + 3x$.

Solution.
$$\begin{array}{r} 2x^2 + 3x \\ + \ x^2 \qquad - 6 \\ \hline 3x^2 + 3x - 6 \end{array}$$

Combine like terms:

$$2x^2 + 3x + x^2 - 6 = 3x^2 + 3x - 6 \qquad \blacksquare$$

Example 2   Subtract $2 + x$ from $5 - x$.

Solution. $(5 - x) - (2 + x)$

Remove parentheses: $5 - x - 2 - x$

Combine like terms: $3 - 2x$

$$(5 - x) - (2 + x) = 3 - 2x \qquad \blacksquare$$

MINI-PRACTICE
(Use this space for your work.)

(a) Add $5a^3 + b$ to $a^3 - 2b$.   *Answer:* _____

(b) Subtract $a - b$ from $2a + b$.   *Answer:* _____

(c) Subtract $x^2 + 2$ from $6x^2 - x$.   *Answer:* _____

IN YOUR OWN WORDS   explain when we must use parentheses in subtraction of polynomials: _____

_____

_____

Answer:

When we subtract polynomials we must enclose them in parentheses so that later we can distribute the minus sign over the operations in the second polynomial.

Practice Exercises

In Exercises 1–30, simplify.

1.   Add $a^2 + b^2 + c^2$ to $6a^2 - 4b^2$.   *Answer:* _____

2.   Subtract $x + y$ from $x - y$.   *Answer:* _____

3.   Add $-x - y$ to $2x - y$.   *Answer:* _____

4.   Subtract $-x - y$ from $2x - y$.   *Answer:* _____

5.   From $-x - y$ take $2x - y$.   *Answer:* _____

6.   From $-xy - 4xz + 2yz$ take $4xz - xy$.   *Answer:* _____

7.   Add $7x - 4y$ to the sum of $3x$ and $4y$.   *Answer:* _____

8.   Subtract $6x - 4y + 3$ from $x + y + 4$.   *Answer:* _____

9.   From the sum of $2a - 3b$ and $-6a + 3b$ subtract $12x - 8b$.   *Answer:* _____

10.   From the sum of $5x$ and $-3y$ subtract the difference of $5x$ minus $-3y$.   *Answer:* _____

11.   From $5.2xy$ take $3.2xy$.   *Answer:* _____

12.   Subtract $8.7a^2b^2 - ab$ from $9.7a^2b^2 + 5ab$.   *Answer:* _____

13.   From $\frac{7}{4}pq$ subtract $\frac{3}{2}pq$.   *Answer:* _____

14.   Subtract $\frac{6}{5}cd - \frac{3}{2}st$ from $\frac{9}{5}cd - \frac{7}{2}st$.   *Answer:* _____

15.   Add $\frac{3}{8}xy$ to the difference of $\frac{5}{8}xy$ and xy (in this order).   *Answer:* _____

16.   Add $\frac{2}{3}a^2b^2$ to the difference of $\frac{4}{3}a^2b^2$ minus a^2b^2.   *Answer:* _____

Answers to Mini-Practice:
(a) $6a^3 - b$ (b) $a + 2b$ (c) $5x^2 - x - 2$

17. Subtract r^3s^3 from the sum of $-5r^3s^3$ and $7r^3s^3$.   *Answer:* _____

18. Subtract $3.5c^2d$ from the sum of $2.3c^2d$ and $3.7c^2d$.   *Answer:* _____

19. From the sum of abc and $-6abc$ subtract $10abc$.   *Answer:* _____

20. From the sum of $r^2s^2t^3$ and $4r^2s^2t^3$ subtract $-5r^2s^2t^3$.   *Answer:* _____

21. From $5u^2p^3$ minus $-2u^2p^3$, take u^2p^3.   *Answer:* _____

22. From the difference of $9x^2y^3$ and $-x^2y^3$ (in this order), take $2x^2y^3$.   *Answer:* _____

23. From the sum of $6p^2q^4$ and $-3p^2q^4$ subtract $-5p^2q^4$ minus $4p^2q^4$.   *Answer:* _____

24. From the sum of $3x^2y$ and $-8x^2y$ subtract $-x^2y$ minus $4x^2y$.   *Answer:* _____

25. To $5.3a^3b^5$ minus $0.3a^3b^5$ add the difference of $6.2a^3b^5$ and $-4.8a^3b^5$ (in this order).

    Answer: _____

26. To $9.6m^4p^2$ minus $8.9m^4p^2$ add $2.7m^4p^2$ minus $-3.8m^4p^2$.   *Answer:* _____

27. From the sum of $\frac{5}{3}f^2g$ and $-\frac{8}{3}f^2g$ subtract the sum of $\frac{4}{3}f^2g$ and $-\frac{7}{3}f^2g$.

    Answer: _____

28. From the sum of $\frac{3}{2}d^3m^4$ and $\frac{5}{4}d^3m^4$ subtract the sum of $-\frac{5}{2}d^3m^4$ and $\frac{3}{4}d^3m^4$.

    Answer: _____

29. Subtract $3y^2 - 5y - 9$ from the sum of $6 + 9y - y^2$ and $7y^2$.   *Answer:* _____

30. Subtract $8ab^2c^3$ from the sum of $b^2c^3 - 4ab^2c^3$ and $9ab^2c^3$.   *Answer:* _____

Summary

1. In translations, parentheses must be used when subtracting a polynomial containing two or more terms.

2. All parentheses are removed in expressions by the use of the distributive law.

3. Like terms are combined.

Review Practice Exercises

In Exercises 1–30, simplify.

1. $3n + 2(n + 5)$ *Answer:* _____   2. $2x + 5(6 - x)$ *Answer:* _____

3. $-7u + 5(2u - 3)$ *Answer:* _____   4. $5p - 7(3 - p)$ *Answer:* _____

5. $5pq - 7p(q - 4p)$ *Answer:* _____   6. $2(x - 2y) - 3y$ *Answer:* _____

7. $-3(4x + 2y) - 2x + y$ *Answer:* _____   8. $7x - (-3x - 4)$ *Answer:* _____

9. $3(a - b) + 2a - 5b$ *Answer:* _____   10. $3a + 4b - (5a + 4b)$ *Answer:* _____

11. $-3(2x + 4y + 2) - 7(-3x - 2y)$ Answer: _____

12. $x(x^2 - 3x + 2) - 4(x^3 + 3x^2 - 2x)$ Answer: _____

13. $3x - 3[2(x - 4) - 8]$ Answer: _____   14. $6 - \{2y - [3y - (5 - y)]\}$ Answer: _____

15. $3x^2y(2x - 1)$ Answer: _____   16. $9ab^2(ab + 1)$ Answer: _____

17. $-4xy^3(x^2 - xy)$ Answer: _____   18. $6x^2 + 7x(x^2 + 3x)$ Answer: _____

19. $2x^3 - 7x(x^2 - 4)$ Answer: _____   20. $10p^2q^2 - 3pq(5p - 6pq)$ Answer: _____

21. $3[2x - (x + y)]$ Answer: _____   22. $8 - 5[x + 2 - 3(x - 2)]$ Answer: _____

23. $5[3x - 4(x + 2y)]$ Answer: _____   24. $16 - 3[a + 3 - 4(a - 1)]$ Answer: _____

25. $3 - 4[5(x - y) + 6x]$ Answer: _____   26. $10 - 6[7x - 6(2x - y)]$ Answer: _____

27. $5x - 2[x - 3(x + 5)]$ Answer: _____

28. $r - \{r - [2r - 5(r + 1) - 3]\}$ Answer: _____

29. $7y - 3[9 - 2(2y + 1) - 3]$ Answer: _____

30. $-3(2x + 2) - (3x - 4)$ Answer: _____

31. Subtract $4x + y$ from $4x - y$.   Answer: _____

32. Subtract $-a - b$ from $5ab + 3a - 4b$.   Answer: _____

33. From $-ab - 4ac + 2bc$ take $5ac - 6bc$.   Answer: _____

34. From $7x^2y - 4xy^2 + 6x^2y^2$ take $10x^2y - 4x^2y^2 + 3xy^2$.   Answer: _____

35. Add $5c^4k^2$ to the sum of $-3c^4k^2$ and $-2c^4k^2$.   Answer: _____

36. Add $6b^2d^5 + 3b^5d^2$ to the sum of $4b^5d^2$ and $-5b^2d^5$.   Answer: _____

37. To the difference of $-ab$ and $4\,abc$ (in this order) add the sum of $-5abc$ and $9\,abc$.   Answer: _____

38. To $-4xy$ minus $6xyz$ add the sum of $2xyz$ and $-5xyz$.   Answer: _____

39. Subtract the sum of m^4p^3 and $4m^4p^3$ from $-5m^4p^3$.   Answer: _____

40. Subtract the sum of $-2f^2g^3$ and $5f^2g^3$ from $-10f^2g^3$.   Answer: _____

41. The difference of $3r^6t^5$ and $-2r^6t^5$ (in this order) is subtracted from the sum of $8r^6t^5$ and $-r^6t^5$.

   Answer: _____

42. The difference of $5x^2z^3$ minus $-8x^2z^3$ is subtracted from the sum of x^2z^3 and $-3x^2z^3$. Answer: _____

If you have questions about this chapter, write them here: _____

_____

_____

_____

_____

PART IV

Equations

9 Equalities and Equations

Remember to

* *BEGIN BY ANALYZING A PROBLEM*
* *LEARN THE NEW VOCABULARY*
* *DO THE LEARNING ACTIVITIES*
* *ANSWER ALL QUESTIONS IN WRITING*

Equations are mathematical statements asserting that various quantities are equal. You will learn how to "solve" certain equations and how to distinguish among different types of equations.

9.1 EQUALITIES

Write the symbol " = " in words. _____

Most likely you know that " = " is an *equals sign.*

The idea of using a symbol for "equals" occurred around 1000 B.C. when the Egyptian scribe Ahmes used it. But the equals sign as we know it was first used by the Englishman Robert Recorde. He published the first English algebra book in 1557: "I will use a pair of paralleles of one length, thus:

———————
———————

because no two things can be more equal." His sign was longer than the one we use nowadays, but the meaning is the same as ours: it conveys *sameness.*

We have just said that " = " stands for "equals," but there are other words and expressions that also can be replaced with " = ":

Translate 2 + 3 = 5 into English: _____

205

The Math Lab

1. **This is a place to get help**
   - With homework.
   - With topics you find difficult.
   - With work you've missed due to absence.

2. **Where and when?**
   - Find out where the math lab is located.
   - Find out when it is open.
   - Keep the schedule handy.

3. **Attend regularly.**
   - Pick out several hours each week to attend.
   - You can work with the same tutors at these hours.
   - Get to know the tutors. They will give you special attention if they see you trying.

4. **Also at the math lab:**
   - There may be programmed instruction.
   - There may be other workbooks for you to use.

Write in symbols:

   1. Two plus three is the same as five. _____

   2. Two plus three has the same value as five. _____

   3. Adding two and three, the result is five. _____

As you undoubtedly saw,

$$\left.\begin{array}{l}\text{equals}\\\text{is}\\\text{is the same as}\\\text{the result is}\\\text{has the same value as}\end{array}\right\} \text{can be represented by} =$$

When we say that things are equal, we mean that they have the same value or stand for the same thing. Two quantities of equal value do not have to look the same. Think, for example, about the different ways you can pay when you buy something that costs $1. You can pay with a dollar bill or in several other ways.

Give examples of how you can combine coins to pay for the one-dollar item: _____

_____

_____

You may have said ten dimes, for example, or four quarters, or three quarters, two dimes, and one nickel. We use the equals sign to show this:

$$\$1 = 4 \times 25¢$$
$$\$1 = 10 \times 10¢$$

and so on. The **value** of all these combinations is still $1, but the coins look different.

Statements that two things are equal are called **equalities**. Here is an example of the use of the equals sign.

$$2(3 - 1)^2 = 2(2)^2 = 2 \cdot 4 = 8$$

Now try the following exercises.

$$3(-2 - 1)^2 = \underline{\hspace{2cm}} = \underline{\hspace{2cm}} = 27$$

In our earlier example of how to pay for the $1 item, we could have used equalities to show the possible combinations of coins. Since 100 cents is equal to a dollar, we can show different combinations.

Complete the blanks:

$$100 = 50 + \underline{\hspace{0.7cm}} = 50 + \underline{\hspace{0.7cm}} (25) = \underline{\hspace{0.7cm}} (10) = \underline{\hspace{0.7cm}} (10) + \underline{\hspace{0.7cm}} (5) = \underline{\hspace{0.7cm}} (5)$$

Example 1   Give four equalities for 28.

Solution.

$$28 = 14 + 14 = 4(7) = 2(14) = 7 + 7 + 7 + 7 = 5 + 3 + 10 + 10 \text{ (and so on)}$$

∎

As you probably recall, to simplify an expression means to make it shorter and easier to use. When we simplify, every step gives us a new equality.

The following three polynomials are steps in a simplification process:

$$4x - 3(x + 5)$$
$$4x - 3x - 15 \qquad \text{(Distribute } -3.)$$
$$x - 15 \qquad \text{(Combine like terms.)}$$

These three equalities should have the same value for any specific value of x. To see that this is true, let $x = 1$. Then we will have

$$
\begin{array}{ccc}
4x - 3(x + 5) \longrightarrow & 4x - 3x - 15 \longrightarrow & x - 15 \\
\downarrow & \downarrow & \downarrow \\
4 - 3(1 + 5) & 4 - 3 - 15 & 1 - 15 \\
\downarrow & \downarrow & \downarrow \\
4 - 3(6) & 1 - 15 & -14 \\
\downarrow & \downarrow & \\
4 - 18 & -14 & \\
\downarrow & & \\
-14 & &
\end{array}
$$

MINI-PRACTICE
(Use this space for your work.)

(a) Give three equalities for 32.   *Answer:*_____

(b) Simplify $2x - 8(x - 3)$ as far as possible.   *Answer:*_____

Show that the following polynomials are equal when $x = -1$:

(c) $2x - 8(x - 3)$ *Answer:* _____

(d) $2x - 8x + 24$ *Answer:* _____

(e) $-6x + 24$ *Answer:* _____

IN YOUR OWN WORDS   give a definition of "equalities." _____

_____

_____

Answer:

   Equalities are relationships among quantities that have the same value or mean the same thing.

Practice Exercises

1. Give four equalities for 25.   *Answer:* _____

2. Give five equalities for $2.50.   *Answer:* _____

3. Give three equalities for 10.   *Answer:* _____

4. Give four equalities for 1.5.   *Answer:* _____

5. Simplify $5 - 3(x + 2)$ as far as possible and show that each expression in the simplification process has the same value when $x = 2$. (See Mini-Practice Exercises (c)–(e).)

    Answer: _____

6. Simplify $2x + 4(x - 5)$ as far as possible and show that both the original expression and the simplified one have the same value when $x = -3$.

    Answer: _____

7. Simplify $4x - 3x(2 - x)$ as far as possible and show that both the original expression and the simplified one have the same value when $x = 1$.

    Answer: _____

8. In Practice Exercise 7, show that both the original expression and the simplified one have the same value when $x = -1$.

    Answer: _____

Answers to Mini-Practice:

(a) $2(16) = 16 + 16 = 4(8) = 2^5$. (There are many other answers.)   (b) $-6x + 24$

(c) $2(-1) - 8(-1 - 3) = 2(-1) - 8(-4) = -2 + 32 = 30$

(d) $2(-1) - 8(-1) + 24 = -2 + 8 + 24 = 6 + 24 = 30$ (e) $-6(-1) + 24 = 6 + 24 = 30$

9. Simplify $-5y - 3y(y - 5) + 2y$ as far as possible and show that both the original expression and the simplified one have the same value when $y = -2$.

   Answer: _____

10. Simplify $4ab^2 - b(5ab - 3a) + 2ab$ as far as possible and show that both the original expression and the simplified one have the same value when $a = -1$ and $b = 2$.

    Answer: _____

9.2 EQUATIONS

In the previous activities we were able to line up quantities that are equal to one another. In mathematics we often work with two given equal expressions, linked by an equals sign.

You might have seen something like

$$x + 2 = 5$$

before.

What is the value of x? _____

You probably said x is 3, and that is true. When two expressions are linked by an equals sign, they form an **equation**; we often use equations to find out what number a variable stands for.

> In mathematics, when we have
>
> 1ST EXPRESSION = 2ND EXPRESSION
>
> (that is, when we link two expressions by an equals sign) we have an
>
> EQUATION

Since an equation represents the equality of two expressions,

$$2 + 3 = 1 + 4$$

is an equation. Other examples of equations are:

$$2x + 3 = 7$$
$$x + y = 5$$
$$x^2 + 3 = 7 + 2$$
$$x^3 - 1 = -9$$
$$5^2 - 1 = 24$$
$$x + y = w$$

Encircle the symbol that is common to all of these examples. If you encircled x, look again. Not all equations contain letters, but all equations contain an equals sign!

As mentioned earlier, we often use equations to find the value of a variable. However, it is possible to have equations without variables. Also, when we have an equation with more than one variable, we usually need other equations to determine what the various letters stand for.

In the equation

$$2 + 3 = 1 + 4$$

the only thing we can do is add the numbers. The value of both expressions is known.

Why can we say that the formula $A = l \cdot w$ (area of a rectangle) is an equation? _____

You probably wrote "because we have two expressions (A and $l \cdot w$) with an equals sign between them."

As this equation stands, we cannot do anything further with it. However, we might get additional information about what some of the letters stand for. For example, what is the value of A if we are told that $l = 5$ and $w = 2$?

$$A = l \cdot w$$
$$? = 5 \cdot 2$$

Therefore,

$$\text{Area} = A = 10$$

The length, l, of the rectangle can be "unknown." What is the length of a rectangle whose area is 12 and whose width is 2?

$$A = l \cdot w$$
$$12 = ? \cdot 2$$

Therefore, $l = 6$

In mathematics we use equations mostly to find the value of something we don't know. For example, "What number added to 3 gives a result of 5?" can be written as an equation:

$$? + 3 = 5 \quad \text{or} \quad n + 3 = 5 \quad \text{or} \quad x + 3 = 5$$

or just about anything indicating

an unknown quantity + 3 = 5

The unknown number is obviously 2. Usually, single letters are preferred to question marks.

The French mathematician Vieta started to use letters to represent numbers in the sixteenth century; later the French philosopher-mathematician René Descartes started using the first letters of the alphabet ($a, b, c, \ldots$) to stand for known numbers and the last letters ($\ldots, x, y, z$) to stand for unknowns.

Is $x - 4 = 5 + 3$ an equation? _____

The right side of the equation, $5 + 3$, is equal to 8, so we get a new equation,

$$x - 4 = 8$$

"Truth" in Equations   Complete the following table:

x	$x + 2$
0	$0 + 2$
5	
7	
9	

What would happen if we use the expression $x + 2$ from the table in an equation such as $x + 2 = 9$? You can replace x with any number, but only one value of x makes this equation true.

Which value of x makes the equation true? _____

As you saw, only $x = 7$ is true for $x + 2 = 9$.

Use the values of x that have been selected for you in the equation

$$x + 2 = x + 5 - 3$$

and make a check if you get a true statement:

x	$x + 2 = x + 5 - 3$	*Check, if true*
0	$0 + 2 \overset{?}{=} 0 + 5 - 3$	✓
−1		
2		
31		
−2		
3		
1		

This equation is true for all values of x.

Explain why this is so: _____

_____

This equation is true because the left side of the equation is equal to the right side for *every* value of x. We call this kind of equation an **identity**.

Identities are equations that are true for all values of the variables.

The next equation we consider is

$$x + 2 = x + 5$$

x	$x + 2 = x + 5$	*Check, if true*
0	$0 + 2 \overset{?}{=} 0 + 5$	
5		
–1		
–4		

Did you make any checks? _____

Why is the equation false for all values of x? _____

_____

As you probably saw, this is not a true equation: a number plus two can never equal the same number plus five. In fact, it should be written as

$$x + 2 \neq x + 5$$ (Recall that " $\neq$ " denotes "is *not* equal to.")

Mathematical symbols have been used since 1000 B.C., and algebra itself started to develop around A.D. 300 when Diophantus, "the father of algebra," used symbols and also equations.

Equations are informally used in "number tricks" and "brain-teasers." For instance, "Take a number, add three. The answer is eight."

What is the number? _____

What do you do to find that number? _____

_____

You probably said that you subtracted three from eight and got five, which is correct. We *undo* addition by subtraction.

In the example,

A number is multiplied by three and the product is fifteen

What is the number? _____

How do you find the number? _____

You probably said that you divided fifteen by three and got five.

Operations that "undo" each other are called **inverse** operations.

Addition and subtraction are inverse operations. Multiplication and division are inverse operations.

Take a number and subtract five. The answer is two.

What is the number? _____

The inverse operation is addition.

$$5 + 2 = 7$$
$$7 - 5 = 2$$

The next example is more complicated. If you start from the answer and go backwards, you can probably get the solution. Remember that these examples deal with number tricks and are included here to make you understand operations and their inverses.

Try to find this number:

_____	?
_____	÷ 9
_____	− 2
_____	× 4
_____	= 20

Did you get the number 63?

If you start with 20 and go backwards, you can solve the problem. Instead of multiplying, you divide. We do the "inverse" operation of what was specified in the problem:

$$20 \div 4 = 5$$
$$5 + 2 = 7$$
$$7 \times 9 = 63$$

Example 1 Take a number, multiply by 3, add 4, and the answer is 13. What is the number?

Solution. Begin with the answer, and work backwards.

$$13 - 4 = 9$$
$$9 \div 3 = 3$$

The number is 3.

To find out if this is true, "plug in" the number 3 in the original statement:

$$3 \times 3 + 4 = 9 + 4 = 13$$

We have *checked* the solution and it was correct.   ■

Example 2 Take a number, add 5, divide the sum by 8, multiply by 3, and subtract 5. If the answer is 1, what is the unknown number?

Solution.   Analyze the problem:

Take a number.   _____

Add 5.   _____

Divide by 8.   _____

Multiply by 3.   _____

Subtract 5.   _____

The answer is 1.   _____

Now we go in the opposite direction. Start with 1.

What to do:	We now have:
Add 5. (Addition is the inverse of subtraction.)	6
Divide by 3 (Division is the inverse of multiplication.)	2
Multiply 2 by 8.	16
Subtract 5.	11

Is 11 the correct answer? You will find that it is if you do the original procedure again, starting with the number 11.   ■

MINI-PRACTICE

(Use this space for your work.)

(a) Take a number, multiply by 5; the answer is 10. Find the number.
(b) Take a number, multiply by 4, add 6; the answer is 14. Find the number.
(c) Take a number, divide by 5, subtract 3; the answer is 1. Find the number.

Answers: (a) _____   (b) _____   (c) _____

IN YOUR OWN WORDS   explain how you can find an unknown number when operations have been performed on the number, and you know the final answer: _____

_____

_____

Answer:

Start with the answer and do the inverse operations. You will end up with the unknown number.

Practice Exercises

1. Take a number, divide by five, and add three. If the answer is five, find the number.

    Answer: _____

2. Take a number, add seven, and divide the sum by five. If the answer is three, find the number.

    Answer: _____

_____

Answers to Mini-Practice:

(a) 2   (b) 2   (c) 20

3. When two is added to a number and the sum is multiplied by six, the product is thirty. Find the number.

   Answer: _____

4. Five is added to the product of a number and three. If the sum is fourteen, find the number.

   Answer: _____

5. Four is subtracted from a number, and the difference is multiplied by three. If the product is eighteen, what is the number?

   Answer: _____

6. When negative three is added to a number and the sum is multiplied by four, the product is twelve. Find the number.

   Answer: _____

7. If five is subtracted from a number and the difference is multiplied by three, the product is twenty-one. Find the number.

   Answer: _____

8. If negative four is added to twice a number and the sum is divided by two, the result is five. Find the number.

   Answer: _____

9. When nine is added to three times a number and the sum is divided by negative two, the result is six. Find the number.

   Answer: _____

10. If seven multiplies the sum of a number and two, the product is fourteen. Find the number.

    Answer: _____

11. If negative two divides the difference of twice a number minus eight, the quotient is four. Find the number.

    Answer: _____

12. If the difference of twice a number minus eight is divided by negative two, the quotient is four. Find the number.

    Answer: _____

9.3 CLASSIFICATION OF EQUATIONS

We have seen examples of equations containing only numbers—for example, $2 + 3 = 1 + 4$. They are called **numerical equations**. Other types of equations contain variables—for example, $x + 3 = 5$ or $x - y = z$; we simply call them **equations**.

How many variables does each of the following equations contain?

Equation	Number of Variables
1. $y + 5 = 2$	_____
2. $x + y = w$	_____
3. $2x + 5 = 5x - 7$	_____

$$4.\ 5x - 2z = 10$$ _____

$$5.\ x^2 + 8x - 9 = 0$$ _____

Equations 1, 3, and 5 contain only one variable. Recall that x stands for the same number, even if it is on both sides of the equals sign or if it is raised to a power. Equation 2 contains three variables and equation 4 contains two variables.

MINI-PRACTICE
(Use this space for your work.)

Determine the number of variables for the following equations:

(a) $x^2 + 4 = 5$ (b) $x - 3x + 3 = 0$ (c) $x^2 + 4y - 9 = 0$

Answers: (a) _____   (b) _____   (c) _____

Practice Exercises

In Exercises 1–10, determine the number of variables in each of the following equations.

1. $x^4 - 3x^3 + 2x^2 - 2 = 0$ *Answer:* _____   2. $2x^2 + 3x + 8 = 0$ *Answer:* _____

3. $y = -x + 5$ *Answer:* _____   4. $x + y^4 = 6$ *Answer:* _____

5. $-x = 2$ *Answer:* _____   6. $x^2 + y^2 + 4$ *Answer:* _____

7. $y = x^3 + z^3$ *Answer:* _____   8. $a^2 = b^2 + c^2$ *Answer:* _____

9. $(x - 2)^2 + (y - 3)^2 = r^2$ *Answer:* _____   10. $P = 2l + 2w + x^2$ *Answer:* _____

> ### SUMMARY
>
> **1.** An equality is a relationship among quantities that have the same value or that mean the same thing.
>
> **2.** An equation is a statement that contains two expressions separated by an equals sign.

Review Practice Exercises

1. Give four equalities for 50.   *Answer:* _____

2. Give three equalities for 3.5.   *Answer:* _____

_____

Answers to Mini-Practice:
(a) 1   (b) 1   (c) 2

3. Simplify $10 - 5(x + 3)$ as far as possible and show that each expression in the simplification process has the same value when $x = -1$.

    Answer: _____

4. Simplify $12 - 2x(3 - x)$ as far as possible and show that each expression in the simplification process has the same value when $x = 2$.

    Answer: _____

5. Simplify $8x^2 - 3x(x + 2)$ as far as possible and show that each expression in the simplification process has the same value when $x = -2$.

    Answer: _____

6. Simplify $-3xy^2 + x^2y - 2xy(4y - 5x)$ as far as possible and show that each expression in the simplification process has the same value when $x = -1$ and $y = 2$.

    Answer: _____

7. Take a number, add five, and multiply the sum by three. If the answer is twenty-four, find the number.

    Answer: _____

8. When six is subtracted from twice a number, the result is ten. Find the number.   *Answer:* _____

9. Twelve is added to the product of five and a number. If the sum is twenty-two, what is the number?

    Answer: _____

10. Five is added to the quotient of a number and two. If the result is negative three, what is the number?

    Answer: _____

11. When three is subtracted from the product of a number and five, the difference is twelve. Find the number.

    Answer: _____

12. A number is subtracted from 10 and the difference is divided by two to give a result of four. Find the number.

    Answer: _____

In Exercises 13–18, determine the number of variables in each of the following equations.

13. $x^3 - 2x^2 + 3x = 4$ *Answer:* _____   14. $x^2 - y^2 = 1$ *Answer:* _____

15. $z = y - 2x$ *Answer:* _____   16. $I = PRT$ *Answer:* _____

17. $A = \dfrac{BH}{2}$ *Answer:* _____   18. $(x - 1)^2 - (y + 2)^2 = 1$ *Answer:* _____

If you have questions about this chapter, write them here: _____

_____

_____

_____

_____

_____

10 Solving Equations

10.1 METHODS OF SOLUTION

To **solve** an equation in one variable means to find all numbers that when substituted for that variable yield a true statement.

Different methods are used for solving different types of equations. We assume that the equation contains one variable and that only the first power of that variable occurs. Examples of such equations are

$$x + 4 = 10$$

$$2y - 3 = y + 1$$

and

$$\frac{t}{5} = 2t + 3$$

There are essentially three methods to solve equations: "by inspection," by trial and error, and by algebraic operations.

Solving by Inspection   The "by inspection" method can be called the "light bulb" approach. You look at the equation and try to come up with the answer.

Solve these equations "by inspection":

$$2n = 6 \underline{\hspace{3cm}}$$

Remember that $2n$ is the same as "two times a number." Here n stands for 3 because $2(3) = 6$.

$$n \div 3 = 12 \underline{\hspace{3cm}}$$

The variable n must stand for 36 in this equation because $36 \div 3 = 12$.

$$n - 5 = 10 \underline{\hspace{3cm}}$$

Here n is 15 because $15 - 5 = 10$.

Solving by Trial and Error

Closely related to the "by inspection" method is the method of guessing ("trial and error"). Try some numbers and see if they fit. Computers sometimes use this method: a number is picked and if it does not fit, the computer is instructed to try another number.

Try to figure out what x stands for in the equation

$$2x + 3 = 7$$

by guessing the answer. It is often helpful to restate the equation in plain English so that what you are looking for becomes clear.

Translate $2x + 3 = 7$ into English: $\underline{\hspace{4cm}}$

$\underline{\hspace{8cm}}$

You probably wrote something like: take a number, double it, add three, and the answer is seven.

Guess a number you think might fit. $\underline{\hspace{2cm}}$

You probably guessed a fairly small number, perhaps 3. Does 3 fit?

$$2(3) + 3 = 6 + 3 = 9$$

No, 3 gives an answer that is too big. We should then try a number smaller than 3. How about 1?

$$2(1) + 3 = \underline{\hspace{2cm}}$$

Too small! 3 was too big and 1 was too small.

Therefore, the next guess should be $\underline{\hspace{2cm}}$

Does it fit? $2(\quad) + 3 = \underline{\hspace{2cm}}$

The correct solution is 2 because

$$2(2) + 3 = 4 + 3 = 7$$

Here is another example to try:

$$-3y + 1 = -8$$

Translate this equation into words: $\underline{\hspace{4cm}}$

$\underline{\hspace{8cm}}$

Working with Classmates

1. Find classmates with whom you can work.
- Ask them your questions.
- You may get valuable help from a classmate who may recognize your errors.
- Teach your friend when you can.
- This will improve your understanding.

2. Set aside time to do homework together
- At school.
- At home.
- In the math lab.

3. Exchange phone numbers.
- Call each other with questions.
- Discuss the work.
- Call if you've missed class, so you don't fall behind.

You are supposed to find a number, multiply that number by –3, add 1, and get a result of –8.

Select a number arbitrarily and substitute it into the equation. Use as many guesses as you need to get the answer –8 on the right side.

$$-3(\underline{\quad}) + 1 = \underline{\qquad}$$

$$-3(\underline{\quad}) + 1 = \underline{\qquad}$$

$$-3(\underline{\quad}) + 1 = \underline{\qquad}$$

$$-3(\underline{\quad}) + 1 = \underline{\qquad}$$

$$-3(\underline{\quad}) + 1 = \underline{\qquad}$$

The number that fits the equation is called the **solution** of the equation. In this case, the correct solution is 3.

How do you know that your solution is the right one? _____

_____

We should always *check* the solutions of our equations. To check a solution, we substitute it for the variable in the original equation. We must make sure we obtain a true statement.

In this case we said that 3 was the correct solution. We could also have said that 3 was a **satisfactory** answer.

Why is it *satisfactory*? _____

When we get a solution that gives a true statement we say in mathematics that

the solution of an equation satisfies the equation

Solve $2(x - 1) = 10$ in two steps:

What number can replace ( )? _____

Since $2(5) = 10$, $x - 1$, the expression within parentheses, is also equal to 5, and we have to solve the equation $x - 1 = 5$.

What does x stand for? _____

Take the number you found and replace x with it:

$$2[(\underline{\quad}) - 1] \stackrel{?}{=} 10$$

Did you get a true statement? _____

If not, try again! The correct solution is 6 because

$$2(6 - 1) = 2(5) = 10$$

Solving by Algebraic Operations

Not all equations can be solved easily by inspection. Consider the equation

$$\frac{x}{2} - 1 = \frac{3}{5}$$

It is hard to guess what x stands for; we suspect that it is a fraction, and it could take a long time to guess the answer. Therefore, we need a method for solving equations when we cannot solve them easily by inspection or by trial and error.

To solve an equation is to answer the question: What is x equal to? In other words, we must be able to write

x equals a number   ($x = \underline{\quad}$)

that is, x must be all alone on one side of the equation. The procedure for getting x alone is called **isolating the variable**.

Isolating the Variable.   Recall that every equation has three parts: two algebraic expressions linked by an equals sign.

As an example, let us work with the equation

$$x + 3 = 5$$

To isolate the variable means to remove "+3" from the left side so that "x" is *by itself*.

How do we isolate x on the left-hand side? Before we can answer that question, we have to consider an important property of equations:

Start with $3 + 2 = 5$. Subtract 2 from each side.

Do we still have equality? _____

Start with $1 + 1 = 2$. Add 6 to each side.

Do we still have equality? _____

Start with $6 \times 8 = 48$. Divide both sides by 8.

Do we still have equality? _____

Start with $9 = 3 \times 3$. Multiply each side by 2.

Do we still have equality? _____

State what you think we can do with equations and still keep their equality: _____

_____

_____

You probably realized that you can perform a mathematical operation on either side of the equation as long as you also do it on the other side.

This can be illustrated with a picture of a balance:

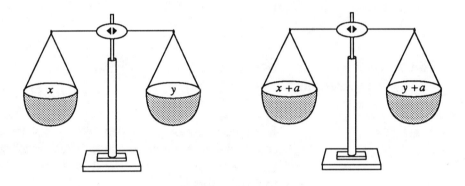

You preserve the equality of the two sides if you add the same or equal quantities to both sides. The Greek mathematicians used the axiom: "If equals be added to equals the wholes are equal."

We can subtract equal amounts from both sides and still preserve equality. We can multiply both sides by the same number or divide both sides by the same *nonzero* number and still preserve equality.

Consider the following equations:

$x + 3 = 15$	$x - 3 = 15$
$3x = 15$	$\dfrac{x}{3} = 15$

Each equation contains just one operation. In each case perform the inverse operation on each side. You should get:

$x + 3 = 15$ $ -3 -3$ $x = 12$	$x - 3 = 15$ $ +3 +3$ $x = 18$
$\dfrac{3x}{3} = \dfrac{15}{3}$ $x = 5$	$(3)\dfrac{x}{3} = 15(3)$ $x = 45$

Check if the solutions are right:

$12 + 3 = 15 \ \surd$	$18 - 3 = 15 \ \surd$
$3(5) = 15 \quad \surd$	$\dfrac{45}{3} = 15 \quad \surd$

Complete the following table:

Equation	Operation	Inverse Operation
$x + 3 = 15$	Addition	Subtraction
$x - 3 = 15$	_____	_____
$3x = 15$	_____	_____
$\dfrac{x}{3} = 15$	_____	_____

Example 1 Solve the equation $5y = 30$.

Solution.

Operation: Multiplication
Inverse Operation: Division

$$\frac{5y}{5} = \frac{30}{5}$$
$$y = 6$$

Check. $5 \times 6 \overset{?}{=} 30$
$ 30 = 30$

Example 2   Solve the equation $\frac{n}{7} = 14$.

Solution.

Operation: Division (as indicated by the fraction bar)
Inverse Operation: Multiplication

$$7 \cdot \frac{n}{7} = 7 \cdot 14$$

$$\frac{7}{1} \cdot \frac{n}{7} = 7 \cdot 14$$

$$n = 98$$

Check. $\frac{98}{7} \overset{?}{=} 14$

$$14 = 14$$ ■

MINI-PRACTICE
(Use this space for your work.)

Solve the following equations:

(a) $x - 7 = 14$ (b) $y + 8 = -3$ (c) $3m = 6$ (d) $\frac{b}{7} = 4$

Answers: (a) _____   (b) _____   (c) _____   (d) _____

IN YOUR OWN WORDS   explain how to solve an equation when the variable is related to a number through one operation only: _____

_____

_____

Answer:

We remove terms from one side of the equation by performing the inverse (opposite) operation. We remove coefficients from the variable by performing the inverse (reciprocal) operation.

Practice Exercises

Solve the following equations by isolating the variable.

1. $x + 7 = 11$ Answer: _____   2. $x + 5 = 8$ Answer: _____   3. $x - 12 = 15$ Answer: _____

4. $x - 31 = 50$ Answer: _____   5. $5 = x - 7$ Answer: _____   6. $48 = x + 42$ Answer: _____

7. $4x = 28$ Answer: _____   8. $9x = 36$ Answer: _____   9. $\frac{x}{2} = 7$ Answer: _____

Answers to Mini-Practice:

(a) 21   (b) –11   (c) 2   (d) 28

10. $\frac{x}{6} = 3$ *Answer:* _____   **11.** $x + 8 = 12$ *Answer:* _____   **12.** $\frac{x}{2} = 5$ *Answer:* _____

13. $x - 2 = 5$ *Answer:* _____   **14.** $3x = 21$ *Answer:* _____   **15.** $x + 5 = 6$ *Answer:* _____

16. $\frac{x}{3} = 1$ *Answer:* _____   **17.** $5x = 5$ *Answer:* _____   **18.** $\frac{x}{5} = 3$ *Answer:* _____

19. $16x = 4$ *Answer:* _____   **20.** $x - 9 = 9$ *Answer:* _____   **21.** $x + 10 = 3$ *Answer:* _____

22. $x - 1 = 1$ *Answer:* _____   **23.** $3x = -9$ *Answer:* _____   **24.** $x - 4 = -2$ *Answer:* _____

25. $\frac{x}{2} = 10$ *Answer:* _____   **26.** $15 = x + 2$ *Answer:* _____   **27.** $18 = 6x$ *Answer:* _____

28. $\frac{x}{4} = 4$ *Answer:* _____   **29.** $13 = x - 14$ *Answer:* _____   **30.** $x + 7 = -3$ *Answer:* _____

10.2 DIFFERENT WAYS TO ISOLATE THE VARIABLE

The equation $x + 3 = 5$ can, of course, be solved by inspection (and should be!). But to illustrate how we can isolate x by using mathematical operations, we write:

$$
\begin{array}{rcl}
x + \quad 3 & = & 5 \qquad \text{(Original equation)} \\
+ (-3) & = & + (-3) \qquad \text{(Equals added to equals)} \\
\hline
x & = & 2
\end{array}
$$

Here, we write out the addition vertically. Instead of adding -3 to each side, we can simply subtract 3 from both sides. We can also write subtraction horizontally:

$$x + 3 = 5$$
$$x + 3 - 3 = 5 - 3$$
$$x = 2$$

Some people move terms from one side of an equation to the other by performing the opposite operation:

$$x + 3 = 5$$
$$x \quad = 5 - 3$$

When a term is moved to the opposite side, we write its inverse instead. Each time we move a number, we should simplify the new expressions, if possible.

Equations with Two Operations

According to the rules for the order of operations, which operations are performed first—multiplication/division or addition/subtraction?

_____

When we solve equations, we "undo" operations by performing the inverse operation. We "undo" the operations *in the reverse order in which they are performed.*

In the equation $2x + 3 = 15$, there are two operations. Which ones?

_____ and _____

Which operation should be "undone" first? _____

We first remove $+ 3$ by subtracting 3 from both sides of the equation:

$$\begin{array}{rcl} 2x + 3 & = & 15 \\ -3 & & -3 \\ \hline 2x & = & 12 \end{array}$$

Divide both sides by 2:

$$\frac{2x}{2} = \frac{12}{2}$$

$$x = 6$$

Check. $\quad 2(6) + 3 \overset{?}{=} 15$

$\qquad\quad 12 \ + 3 = 15 \quad \checkmark$

Note that in the equation $3 + 2x = 15$, addition is performed *after* multiplication, even though $+$ is written first. Thus, we solve this as we did the equation $2x + 3 = 15$.

Example 1   Solve the equation $\frac{x}{3} - 5 = 1$.

Solution.

Two operations: Division and subtraction
"Undo" subtraction, which is the *last* operation done. Add 5 to both sides:

$$\begin{array}{rcl} \frac{x}{3} - 5 & = & 1 \\ +5 & & +5 \\ \hline \frac{x}{3} & = & 6 \end{array}$$

"Undo" division. Multiply by 3:

$$3 \cdot \frac{x}{3} = 3(6)$$

$$x = 18$$

Check. $\quad \frac{18}{3} - 5 \overset{?}{=} 1$

$\qquad\quad 6 - 5 = 1 \quad \checkmark$ ∎

Example 2   Solve the equation $-x + 2 = 5$.

Solution.

Two operations: Multiplication (by -1) and addition. Subtract 2:

$$\begin{array}{rcl} -x + 2 & = & 5 \\ -2 & & -2 \\ \hline -x & = & 3 \end{array}$$

Divide by –1:

$$\frac{-x}{-1} = \frac{3}{-1}$$
$$x = -3$$

Check. $-(-3) + 2 \overset{?}{=} 5$
$+3 \ + 2 = 5$ $\checkmark$

From the equation

$$-x = 3$$

in the solution of Example 2, we could have also proceeded by *multiplying* both sides by –1. Thus,

$$(-x)(-1) = 3(-1)$$
$$x = -3$$ ■

MINI-PRACTICE
(Use this space for your work.)

Solve the following equations:

(a) $3x + 1 = 10$ (b) $2x - 3 = 9$ (c) $\frac{x}{2} + 2 = 8$ (d) $\frac{x}{3} - 2 = 5$

Answers   (a) _____   (b) _____   (c) _____   (d) _____

IN YOUR OWN WORDS

explain how to solve equations with both multiplication/division and addition/subtraction: _____

_____

_____

Answer:

   Remove the constant term by performing the inverse operation of addition or subtraction. Remove the coefficient by performing the inverse operation of multiplication or division.

Practice Exercises

Solve the following equations.

1. $8R - 10 = -40$ *Answer:* _____
2. $2x - 15 = 3$ *Answer:* _____
3. $5m - 1 = 2$ *Answer:* _____
4. $5y + 30 = 10$ *Answer:* _____
5. $4y - 10 = 3$ *Answer:* _____
6. $-5m - 7 = -1$ *Answer:* _____
7. $22 - F = 2$ *Answer:* _____
8. $5T - 5 = 5$ *Answer:* _____
9. $36y - 8 = 10$ *Answer:* _____
10. $2x - 1 = 0$ *Answer:* _____
11. $2x + 3 = 9$ *Answer:* _____
12. $\frac{x}{2} + 4 = 5$ *Answer:* _____

Answers to Mini-Practice:

(a) 3   (b) 6   (c) 12   (d) 21

13. $\dfrac{x}{5} - 3 = 4$ *Answer:* _____

14. $5x + 4 = 24$ *Answer:* _____

15. $\dfrac{x}{3} - 1 = 2$ *Answer:* _____

16. $4x - 3 = 5$ *Answer:* _____

17. $\dfrac{x}{7} + 2 = 5$ *Answer:* _____

18. $8x - 1 = 3$ *Answer:* _____

19. $3x + 1 = 2$ *Answer:* _____

20. $3 + \dfrac{x}{8} = 5$ *Answer:* _____

21. $-2x - 5 = 3$ *Answer:* _____

22. $\dfrac{x}{6} - 2 = -1$ *Answer:* _____

23. $\dfrac{x}{9} + 5 = 2$ *Answer:* _____

24. $-7x + 3 = 10$ *Answer:* _____

25. $16 - 5m = 1$ *Answer:* _____

26. $2 + \dfrac{x}{3} = 8$ *Answer:* _____

27. $3 - \dfrac{x}{7} = 2$ *Answer:* _____

28. $12p - 3 = -1$ *Answer:* _____

29. $5 - \dfrac{x}{4} = 1$ *Answer:* _____

30. $3x + 5 = 2$ *Answer:* _____

10.3 EQUATIONS THAT CAN BE SIMPLIFIED

Simplify the expression on the left side of the equation:

$$2x + 8 - 5 = 15$$

_____

Combine like terms on each side of the following equations:

$$2y + 3y - 10 = 5 - 2 - 6$$

_____ = _____

$$-6m + 3m = 7 - 8 + 1$$

_____ = _____

$$-3 + 3 = 5x + 6x$$

_____ = _____

If on one side the like terms add up to 0, we **MUST** show this by writing "0." Go back over the preceding examples and check that you have two expressions (one might be zero) separated by an equals sign.

Example 1   Solve the equation $5t - 4t + 2 = 5 - 10$.

Solution. $5t - 4t + 2 = 5 - 10$

Simplify both sides:

$$t + 2 = -5$$

Add –2:
$$\frac{-2 \quad -2}{t \quad = -7}$$

Check. $5(-7) - 4(-7) + 2 \overset{?}{=} 5 - 10$
$$-35 + 28 + 2 \overset{?}{=} -5$$
$$-35 + 30 = -5 \quad \checkmark$$

■

Example 2 Solve the equation $1 - 3 + 2 = 2p - 5p - 9$.

Solution. $\qquad 1 - 3 + 2 = 2p - 5p - 9$

Simplify both sides:
$$0 = -3p - 9$$
Add 9:
$$\frac{9 \qquad + 9}{9 = -3p}$$

Divide by –3:
$$-3 = p$$

Check. $1 - 3 + 2 \overset{?}{=} 2(-3) - 5(-3) - 9$
$$0 \overset{?}{=} -6 + 15 - 9$$
$$0 = 0 \quad \checkmark$$

■

MINI-PRACTICE
(Use this space for your work.)

Simplify each expression and then solve the equation:

(a) $y + (y + 2) = 10$ (b) $2M = -(10 + 2)$ (c) $-1 = y - (2y - 10)$

(d) $2(5c - 3c) = 8$ (e) $2a - 3a = 1$

Answers: (a) _____   (b) _____   (c) _____   (d) _____   (e) _____

IN YOUR OWN WORDS explain how to solve equations containing expressions that can be simplified:

_____

_____

_____

Answer:

   Simplify each side as far as possible. This means that we end up with one term (possibly equal to 0) on each side.

Summary of the steps:

Step 1. Simplify each side as far as possible.
Step 2. Isolate the term containing the variable on one side of the equation.
Step 3. Isolate the variable.
Step 4. Check your solution.

_____

Answers to Mini-Practice:
(a) 4   (b) –6   (c) 11   (d) 2   (e) –1

Practice Exercises

Solve the following equations.

1. $2x + 1 + x = 10$ Answer: _____
2. $x + 2x + 4 = 19$ Answer: _____
3. $7x - 2 - 5x - 3 = 5$ Answer: _____
4. $5x + 7 - 3x + 5 = 16$ Answer: _____
5. $5x + 2 + 2x = 16$ Answer: _____
6. $3x + 4x + 9 = 30$ Answer: _____
7. $8x - 5 + 2x + 10 = 15$ Answer: _____
8. $x + 4 + 7x - 11 = 9$ Answer: _____
9. $9x - 8 - 11x + 2 = 4$ Answer: _____
10. $5(x + 3) + 4(x - 1) = 56$ Answer: _____
11. $3(x + 2) - (x - 4) = 14$ Answer: _____
12. $4(x + 3) - 2(x + 10) = 0$ Answer: _____
13. $2(x - 7) - (3x + 4) + 24 = 2$ Answer: _____
14. $12 + 3(2x + 3) = 33$ Answer: _____
15. $3(x - 1) - (x - 10) - 4x = 5$ Answer: _____
16. $2(x + 5) + 4(x - 3) = 10$ Answer: _____
17. $5(x - 2) + 6(x + 4) = 47$ Answer: _____
18. $3(x + 1) + 2(x + 4) = 10$ Answer: _____
19. $8(x + 2) - 3(x + 5) = 11$ Answer: _____
20. $4(x - 9) - 5(x - 3) = 0$ Answer: _____
21. $5(4 - x) - 7(x - 4) = 24$ Answer: _____
22. $2(-3 - x) - 6(3 - x) = 4$ Answer: _____
23. $3(x + 6) + 2(4 - x) = 30$ Answer: _____
24. $8 + 2(3x - 5) = 10$ Answer: _____
25. $7(x - 3) - (2x + 5) - 13 = 1$ Answer: _____
26. $10 - 9(2x - 3) = 1$ Answer: _____
27. $2(x - 4) - (x - 8) + 2x = 9$ Answer: _____
28. $6(3 - x) + 2(x - 4) + 5x = 0$ Answer: _____
29. $5(3x + 7) - 8(3 - x) - 10x = -2$ Answer: _____
30. $10(3 - 5x) - 4(2x - 7) - 13x = -84$ Answer: _____

10.4 EQUATIONS WITH DECIMALS

When decimals occur in equations, it is usually best to multiply both sides of the equation by a multiple of 10 (10, 100, 1000, etc.) to eliminate decimals. The equations will then involve only whole numbers.

Example 1   Solve the equation $36 = -2.5y + 1$.

Solution. $36 = -2.5y + 1$

Step 1: Multiply both sides by 10 to eliminate decimals. By using the distributive law on the right side, each term is multiplied by 10:

$$360 = -25y + 10$$

Step 2: Isolate the term with the variable:

$$360 - 10 = -25y$$
$$350 = -25y$$

Step 3: Isolate the variable:

$$\frac{350}{-25} = y$$

$$-14 = y$$

Check. $36 \stackrel{?}{=} -2.5(-14) + 1$

$36 = 35 + 1$ $\sqrt{}$ ∎

Example 2 Solve the equation $3x + 13 - x + 14.5 + 3x - x = 45.5$.

Solution. $3x + 13 - x + 14.5 + 3x - x = 45.5$

Multiply both sides by 10 to eliminate decimals.

$$30x + 130 - 10x + 145 + 30x - 10x = 455$$

Simplify:

$$40x + 275 = 455$$

Add -275:

$$\underline{\quad -275 \quad -275}$$

$$40x \quad = \quad 180$$

Divide by 40:

$$x = 4.5$$

Check. $3(4.5) + 13 - 4.5 + 14.5 + 3(4.5) - 4.5 \stackrel{?}{=} 45.5$

$45.5 = 45.5$ $\sqrt{}$ ∎

MINI-PRACTICE   Solve the following equations:
(Use this space for your work.)

  (a) $0.5x - 1 = 2$ (b) $2.5y + 5 = 0$ (c) $1 + 3.5a = -2.5$

  (d) $2.1a - 1 = 3.2$

Answers: (a) _____   (b) _____   (c) _____   (d) _____

IN YOUR OWN WORDS   explain how we solve equations that contain decimals: _____

_____

_____

Answer:

  To solve an equation with decimals, multiply both sides of the equation by 10, 100, 1000, etc., to eliminate decimals. Then apply the previous methods to the resulting equation.

_____

Answers to Mini-Practice:

(a) 6   (b) –2   (c) –1   (d) 2

Practice Exercises

Solve the following equations.

1. $d + 1.3619 = 2.0148$ *Answer:* _____
2. $w + 2.932 = 4.801$ *Answer:* _____
3. $-0.813 + x = -1.098$ *Answer:* _____
4. $-1.926 + t = -1.042$ *Answer:* _____
5. $x + 2.45 = 3.91$ *Answer:* _____
6. $y + 0.829 = 1.102$ *Answer:* _____
7. $z - 3.72 = 2.459$ *Answer:* _____
8. $1.25 - a = 0.75$ *Answer:* _____
9. $-0.593 + b = -0.286$ *Answer:* _____
10. $6.284 - x = 4.937$ *Answer:* _____
11. $5.237 = -2.014 + 2x$ *Answer:* _____
12. $3.47a = 7.1482$ *Answer:* _____
13. $3x + 1.45 = 2.44$ *Answer:* _____
14. $1.2x - 3.9 = 2.1$ *Answer:* _____
15. $0.23a + 1.43 = 2.81$ *Answer:* _____
16. $5.21y = -1.563$ *Answer:* _____
17. $-0.42z = 2.94$ *Answer:* _____
18. $1.27 - 3.5b = -2.93$ *Answer:* _____
19. $-6.54 = 1.48y - 3.062$ *Answer:* _____
20. $1.25x - 4 = 1$ *Answer:* _____
21. $2x + 5 - 3x + 7.2 + 5x = 20.2$ *Answer:* _____
22. $11.3 - 2x + 4.6 + 5x = 19.5$ *Answer:* _____
23. $5y - 9.3 + 2y + 5.4 - 3y = 4.5$ *Answer:* _____
24. $-5.3 = 1.5a - 3.8 - 4.1a + 1.1$ *Answer:* _____
25. $2x + 18.7 - 3.3x - 16.75 = 0$ *Answer:* _____
26. $-4.7x + 2.3 - 5.3x + 4 = 26.3$ *Answer:* _____
27. $7.3x - 2 - 5.1x - 3 = 8.2$ *Answer:* _____
28. $5.1p + 7.8 - 3.9p + 5.2 = 12.88$ *Answer:* _____
29. $12y - 6.3 + 2.5y - 3.8 = 33.4$ *Answer:* _____
30. $3.25 - 11.7x + 14.6 - 4.83x = -64.8$ *Answer:* _____

10.5 EQUATIONS WITH FRACTIONS

In equations with fractions, multiply each term of the equation by the least common denominator of all of the fractions. Then follow these steps:

Step 1. Simplify each side as far as possible.
Step 2. Isolate the term containing the variable on one side of the equation.
Step 3. Isolate the variable.
Step 4. Check your solution.

Example 1   Solve the equation $\frac{x}{2} - 1 = \frac{3}{5}$.

Solution.   The least common denominator of 2 and 5 is 10. Multiply each term by 10.

$$10 \cdot \frac{x}{2} - 10 \cdot 1 = 10 \cdot \frac{3}{5}$$

$$5x - 10 \quad = \quad 6$$

$$5x \qquad\quad = 16$$

$$x = \frac{16}{5}$$

$$x = 3.2$$

Check. $\dfrac{3.2}{2} - 1 \overset{?}{=} \dfrac{3}{5}$

$1.6 - 1 = 0.6$ $\checkmark$

The equation $\dfrac{x}{2} - 1 = \dfrac{3}{5}$ can also be solved the following way:

$$\frac{x}{2} - 1 = \frac{3}{5}$$

$$\underline{\quad +1 \qquad +1 \left(\text{or} + \frac{5}{5}\right)}$$

$$\frac{x}{2} \quad = \frac{8}{5}$$

$$2 \cdot \frac{x}{2} = 2 \cdot \frac{8}{5}$$

$$x = \frac{16}{5}$$

$$x = 3.2 \qquad\qquad\qquad \blacksquare$$

Example 2   Solve the equation $\dfrac{x}{3} - \dfrac{x+2}{4} = \dfrac{1}{12}$.

Solution.   The least common denominator of 3, 4, and 12 is 12. Multiply both sides of the equation by 12:

$$12 \cdot \frac{x}{3} - 12 \cdot \frac{x+2}{4} = 12 \cdot \frac{1}{12}$$

Simplify. One of the numerators contains two terms. Place parentheses around this numerator so that the minus sign applies to both terms.

$$4x - 3(x + 2) = 1$$

$$4x - 3x - 6 = 1$$

$$x - 6 = 1$$

$$x = 7$$

Check. $\dfrac{7}{3} - \dfrac{9}{4} \overset{?}{=} \dfrac{1}{12}$

$$\frac{28}{12} - \frac{27}{12} = \frac{1}{12} \quad \checkmark \qquad\qquad \blacksquare$$

MINI-PRACTICE
(Use this space for your work.)

Solve the following equations:

(a) $\frac{x}{3} - 1 = 8$ (b) $x - \frac{x}{3} + \frac{x}{5} = 26$ (c) $\frac{5y - 4}{8} = 2$

(d) $\frac{x - 4}{2} - \frac{x - 6}{4} = 2$

Answers: (a) _____   (b) _____   (c) _____   (d) _____

IN YOUR OWN WORDS   explain how equations containing fractions are simplified. _____

_____

_____

Answer:

Multiply each term by the least common denominator. Be careful when a numerator has more than one term and is preceded by a minus sign. Keep parentheses around the numerator and "distribute the minus sign."

Practice Exercises

Solve the following equations.

1. $\frac{x}{3} + 2 = 4$ *Answer:* _____

2. $\frac{y}{3} - 1 = 2$ *Answer:* _____

3. $\frac{t}{6} + 2 = \frac{5}{2}$ *Answer:* _____

4. $\frac{3x}{4} - 5 = 1$ *Answer:* _____

5. $\frac{4y}{5} + 3 = 7$ *Answer:* _____

6. $\frac{3 - x}{5} = 2$ *Answer:* _____

7. $\frac{2y}{3} - 3 = 9$ *Answer:* _____

8. $\frac{5 - u}{2} = 9$ *Answer:* _____

9. $\frac{4x - 5}{3} + 2 = 7$ *Answer:* _____

10. $\frac{3x - 7}{5} = 4$ *Answer:* _____

11. $\frac{3y}{5} + 26 = 5$ *Answer:* _____

12. $9 - \frac{2t}{3} = 5$ *Answer:* _____

13. $\frac{2x - 7}{5} + 1 = 4$ *Answer:* _____

14. $\frac{t}{5} = \frac{t - 6}{4}$ *Answer:* _____

15. $\frac{x}{2} = \frac{x + 3}{3}$ *Answer:* _____

16. $\frac{x}{2} - \frac{x}{3} = 1$ *Answer:* _____

Answers to Mini-Practice:

(a) 27   (b) 30   (c) 4   (d) 10

17. $\dfrac{x}{3} - \dfrac{x}{9} = 2$ *Answer:* _____

18. $\dfrac{5y + 11}{3} = 7$ *Answer:* _____

19. $\dfrac{z - 5}{4} + 1 = \dfrac{7}{2}$ *Answer:* _____

20. $4 - \dfrac{x - 3}{5} = 3$ *Answer:* _____

21. $\dfrac{x}{2} + \dfrac{x}{3} = 15$ *Answer:* _____

22. $\dfrac{2x + 1}{3} - \dfrac{3x + 2}{6} = 5$ *Answer:* _____

23. $\dfrac{x}{5} - \dfrac{x}{10} = 2$ *Answer:* _____

24. $\dfrac{x}{5} - \dfrac{x}{2} = 3$ *Answer:* _____

25. $\dfrac{x + 5}{4} - \dfrac{x + 3}{8} = \dfrac{7}{4}$ *Answer:* _____

26. $\dfrac{x + 1}{3} - \dfrac{x + 2}{6} = 3$ *Answer:* _____

27. $\dfrac{x}{5} + \dfrac{x + 3}{15} = 1$ *Answer:* _____

28. $\dfrac{x - 4}{4} - \dfrac{x + 6}{8} = 2$ *Answer:* _____

29. $\dfrac{2y + 3}{10} + \dfrac{y - 5}{15} = \dfrac{1}{2}$ *Answer:* _____

30. $\dfrac{3x - 1}{12} - \dfrac{5x + 1}{16} = \dfrac{7}{24}$ *Answer:* _____

10.6 VARIABLES ON BOTH SIDES OF THE EQUATION

So far, we have seen equations in which the term containing the variable is on only one side of the equation. But frequently, terms containing the variable are on *both* sides of the equation, as in the following example.

$$2x + 6 = 5x$$

How can we "combine like terms" when the terms are on opposite sides of the equation? _____

You probably realized that we have to move one of the variable terms to the other side. It doesn't matter which term we move.

How can you remove $2x$ from the left side? _____

As in all equations, we are allowed to add anything we want to both sides of the equation. We add $-2x$ (or subtract $2x$). We get:

$$
\begin{array}{rcl}
2x + 6 &=& 5x \\
-2x & & -2x \\
\hline
6 &=& 3x \\
x &=& 2
\end{array}
$$

Check. $2(2) + 6 \overset{?}{=} 5(2)$
           $4 + 6 = 10$ $\surd$

You could have added $-5x$ to both sides instead:

$$
\begin{array}{rcl}
2x + 6 &=& 5x \\
-5x & & -5x \\
\hline
-3x + 6 &=& 0
\end{array}
$$

Notice that if you are left with a zero on the right side, you must write that zero! Remember that an equation always has three parts: two algebraic expressions that are connected by an equals sign.

$$
\begin{array}{rcr}
-3x + 6 &=& 0 \\
-6 && -6 \\
\hline
-3x &=& -6
\end{array}
$$

Divide by -3:

$$x = 2$$

Example 1   Solve the equation $10x = 7 + 2x$.

Solution.

$$
\begin{array}{rcr}
10x &=& 7 + 2x \\
-2x && -2x \\
\hline
8x &=& 7
\end{array}
$$

Add $-2x$:

Divide by 8:

$$x = \frac{7}{8}$$

Check. $10 \cdot \dfrac{7}{8} \overset{?}{=} 7 + 2 \cdot \dfrac{7}{8}$

$5 \cdot \dfrac{7}{4} \overset{?}{=} \dfrac{28}{4} + \dfrac{7}{4}$

$\dfrac{35}{4} = \dfrac{35}{4}$ $\checkmark$ ∎

Example 2   Solve the equation $\dfrac{x+8}{3} - \dfrac{x-4}{5} = x$.

Solution.   Multiply both sides by 15 (the least common denominator).

$$15\left(\frac{x+8}{3}\right) - 15\left(\frac{x-4}{5}\right) = 15x$$

$$5(x+8) - 3(x-4) = 15x$$

$$5x + 40 - 3x + 12 = 15x$$

$$
\begin{array}{rcr}
2x + 52 &=& 15x \\
-2x && -2x \\
\hline
52 &=& 13x \\
x &=& 4
\end{array}
$$

Check. $\dfrac{4+8}{3} - \dfrac{4-4}{5} \overset{?}{=} 4$

$4 - 0 = 4$ $\checkmark$ ∎

MINI-PRACTICE
(Use this space for your work.)

Solve the following equations:

(a) $5c = 28 + c$ (b) $8r + 10 = 3r - 40$ (c) $5m - 1 = -m + 2$

(d) $4x = -6 + 2x + 3x$ (e) $5y + 30 = 10y$

Answers: (a) _____   (b) _____   (c) _____   (d) _____   (e) _____

IN YOUR OWN WORDS   explain the method we use to solve equations with variable terms on both sides of the equation. _____

_____

_____

Answer:

As with all equations, we first simplify as far as possible. When we have gotten rid of all denominators and have at the most two (unlike) terms on each side of the equation, we transfer one variable term from one side to the other by using the inverse operation.

Practice Exercises

Solve the following equations.

1. $2F = 36 + 10F$ *Answer:* _____
2. $4y - 10 = 5y + 3$ *Answer:* _____
3. $3m - 5m - 7 = 2m - 1$ *Answer:* _____
4. $22 - y = 10y + 2$ *Answer:* _____
5. $5T - 5 = -5T - 5$ *Answer:* _____
6. $36y = 10y + 8$ *Answer:* _____
7. $3y - 5 = 6y + 4$ *Answer:* _____
8. $10x + 4 = 8x - 2$ *Answer:* _____
9. $y + 3(y + 2) = 10y$ *Answer:* _____
10. $M + 2(M - 1) = 5$ *Answer:* _____
11. $2x - (3x - 1) = 0$ *Answer:* _____
12. $-1 + y = y - (2y - 10)$ *Answer:* _____
13. $5x - (2x + 1) = x + 7$ *Answer:* _____
14. $8 - 3(2 - x) = 2x + 5$ *Answer:* _____
15. $5p + 9 = 10 - 3p$ *Answer:* _____
16. $3(x + 4) = 6(x - 3)$ *Answer:* _____
17. $8(1 - y) = 24(y - 3)$ *Answer:* _____
18. $-4 - 2p = p - (3p + 5)$ *Answer:* _____
19. $3a - (5a - 4) = a + 2$ *Answer:* _____
20. $1.5x - 3.2 = 0.4 + 0.3x$ *Answer:* _____
21. $2y - 3 = -2y + 3$ *Answer:* _____
22. $4z - 5 = 10 - 6(z - 1)$ *Answer:* _____
23. $15t + 4 = 3t + 6$ *Answer:* _____
24. $7(x - 3) - 5(2x + 1) = 10 - 8(x - 0.5)$ *Answer:* _____
25. $8y - 9(2 + y) = 15 - 11(y + 7)$ *Answer:* _____
26. $\dfrac{x - 5}{2} - \dfrac{x - 3}{4} = \dfrac{x - 7}{10}$ *Answer:* _____
27. $\dfrac{a + 1}{3} + \dfrac{a + 3}{4} = \dfrac{a + 3}{2}$ *Answer:* _____
28. $\dfrac{x + 10}{4} - \dfrac{x - 10}{6} = \dfrac{x + 10}{2}$ *Answer:* _____

Answers to Mini-Practice:

(a) 7   (b) –10   (c) $\frac{1}{2}$ (d) 6   (e) 6

29. $\dfrac{2x-3}{10} + \dfrac{x+2}{15} = \dfrac{x+40}{20}$ *Answer:* _____

30. $\dfrac{a+3}{12} - \dfrac{a-2}{8} = \dfrac{9-a}{6}$ *Answer:* _____

SUMMARY

The four major steps in solving any equation in one variable in which only the first power of that variable occurs are:

Step 1. Simplify each side as far as possible.
Step 2. Isolate the term containing the variable on one side of the equation.
Step 3. Isolate the variable.
Step 4. Check your solution.

This sequence of steps for solving linear equations is illustrated in the flowchart on the following page.

Review Practice Exercises

Solve the following equations.

1. $x - 9 = 13$ *Answer:* _____

2. $y + 5 = 10$ *Answer:* _____

3. $z - 3 = -12$ *Answer:* _____

4. $a + 3 = -1$ *Answer:* _____

5. $b - 5 = 0$ *Answer:* _____

6. $n + 7 = 4$ *Answer:* _____

7. $-5 = -7 + c$ *Answer:* _____

8. $-4 = d - 8$ *Answer:* _____

9. $3 = y + 12$ *Answer:* _____

10. $s + 12 = 8$ *Answer:* _____

11. $3r = 15$ *Answer:* _____

12. $-4x = 36$ *Answer:* _____

13. $21 = 7y$ *Answer:* _____

14. $-5a = 20$ *Answer:* _____

15. $-b = 10$ *Answer:* _____

16. $100m = 0$ *Answer:* _____

17. $\dfrac{n}{3} = 12$ *Answer:* _____

18. $\dfrac{-4x}{8} = -28$ *Answer:* _____

19. $40 = \dfrac{y}{4}$ *Answer:* _____

20. $-6 = \dfrac{a}{2}$ *Answer:* _____

21. $\dfrac{b}{3} = 7$ *Answer:* _____

22. $\dfrac{R}{10} = 0$ *Answer:* _____

23. $\dfrac{T}{5} = -1$ *Answer:* _____

24. $\dfrac{M}{-2} = -3$ *Answer:* _____

Flowchart for Solving Linear Equations

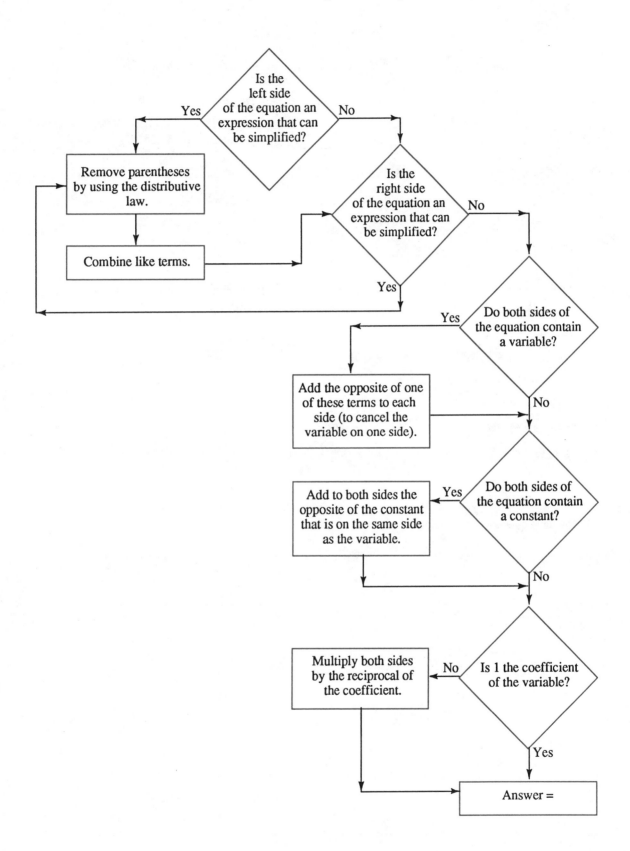

25. $\dfrac{x}{-5} = 1$ *Answer:* _____

26. $15 = \dfrac{y}{3}$ *Answer:* _____

27. $8x - 10 = -1$ *Answer:* _____

28. $-2y + 3y = -y + 1$ *Answer:* _____

29. $4a + 2 - 5 = 3a - 3$ *Answer:* _____

30. $\dfrac{x}{3} - 6 = \dfrac{1}{2}$ *Answer:* _____

31. $2.5y = 3.5y + 4$ *Answer:* _____

32. $\dfrac{x}{2} - \dfrac{x+2}{5} = 2$ *Answer:* _____

33. $\dfrac{x-4}{2} - 2 = \dfrac{x-7}{4}$ *Answer:* _____

34. $\dfrac{x+8}{3} - \dfrac{x-4}{5} = x$ *Answer:* _____

35. $8x - (4x + 3) = -1$ *Answer:* _____

36. $4(3n - 2) = 0$ *Answer:* _____

37. $-2(5c + 3) = 30$ *Answer:* _____

38. $-(6a + 2) = 5$ *Answer:* _____

39. $-3(b + 2) = 5(7b - 1)$ *Answer:* _____

40. $20 - 5c(-2) = -10$ *Answer:* _____

41. $7x - (3x + 2) = x - (2x + 1)$ *Answer:* _____

42. $\dfrac{11x}{3} + 0.9 = 2$ *Answer:* _____

43. $2(x - 7) - (3x + 4) + 24 = 2$ *Answer:* _____

44. $27 - 3(x + 1) = 10x - 3(x - 4)$ *Answer:* _____

45. $12 + 3(2x - 3) = 5(2x - 7) - 2(5x - 20)$ *Answer:* _____

46. $5(x - 7) + 4(x + 9) = 2(2x + 21) + 6(x - 6.5)$ *Answer:* _____

47. $5x - 3(1 - 2x) = 7 + 3(2x - 3) + x$ *Answer:* _____

48. $2(2 - 3x) - 2x = 2(1 - 2x)$ *Answer:* _____

49. $88 - 3(x - 6) - 4(x - 1) - 7(2x - 3) + 4x + 5 = 0$ *Answer:* _____

50. $5x - 2(3 - x) + 11 - 6(x + 2) - 13 - 3(x - 1) - (x - 2) = 0$ *Answer:* _____

If you have questions about this chapter, write them here: _____

_____

_____

_____

_____

_____

11 Applications of Equations

Remember to

* *SOLVE LITERAL EQUATIONS THE SAME WAY AS YOU SOLVE EQUATIONS IN ONE VARIABLE*
* *COMBINE ONLY LIKE TERMS*
* *REVIEW THE FORMULAS OF CHAPTER 7*
* *ASK QUESTIONS WHEN YOU DON'T UNDERSTAND*

Next, you will learn about equations with several variables. You will also work with formulas you have used earlier in the book. Refer to page 193 for a listing of these formulas.

11.1 LITERAL EQUATIONS

Why is

$$x + y = z$$

an equation? _____

Two expressions linked by an equals sign form an equation. We call an equation that has more than one variable a **literal equation**. Scientific and mathematical formulas can be thought of as literal equations. A literal equation can be solved for any of the variables in terms of the remaining variables. To do this, we apply the same rules for solving equations that you have already learned. For example, we can solve the equation

$$x + y = z$$

for x.

What would you do if y were a known number? _____

Preparing for Tests

1. **Two weeks before:**
   - Begin going over past homework assignments.
   - Never leave a question about math unanswered.
   - Ask your instructor for copies of old exams. Work out the problems.
   - Read your notes and the text again.
   - With your book closed, think about the concepts and methods learned.

2. **The night before:**
   - Rework some exercises from each topic.
   - Get enough sleep. Do not stay up all night cramming.
   - Eat enough so you are not hungry during the exam.
   - Approach the exam confidently and with a clear mind.

Now, subtract y from both sides of the equation:

$$\begin{array}{rcl} x+y &=& z \\ -y & & -y \\ \hline x & =& z-y \end{array}$$

Why can we not simplify $z - y$? _____

You probably realized that z and y are not like terms. Remember that we can only simplify like terms when adding or subtracting. So, the solution is $x = z - y$.

We can check the solution as follows: Replace x with $z - y$ in the given equation and obtain

$$\underbrace{z - y + y}_{0} \overset{?}{=} z$$

$$z = z \quad \sqrt{}$$

Example 1   Solve $x - y = 3$ for y.

Solution.

First, subtract x from both sides:

$$\begin{array}{rcl} x-y &=& 3 \\ -x & & -x \\ \hline -y &=& 3-x \end{array}$$

Second, multiply both sides by -1:

$$y = (-1)(3 - x)$$

Third, distribute the minus sign:

$$y = -3 + x$$

or

$$y = x - 3$$

Check. $x - (x - 3) \stackrel{?}{=} 3$
$x - x + 3 = 3$ √

Example 2   Solve $x + 2y = 4$ for y.

Solution.

Subtract x from both sides:

$$\begin{array}{r} x + 2y = 4 \\ \underline{-x \qquad -x} \\ 2y = 4 - x \end{array}$$

Divide both sides by 2:

$$y = \frac{4 - x}{2}$$

or

$$y = 2 - \frac{x}{2}$$

MINI-PRACTICE
(Use this space for your work.)

Solve each of the following literal equations for the indicated variable.

(a) $a + x = 1$ for x (b) $ax = 1$ for x (c) $6b - x = 2a$ for x

(d) $4y - 10 = 6z$ for y (e) $rx + 2d = 7t$ for x

Answers: (a) _____   (b) _____   (c) _____

        (d) _____   (e) _____

IN YOUR OWN WORDS   explain how to solve literal equations. _____

_____

_____

Answer:

     We solve literal equations the same way as we solve equations in one variable. We isolate the variable we want by performing the inverse operations. Only like terms can be combined.

Answers to Mini-Practice:

(a) $1 - a$ (b) $\frac{1}{a}$ (c) $6b - 2a$ (d) $\frac{3z + 5}{2}$ (e) $\frac{7t - 2d}{r}$ $(r \neq 0)$

Practice Exercises

Solve each of the following equations for the indicated variable:

1. $x + y = 10$, solve for x.   *Answer:* _____   2. $m - 3 = R$, solve for m.   *Answer:* _____

3. $ax = 10$, solve for x.   *Answer:* _____   4. $\dfrac{c}{x} = 4$, solve for x.   *Answer:* _____

5. $mx = c$, solve for x.   *Answer:* _____   6. $ty = m$, solve for y.   *Answer:* _____

7. $ln = q$, solve for n.   *Answer:* _____   8. $\dfrac{p}{q} = r$, solve for p.   *Answer:* _____

9. $\dfrac{a}{b} = c$, solve for b.   *Answer:* _____   10. $ex + y = 10$, solve for y.   *Answer:* _____

11. $y - mx = b$, solve for x.   *Answer:* _____   12. $b - 5a = 10$, solve for a.   *Answer:* _____

13. $-n - 5a = 10$, solve for a. *Answer:* _____   14. $-5d = -7 + c$, solve for d.   *Answer:* _____

15. $3R - T = 15$, solve for R. *Answer:* _____   16. $8M - 2N = 5N$, solve for M.   *Answer:* _____

17. $3ABC = -1$, solve for A.   *Answer:* _____   18. $4A + 3B + 5C = 100$, solve for C.   *Answer:* _____

19. $3x + 2y = 6$, solve for y.   *Answer:* _____   20. $4x - y = 3$, solve for y.   *Answer:* _____

21. $x + 6y = 6$, solve for x.   *Answer:* _____   22. $x + 3y = 10$, solve for x.   *Answer:* _____

23. $2x - y = 6$, solve for x.   *Answer:* _____   24. $4x + 3y = 12$, solve for x.   *Answer:* _____

25. $px + qy = t$, solve for x.   *Answer:* _____   26. $px + qy = 10$, solve for y.   *Answer:* _____

27. $4xyz = 6$, solve for x.   *Answer:* _____   28. $3x + 2y - 4z = 10t$, solve for z.   *Answer:* _____

29. $3x + 2y - 4z = 10t$, solve for x.   *Answer:* _____

30. $2p + 3q - 4r = w$, solve for r.   *Answer:* _____

11.2 FORMULAS

A formula, such as $A = lw$, can be considered as a literal equation. This formula expresses the area, A, of a rectangle in terms of the length, l, and the width, w. Sometimes, we would rather feature the length, l, by itself. Thus, we want to solve the equation for l. At other times, we want to express w in terms of A and l.

Solve the equation for l: _____

Solve the equation for w: _____

Example 1   Formula:

$$A = \frac{h(a+b)}{2} \quad \text{(Area of a trapezoid)}$$

Solve for a.

Solution. $A = \dfrac{h(a+b)}{2}$

Multiply both sides by 2: $2A = h(a + b)$

Divide both sides by h: $\qquad \dfrac{2A}{h} = a + b$

Subtract b: $\qquad \dfrac{2A}{h} - b = a$

or

$$a = \frac{2A}{h} - b \qquad \blacksquare$$

Example 2   Formula:

$$C = \frac{5}{9}(F - 32)$$

Solve for F.

Solution.   You may recognize this formula as the one used to convert Fahrenheit temperature to Celsius temperature.

$$C = \frac{5}{9}(F - 32)$$

Multiply by 9:

$$9C = 5(F - 32)$$

Divide by 5:

$$\frac{9C}{5} = F - 32$$

Add 32:

$$\frac{9C}{5} + 32 = F$$

or

$$F = \frac{9C}{5} + 32 \qquad \blacksquare$$

MINI-PRACTICE   Solve each of the following formulas for the given variable.
(Use this space for your work.)

(a) Formula: $A = \dfrac{1}{2}bh$ for b (b) Formula: $P = 2l + 2w$ for w

(c) Formula: $I = Prt$ for t (d) Formula: $V = C - Crt$ for r

Answers: (a) _____   (b)_____   (c) _____   (d) _____

IN YOUR OWN WORDS   explain why formulas are literal equations: _____

_____

_____

Answer:

A formula consists of two or more expressions linked by an equals sign. There are always at least two letters in a formula, so the formula is a *literal equation.*

Answers to Mini-Practice:

(a) $\dfrac{2A}{h}$ (b) $\dfrac{P-2l}{2}$ (c) $\dfrac{I}{Pr}$ (d) $\dfrac{C-V}{Ct}$

Practice Exercises

Use the given formulas to solve for the indicated variable.

1. Formula: $A = \dfrac{h(a+b)}{2}$ for b Answer: _____   2. Formula: $P = 2l + 2w$ for l Answer: _____

3. Formula: $N = R - C$ for C Answer: _____   4. Formula: $A = L + D$ for D Answer: _____

5. Formula: $N = G - D$ for G Answer: _____   6. Formula: $N = G - D$ for D Answer: _____

7. Formula: $M = V(D)$ for D Answer: _____   8. Formula: $V = \dfrac{D}{t}$ for t Answer: _____

9. Formula: $V = \dfrac{D}{t}$ for D Answer: _____   10. Formula: $F = \dfrac{9C}{5} + 32$ for C Answer: _____

11. Formula: $A = \dfrac{1}{2} bh$ for h Answer: _____   12. Formula: $r = \dfrac{1}{3} Ah$ for A Answer: _____

13. Formula: $\dfrac{R}{t} = c - s$ for s Answer: _____   14. Formula: $E = IR$ for R Answer: _____

15. Formula: $p = \dfrac{R - C}{n}$ for R Answer: _____   16. Formula: $PV = nRT$ for T Answer: _____

17. Formula: $A = P + PRT$ for P Answer: _____   18. Formula: $A = Sw + w$ for w Answer: _____

11.3 NUMERICAL APPLICATIONS

Most of the time when we use formulas, we have numerical values for all variables but one. All formulas used here were explained in Chapter 7 in the section titled "Formulas."

Find the width of a rectangle whose length is 5 inches and whose area is 20 square inches.

What is the formula for the area of a rectangle? _____

Substitute all known values in the formula: _____

You should have $20 = 5w$. Solve for w.

Is it true that a rectangle with length 5 inches and width 4 inches has an area of 20 square inches? _____

Since five times four is twenty, it is true.

Example 1   Use the formula

$$C = \frac{5}{9}(F - 32)$$

to solve for F when $C = 35°$.

Solution.

Replace C with 35:

$$35 = \frac{5}{9}(F - 32)$$

Multiply both sides by 9:

$$315 = 5(F - 32)$$

Divide both sides by 5:

$$63 = F - 32$$

Add 32:

$$95 = F$$

We have shown that when $C = 35°$, $F = 95°$.

Check: $35 \overset{?}{=} \dfrac{5}{9}(95 - 32)$

$35 \overset{?}{=} \dfrac{5}{9}(63)$

$35 = 5(7)$ $\sqrt{}$ ■

MINI-PRACTICE
(Use this space for your work.)

Use the following formulas and the given values to find the value of the unknown.

(a) *Net Income:* $N = R - C$ when $C = \$300$ and $N = \$1000$

(b) *Assets:* $A = L + O$ when $A = \$800$ and $L = \$200$

(c) *Net Pay:* $N = G - D$ when $D = \$178.36$ and $N = \$1500$

(d) *Mass:* $M = V(D)$ when $D = 10.4$ and $M = 20.8$

Answers: (a) _____ (b) _____ (c) _____ (d) _____

IN YOUR OWN WORDS

explain how to use formulas with given values to find the one unknown value.

_____

_____

_____

Answer:

You must go back to Chapter 7 where you first learned about using formulas. In previous chapters you evaluated a formula for certain values of the variables. Now you have learned to use a formula as an equation and to solve for the letter that was not given a value in the problem.

Practice Exercises

1. Find the height of a triangle with base equal to 4 inches and area equal to 10 square inches.

   Answer: _____

2. Find the height of a trapezoid with an area of 9 square inches and with bases 2 inches and 4 inches.

   Answer: _____

Answers to Mini-Practice:

(a) $R = \$1300$ (b) $O = \$600$ (c) $G = \$1678.36$ (d) $V = 2$

3. Find the length of a rectangle with a perimeter of 8 inches and a width of 1 inch.

    Answer: _____

4. Find the side of a square with a perimeter of 10 cm.

    Answer: _____

5. Find the owner's equity if assets are $5000 and liabilities are $425.45.

    Answer: _____

6. Find the estimated life of an item if the annual depreciation is $50, the cost is $790, and the residual value is $40.

    Answer: _____

7. Find the cost if the net income is $750 and the revenue is $1100.

    Answer: _____

8. Find the deductions if the net pay is $1740 and the gross pay is $2873.

    Answer: _____

9. Find the distance when the velocity is 80 miles/hour and the time is 3 hours.

    Answer: _____

10. Find the density when the mass is 1.25 grams and the volume is 10 milliliters.

    Answer: _____

11. Find the base of a triangle whose area is 40 cm.^{2} and whose height is 8 cm.

    Answer: _____

12. Find the height of a trapezoid whose area is 50 in.^{2} and whose bases are 12 inches and 8 inches, respectively.

    Answer: _____

13. If the height of a trapezoid is 10 inches, one base is 6 inches, and the area of the trapezoid is 100 square inches, find the other base.

    Answer: _____

14. A man earns $3485 per month. If his deductions amount to $1298, what is his take-home pay?

    Answer: _____

15. The area of a rectangle is 60 in.^{2} What is the length if the width is 5 inches?

    Answer: _____

16. The cost of producing a certain item is $128. If the article is sold for $196, what is the net gain?

    Answer: _____

Summary

Literal equations are equations that contain more than one variable. These equations are solved for one of the variables in terms of the others by isolating this variable, using the rules of Chapter 10.

Formulas are literal equations that often do not contain x, y, or z, but rather, letters referring to words, such as l for length, d for density, etc. A listing of useful formulas is found on page 193.

Numerical applications of formulas refer to the use of formulas as equations in one variable. A suitable formula has to be found, variables should be replaced with the given numerical values, and the resulting equation must be solved for the variable that is left in the equation.

Review Practice Exercises

In Exercises 1–8, solve each literal equation for the given variable.

1. $y = 4 - 5x$ for x *Answer:* _____

2. $y = \dfrac{x + 4}{5}$ for x *Answer:* _____

3. $8x - 3y = -1$ for y *Answer:* _____

4. $3x - \dfrac{y}{3} = 10$ for y *Answer:* _____

5. $2a + b - 4c = d$ for c *Answer:* _____

6. $y = 3x - 8$ for x *Answer:* _____

7. $y = 2 - 7x$ for x *Answer:* _____

8. $y = \dfrac{3x + 1}{4}$ for x *Answer:* _____

In Exercises 9–18, solve each formula for the given variable.

9. $V = A \cdot h$ for h *Answer:* _____

10. $V = \dfrac{A \cdot h}{3}$ for h *Answer:* _____

11. $S = a + b + c$ for b *Answer:* _____

12. $U = R \cdot I$ for R *Answer:* _____

13. $P = A + Akt$ for t *Answer:* _____

14. $ax - by = 16$ for y *Answer:* _____

15. $d = s + at$ for a *Answer:* _____

16. $f = v + at$ for t *Answer:* _____

17. $I = PRT$ for T *Answer:* _____

18. $a = s - sr$ for r *Answer:* _____

In Exercises 19–23, find the correct formula and solve.

19. Find the width of a rectangle when the area is 45 cm.^{2} and the length is 15 cm.

    Answer: _____

20. Find the third side of a triangle with a perimeter of 25 cm. and the other sides of lengths 7.5 cm. and 8.3 cm.

    Answer: _____

21. Find the height of a triangle with area 27 cm.^{2} and base 9 cm.

    Answer: _____

22. Find the volume of a stone that has a density of 5.6 g./ml. and that weighs 140 g.

Answer: _____

23. A car traveled 240.8 miles at a rate of 56 miles per hour. How many hours did the trip take?

Answer: _____

If you have questions about this chapter, write them here: _____

_____

_____

_____

_____

_____

_____

PART V

Word Problems

12 Translations of Equations

Remember to

* **LEARN THE KEY WORDS THAT HELP YOU RECOGNIZE THE BASIC ARITHMETIC OPERATIONS**
* **PAY ATTENTION TO THE DIFFERENCE BETWEEN DIRECT TRANSLATIONS AND REVERSE TRANSLATIONS**
* **REVIEW HOW TO SOLVE EQUATIONS**
* **DO ALL THE LEARNING ACTIVITIES**

In Chapters 4 and 6, you were introduced to translations from mathematics to English, and vice versa. In this chapter you will review these translations and extend your skills to translating English sentences into mathematics. You will also learn to write your own equations.

Problem solving always requires some kind of planning. In mathematics we need to solve many problems containing words. Most problems found in math books or on tests can be translated into equations.

What is an equation? _____

_____

_____

Recall that an equation consists of two expressions that are equal to each other. Therefore, to plan for solving word problems, you need to look for two expressions that are equal.

To help you acquire the skills you need to translate English statements into equations, we will begin by translating equations into English. Then we will translate English into mathematics.

Taking a Test

1. Do the problems you know best first.
- This will give you confidence.
- Save the most difficult problems for last.
- Come back to problems that take a lot of time.

2. Don't panic!
- Take a deep breath when you're stuck.
- Tell yourself that you've done these kinds of problems before. You can do them again.

3. Keep your mind on your test.
- Ignore the sniffling and shuffling of other students.

4. Check all of your work.
- Don't try to finish early, even if other students do.
- Give yourself the full amount of time to do the problems and check your answers.

12.1 TRANSLATING EQUATIONS WITH ONE OPERATION INTO ENGLISH

Translate each of the following equations into English in two different ways:

(a) $x + 4 = 5$ _____

_____

(b) $x - 4 = 5$ _____

_____

(c) $4x = 5$ _____

_____

(d) $x \div 4 = 5$ _____

_____

Here are various translations:

(a) The sum of a number and 4 is five.
Four more than a number is five.
A number plus four is five.
A number increased by four is five.

(b) The difference between a number and four (in this order) is five.
   Four less than a number is five.
   A number minus four is five.
   A number decreased by four is five.
   Four subtracted from a number is five.

(c) The product of four and a number is five.
   Four times a number is five.

(d) A number divided by four is five.
   Four into a number is five.

Be careful when you translate subtraction and division. Remember that the order is important! Review Chapters 4 and 6 if you feel uncertain about this.

Translate the following equations using the words "less than":

(a) $x - 15 = 34$ _____

(b) $x - 8 = 7$ _____

You should have written:

(a) Fifteen less than a number is 34.
(b) Eight less than a number is seven.

Translate the same equations using the words "subtracted from":

(a) $x - 15 = 34$ _____

(b) $x - 8 = 7$ _____

Here the order is again reversed:

(a) Fifteen subtracted from a number is 34.
(b) Eight subtracted from a number is seven.

Example 1   Translate $x - 4 = 29$ into English.

Solution.

$$x \quad - \quad 4 \quad = \quad 29$$

A number minus four equals twenty-nine.

or

$$x \qquad - 4 \quad = \quad 29$$

Four less than

a number   is   twenty-nine.   ■

Example 2   Translate $x \div 4 = 6$ into English.

Solution.

$$x \qquad \div \quad 4 \quad = \quad 6$$

A number divided by four equals six.

or

$$x \quad \div 4 \quad = \quad 6$$

Four into

a number   equals   six.   ■

MINI-PRACTICE

(Use this space for your work.)

Translate each of the following equations into English.

(a) $x + 9 = 23$ (b) $x - 4 = 29$ (c) $4x = 48$ (d) $\frac{6}{x} = 2$

Answers:

(a) _____

(b) _____

(c) _____

(d) _____

Practice Exercises

Translate the following equations into English.

1. $x - 3 = 2$ *Answer:* _____

2. $3 - x = 2$ *Answer:* _____

3. $4 + x = 5$ *Answer:* _____

4. $x + 4 = 5$ *Answer:* _____

5. $5x = 15$ *Answer:* _____

6. $x \cdot 5 = 15$ *Answer:* _____

7. $\frac{x}{3} = 6$ *Answer:* _____

8. $\frac{3}{x} = 6$ *Answer:* _____

9. $x + 5 = 7$ *Answer:* _____

10. $a - 3 = -5$ *Answer:* _____

11. $x - 6 = 10$ *Answer:* _____

12. $9 + x = -3$ *Answer:* _____

13. $2a = -14$ *Answer:* _____

14. $-13x = 6$ *Answer:* _____

15. $20 = 4c$ *Answer:* _____

Answers to Mini-Practice:

There are many possible answers. We are giving you only one answer for each question.

(a) Nine more than a number is 23.   (b) Four less than a number is 29.

(c) Four times a number is 48.   (d) Six divided by a number is 2.

16. $\frac{x}{4} = 3$ *Answer:* _____

17. $\frac{3}{4}y = 9$ *Answer:* _____

18. $-\frac{2}{5}a = 3$ *Answer:* _____

12.2 TRANSLATING EQUATIONS WITH MORE THAN ONE OPERATION INTO ENGLISH

Consider the equation

$$3x + 5 = 17$$

(a) What are the operations on the left side? _____

(b) Translate the equation into English.

_____

In Part (a), $3x$ indicates multiplication (by 3) and + indicates addition. In Part (b), you probably wrote something like:

Three times a number increased by five is seventeen.

Use the space below to write at least two more translations of

$$3x + 5 = 17$$

_____

_____

You might have written statements such as the following:

Five more than three times a number is seventeen.

Three times a number plus five is seventeen.

Consider the equation:

$$\frac{x}{3} + 5 = 12$$

What are the operations on the left side? _____

The fraction bar implies division, and there is also addition.

Translate the equation into English. _____

_____

You probably wrote something like:

If the quotient of a number divided by three is increased by five, the result is twelve.

Example 1   Translate the equation $3x + 5 = 17$ into English.

Solution.

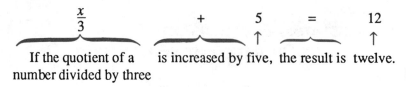

$$3x \qquad + \quad 5 \; = \; 17$$

Three times a number   increased by five is seventeen.   ∎

Example 2   Translate the equation $\frac{x}{3} + 5 = 12$ into English.

Solution.

$$\frac{x}{3} \qquad + \quad 5 \quad = \quad 12$$

If the quotient of a   is increased by five, the result is twelve.
number divided by three   ∎

MINI-PRACTICE   Translate the following equations into English:
(Use this space for your work.)

(a) $5x + 8 = 18$ (b) $3x - 4 = 11$ (c) $\frac{x}{4} + 7 = 16$ (d) $\frac{x}{7} - 4 = 1$

Answers:

(a) _____

(b) _____

(c) _____

(d) _____

Practice Exercises

Translate the following equations into English.

1. $2x + 4 = 6$ *Answer:* _____

2. $2x - 4 = 6$ *Answer:* _____

3. $\frac{x}{2} + 2 = 3$ *Answer:* _____

4. $\frac{2}{x} + 2 = 3$ *Answer:* _____

5. $\frac{x}{3} - 3 = 5$ *Answer:* _____

Answers to Mini-Practice:

(a) 8 more than 5 times a number is 18.   (b) 4 less than 3 times a number is 11.

(c) 7 more than the quotient of a number divided by 4 is 16.

(d) 4 less than the quotient of a number divided by 7 is 1.

(There are other correct answers.)

6. $\dfrac{3}{x} - 3 = 5$ *Answer:* _____

7. $4 - \dfrac{x}{5} = 3$ *Answer:* _____

8. $4 - \dfrac{5}{x} = 3$ *Answer:* _____

9. $3x + 1 = 10$ *Answer:* _____

10. $5m - 6 = 9$ *Answer:* _____

11. $5 = 4x + 9$ *Answer:* _____

12. $2 = 5x + 12$ *Answer:* _____

13. $7 - C = 9$ *Answer:* _____

14. $6 + 2b = 0$ *Answer:* _____

12.3 TRANSLATING ONE-OPERATION STATEMENTS FROM ENGLISH INTO MATHEMATICS

In this section we will reverse the process we used in the preceding section; that is, we will translate an English statement into an algebraic expression.

Apply the following instructions to Sentences 1–5 that follow.

(a) Draw an oval around the phrase we should label as x.

(b) Draw a rectangle around the phrase that tells us what the operation is.

(c) Make an equals sign (=) above the word that represents "equals."

(d) Can this sentence be translated word for word as you read it?

(e) Translate the whole sentence into an equation.

    1. A number increased by nine is forty-five.

       (d) _____

       (e) _____

    2. A number decreased by nine is thirty-three.

       (d) _____

       (e) _____

    3. Nine less than a number is thirty-four.

       (d) _____

       (e) _____

4. Five times a number is eighty-five.

(d) _____

(e) _____

5. A number divided by seven is forty-three.

(d) _____

(e) _____

Answers:

1. (A number) increased by nine is forty-five.

A number increased by nine is forty-five.

$$x \quad + \quad 9 \quad = \quad 45$$

The translation is made word for word from left to right.

2. (A number) decreased by nine is thirty-three.

A number decreased by nine is thirty-three.

$$x \quad - \quad 9 \quad = \quad 33$$

This is another case where the translation is made from left to right and word for word as you read the statement.

3. Nine less than (a number) is thirty-four.

Nine less than
a number   is thirty-four.

$$x \qquad - 9 \qquad = \qquad 34$$

The translation goes right-left-right. Remember "less than" gives us the message to reverse the order.

4. Five times (a number) is eighty-five.

Five times a number is eighty-five.

$$5 \quad \cdot \quad x \quad = \quad 85$$

$$\text{or} \quad 5x = 85$$

The translation is made word for word from left to right.

5. (A number) divided by seven is forty-three.

A number divided by seven is forty-three.

$$x \quad \div \quad 7 \quad = \quad 43$$

The translation is made word for word from left to right.

MINI-PRACTICE
(Use this space for your work.)

Translate into mathematics and solve the equations:

(a) Three more than a number is five.

(b) Three less than a number is five.

(c) Three minus a number is five.

(d) The product of a number and three is five.

(e) The quotient of a number divided by five is three.

Answers: (a) _____   (b) _____

(c) _____   (d) _____

(e) _____

Practice Exercises

In Exercises 1–24, translate into mathematics and solve the equations.

1. A number plus seven is nineteen.   *Answer:* _____

2. A number increased by seventeen is twenty-nine.   *Answer:* _____

3. Eight more than a number is negative twelve.   *Answer:* _____

4. A number minus thirteen is negative four.   *Answer:* _____

5. Twenty-nine decreased by a number is four.   *Answer:* _____

6. Thirteen less than a number is forty-one.   *Answer:* _____

7. A number subtracted from eighteen is seven.   *Answer:* _____

8. A number subtracted from fifty-four is forty-one.   *Answer:* _____

9. Two multiplied by a number is five.   *Answer:* _____

10. Forty-four times a number is one hundred thirty-two.   *Answer:* _____

11. A number divided by two is seventeen.   *Answer:* _____

12. Thirty-two divided by a number is four.   *Answer:* _____

13. The quotient of a number divided by twelve is twenty-one. *Answer:* _____

14. The product of three and a number equals negative thirty. *Answer:* _____

15. The product of negative two and a number equals zero. *Answer:* _____

16. The sum of a number and twelve is ten.   *Answer:* _____

17. The difference between a number and five (in this order) is two.

Answer: _____

18. Two-thirds of a number is six.   *Answer:* _____

19. The quotient of a number divided by seven is the opposite of three.

Answer: _____

Answers to Mini-Practice:

(a) $x + 3 = 5$; $x = 2$ (b) $x - 3 = 5$; $x = 8$ (c) $3 - x = 5$; $x = -2$

(d) $x \cdot 3 = 5$; $x = \dfrac{5}{3}$ (e) $\dfrac{x}{5} = 3$; $x = 15$

20. The sum of a number and thirteen is thirty-one.   *Answer:* _____

21. When a number is multiplied by eleven, the result is one hundred eighty-seven.

Answer: _____

22. When a number is multiplied by five-thirds, the result is twenty.

Answer: _____

23. When a number is divided by four, the result is fifteen.   *Answer:* _____

24. The quotient of fifty divided by a number is five.   *Answer:* _____

12.4 TRANSLATING STATEMENTS WITH MORE THAN ONE OPERATION FROM ENGLISH INTO MATHEMATICS

Apply the following instructions to Sentences 1 and 2 that follow.

(a) Draw an oval around the phrase we should label as x.
(b) Draw a rectangle around the words that tell us what the operations are.
(c) Make an equals sign (=) above the phrase that represents "equals."
(d) Translate the whole sentence into an equation.

1. If three times a number is increased by five the result is eighty-six.

(d) _____

2. If five is subtracted from the quotient of a number and six, the result is one.

(d) _____

Answers:

1. If three |times| (a number) |is increased by| five the result is eighty-six.

(d) $3x + 5 = 86$

2. If five |is subtracted from| the |quotient| of (a number) and six, the result is one.

(d) $\frac{x}{6} - 5 = 1$

Example 1   Translate into an equation: If three times a number is decreased by five, the result is one.

Solution.

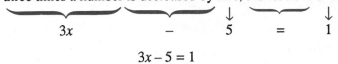

If three times a number is decreased by five, the result is one.

$$3x - 5 = 1$$

$$3x - 5 = 1$$ ■

Example 2   Translate into an equation: If a number divided by three is added to four, the result is nine.

Solution.

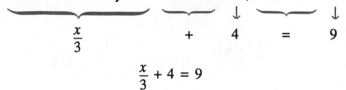

If a number divided by three is added to four, the result is nine.

$$\frac{x}{3} \quad + \quad 4 \quad = \quad 9$$

$$\frac{x}{3} + 4 = 9$$ ■

MINI-PRACTICE
(Use this space for your work.)

Translate the following statements into equations and solve the equations:

(a) Half of a number increased by five is thirty-nine.

(b) Twice a number decreased by thirty is seventy-four.

(c) If the quotient of a number divided by three is increased by eight, the result is thirty-four.

Answers: (a) _____   (b) _____   (c) _____

IN YOUR OWN WORDS

explain which operations we have to be especially careful about when we translate from mathematics into English, and vice versa: _____

_____

_____

_____

_____

Answer:

The operations of subtraction and division are not commutative, so the order in which we subtract or divide is important. A subtraction problem can be translated in direct order by the use of the words "minus," "decreased by," or "the difference of." If we wish to use the words "less than" or "subtracted from," we use a reverse order.

In division "divide by" or "quotient" imply direct order, but "divide into" implies a reverse order.

Answers to Mini-Practice:

(a) $\frac{x}{2} + 5 = 39$; $x = 68$ (b) $2x - 30 = 74$; $x = 52$ (c) $\frac{x}{3} + 8 = 34$; $x = 78$

Practice Exercises

In Exercises 1–20, translate the statements into mathematics and solve the resulting equations.

1. If five times a number is increased by seven, the sum is nine.

   Answer: _____

2. If twenty-one is subtracted from three times a number, the difference is seven.

   Answer: _____

3. Twice a number, increased by negative eight, equals thirty-two.

   Answer: _____

4. If one is subtracted from eight times a number, the difference is eleven.

   Answer: _____

5. One less than one-sixth of a number equals one-half.

   Answer: _____

6. Five times a number increased by fourteen is the same as seven times the number increased by eight.

   Answer: _____

7. Fifteen more than twice a number is the same as five times the number.

   Answer: _____

8. Four less than the quotient of a number divided by five is two.

   Answer: _____

9. If the quotient of eight times a number divided by five is increased by three, the result is seven.

   Answer: _____

10. If two is subtracted from the quotient of twice a number divided by seven, the difference is ten.

    Answer: _____

11. The sum of a number and fourteen is eighty-six.

    Answer: _____

12. When a number is multiplied by thirteen, the result is ninety-one.

    Answer: _____

13. Four-sevenths of a number is forty-four.

    Answer: _____

14. When six is added to a number, the result is fourteen.

    Answer: _____

15. When six is added to seven times a number, the result is sixty-two.

    Answer: _____

16. When nine is subtracted from one-fifth of a number, the result is three.

    Answer: _____

17. When a number is added to four times itself, the resulting number is eighty.

    Answer: _____

18. Seven less than twice a number is equal to eleven.

    Answer: _____

19. Three times the sum of a number and four is fifteen.

    Answer: _____

20. When five times a number is subtracted from twenty, the result is ten.

    Answer: _____

12.5 WRITING EQUATIONS

Let us say that you are getting ready for a picnic and you pack two sodas per person. How many people are going to the picnic if you pack 6 sodas?

How would you write this problem in mathematical notation?

_____

To begin with, we could write our problem this way:

$$2(\text{\# of people}) = 6$$

or this way:

$$2(?) = 6$$

Let's give our unknown quantity, the variable, the letter "P" for "people." Therefore,

$$2(P) = 6$$

or simply,

$$2P = 6$$

Between the coefficient 2 and the variable P there is an *invisible* sign.

Which sign? _____

As you know, one of the conventions of writing mathematics is that between a number and a letter, say between 2 and a in $2a$, or between two letters, say a and b in ab, there is always a multiplication sign taken for granted.

Use n to represent the number and think of the following:

(a) If five is subtracted from a number, the difference is three. What is the number?

(b) If a number is subtracted from 5, the difference is three. What is the number?

Write the equations before you give your answer. Remember that the order in subtraction is essential!

(a) ____ – ____ = 3   (b) ____ – ____ = 3

Answers:

(a) 8   (b) 2

Write as an equation:

How much money can I spend if I have $500 and need to have $200 left? _____

It is helpful to rewrite the sentence in simpler English before you translate it into mathematics:

$$\$500 - \text{how much is } \$200$$

or

$$500 - x = 200$$

Example 1   Translate into an equation:

The temperature went up 10° to 25°.

What was the temperature at the beginning?

Solution.   Let x represent the temperature. The temperature *went up*, so we add.

Equation:

$$x + 10 = 25$$
$$x = 15$$

The temperature at the beginning was 15°, so the temperature went from 15° to 25°.   ■

Example 2   Translate into an equation:

Four cigars cost 72 cents.

How much does one cigar cost?

Solution.   Simplified English: Four times the cost of a cigar is 72¢.

Equation:

$$4x = 72$$
$$x = 18$$

The cost of one cigar is 18¢.   ■

MINI-PRACTICE
(Use this space for your work.)

Translate into equations. Solve the equations and check your solutions:

(a) The sum of five times a number and three times the number is forty. Find the number.

(b) The temperature went down 10° to 20°. What was the temperature at the beginning?

(c) After I buy four notebooks, I have 24¢ left of my $2. How much did I pay for each notebook?

Answers: (a) _____   (b) _____   (c) _____

Practice Exercises

In Exercises 1–21, translate into equations and solve.

1. Multiply a number by two and add four. The result is ten.   *Answer:* _____

2. The sum of twice a number and three times the number is thirty.   *Answer:* _____

3. Three times a number is the same as the number subtracted from twelve.

   Answer: _____

4. Sue had as many nickels as Ron had dimes. Together they had $1.05. How many nickels did Sue have?

   Answer: _____

5. If 1.48 is added to a number, the sum is five times the number. Find the number.

   Answer: _____

6. Grandfather said to eleven-year-old Charlie: "If I divide my age by four and then subtract five from the quotient, I get your age." How old was the grandfather?

   Answer: _____

7. A student had a grade of 85 on a test. What must his grade be on a second test to give him a grade average of 90?

   Answer: _____

8. I think of a number, multiply it by two, and subtract three from the product. Then I multiply the difference by four. The product is forty-four. What is the number?

   Answer: _____

9. One-third of a certain number increased by four is eight.   *Answer:* _____

10. Forty more than a certain number is equal to five times the number.   *Answer:* _____

11. One more than four times a number is equal to nine.   *Answer:* _____

12. Four times six less than a number is sixty.   *Answer:* _____

Answers to Mini-Practice:

(a) $5x + 3x = 40$; $x = 5$ (b) $x - 10 = 20$; $x = 30$ (c) $200 - 4x = 24$; $x = 44$

13. Ten times a number diminished by twice the same number is twenty-two.

                                    Answer: _____

14. Five increased by the product of four and a certain number is twenty-five.

                                    Answer: _____

15. One-half the sum of seven and a number is ten.   *Answer:* _____

16. Five more than three times a given number is the same as twenty.   *Answer:* _____

17. Four is the sum of a certain number and eight less than three times the number.

                                    Answer: _____

18. Seven more than the product of five and a certain number is forty-seven.

                                    Answer: _____

19. When the sum of a number and six is subtracted from fifteen, the result is three.

                                    Answer: _____

20. Twelve minus the product of five and a number equals seven.   *Answer:* _____

21. The sum of three times a number and the number is twelve.   *Answer:* _____

12.6 THE SAME EQUATION REPRESENTING DIFFERENT IDEAS

Translate the following statements into equations:

"A number plus nine is forty-three." _____

"Bob's age increased by nine is forty-three." _____

"Nine degrees more than the current temperature is forty-three degrees." _____

Each statement can be translated by the same equation:

$$x + 9 = 43$$

Example 1   Translate the equation $x - 40 = 60$ into English in three different ways.

Solution.   Three possible answers are:

Forty less than a number is sixty.
How much money do I have if I can spend $40 and still have $60 left?
It was very hot earlier today. The temperature is now 60° after a drop of 40°. How hot was it?   ■

Example 2   Translate each of the following into the same equation:

The sum of three times a number and the number is 60. Find the number.

Ann has three times as much money as Ruth. Together they have $60. How much money does Ruth have?

A father is three times as old as his son. The sum of their ages is 60 years. How old is the son?

Solution.   The equation for each of these is

$$3x + x = 60$$ ∎

MINI-PRACTICE   Translate the following statements into the same equation:

(Use this space for your work.)

(a) Four less than twice a number is ten.

(b) Twice Andrea's age reduced by four is ten years.

(c) If twice the current temperature is decreased by four degrees, the resulting temperature is ten degrees.

Answer: _____

Practice Exercises

In Exercises 1–4, each part gives rise to the same equation. Find this equation.

1. (a) Six less than a number is 2. Find the number.
   (b) The temperature went down 6° to 2°. What was the temperature at the beginning?
   (c) Lisa is six years younger than Paul. If Lisa is two years old, how old is Paul?

   Answer: _____

2. (a) The product of five and a number is 30. What is the number?
   (b) I have $30. How many tapes can I buy if each tape costs $5?
   (c) How long does it take to walk thirty miles if I walk five miles per hour?

   Answer: _____

3. (a) The sum of two numbers is 84. One number is three times the other number. Find the numbers.
   (b) Greg has three times as much money as Roy. Together they have $84. How much money does Roy have?
   (c) Tomatoes cost $3 per pound and potatoes cost $1. How many pounds can I buy for $84 if I want the same number of pounds of each?

   Answer: _____

4. (a) The difference between two positive numbers is 56. One number is 9 times the other. Find the numbers.
   (b) Nancy spent $56 more than Eva did. How much did Eva spend if Nancy spent nine times as much?
   (c) Lisa has nine times as many marbles as Joe. If Joe has 56 less marbles than Lisa, how many does he have?

   Answer: _____

Answer to Mini-Practice:

The equation for each of these is $2x - 4 = 10$.

SUMMARY

1. An equation consists of two expressions separated by an equals sign.

2. The key words that help us recognize the basic operations are the following:

(a) Addition:	the sum of _____ and _____ increased by more than add
(b) Subtraction:	the difference between decreased by less than subtract minus
(c) Multiplication:	the product of times
(d) Division:	the quotient of divide by divided into

3. Subtraction and division are *noncommutative* operations. Care must be taken when expressing the order of these operations.

Review Practice Exercises

In Exercises 1-22, translate each of the following equations into English in at least two different ways. With subtraction problems make sure to practice using "less than."

1. $x + 2 = 17$ *Answer:* _____

_____

2. $2x + 5 = 29$ *Answer:* _____

_____

3. $5x + 7 = 42$ *Answer:* _____

_____

4. $x - 9 = 13$ *Answer:* _____

_____

5. $3x - 9 = 21$ *Answer:* _____

_____

6. $7x - 3 = 25$ *Answer:* _____

_____

7. $\frac{x}{3} + 2 = 5$ *Answer:* _____

_____

8. $\frac{x}{8} - 4 = 2$ *Answer:* _____

_____

9. $10 - \frac{x}{5} = 6$ *Answer:* _____

_____

10. $\frac{6}{x} + 3 = 9$ *Answer:* _____

_____

11. $3x - 5 = 7$ *Answer:* _____

_____

12. $\frac{3x + 7}{2} = 8$ *Answer:* _____

_____

13. $6(x - 5) = 42$ *Answer:* _____

_____

14. $4x + 12 = 32$ *Answer:* _____

_____

15. $\frac{x}{5} - 6 = 3$ *Answer:* _____

_____

16. $2(x - 7) = 6$ *Answer:* _____

_____

17. $5x - 2 = 13$ *Answer:* _____

_____

18. $10 = \frac{x}{2} + 13$ *Answer:* _____

_____

19. $\frac{5x}{4} + 15 = 10$ *Answer:* _____

_____

20. $3\left(\frac{x}{4} + 4\right) = 9$ *Answer:* _____

_____

21. $3(2x + 9) = 9$ *Answer:* _____

_____

22. $\frac{3x - 7}{5} = 4$ *Answer:* _____

_____

In Exercises 23–32, translate the following statements into equations. Solve each equation and check your solution.

23. Seven minus a number is fourteen.   *Answer:* _____

24. A number decreased by two is negative eight.   *Answer:* _____

25. A number minus nine is fifty-one.   *Answer:* _____

26. Four less than a number is twenty-eight.   *Answer:* _____

27. A number subtracted from twenty-five is negative two.   *Answer:* _____

28. Five times a number plus eight is thirty-three.   *Answer:* _____

29. Three less than seven times a number is one hundred nine.   *Answer:* _____

30. Twice a number minus seven is forty-three.   *Answer:* _____

31. Eleven more than the quotient of a number divided by sixteen is thirteen.

                                                                   Answer: _____

32. If the quotient of a number divided by thirty-two is increased by twenty, the result is twenty-six.

                                                                   Answer: _____

If you have questions about this chapter, write them here: _____

_____

_____

_____

_____

_____

13 Word Problems

Remember to

* *APPROACH WORD PROBLEMS CALMLY*
* *READ SLOWLY, MAKE NOTES, AND REREAD*
* *WRITE AN EQUATION, SOLVE IT, AND CHECK YOUR SOLUTION*
* *ASK YOURSELF IF YOUR SOLUTION MAKES SENSE*

In this chapter you will learn how to write equations for several types of word problems. You will learn how to solve percent problems by utilizing proportions and also by using direct translations.

13.1 RATIO AND PROPORTION

In a math lab two calculators are to be used by six students.

> What is the fairest way to distribute the calculators if the students are divided into work groups? _____

The chances are that you said: "There is one calculator for every three students."

In math we say: The **ratio** of calculators to students is 1 to 3.

> How do you write this ratio? _____

A ratio of 1 to 3 is written as 1:3 or as the fraction $\frac{1}{3}$.

> What is the ratio of students to calculators? _____

The correct answer is 3:1 or $\frac{3}{1}$, that is, three students for each calculator.

273

Learning from Mistakes

1. **When you get a problem wrong on an exam:**
   * Find out why.
   * A similar problem might appear on another exam or on the final.

2. **Was it a careless error?**
   * Work slower next time.
   * Check your results more carefully.

3. **Still can't do the problem?**
   * Never ignore a problem you can't do.
   * Get help from your instructor or from classmates.
   * Go back over the problem again and again.
   * Persistence pays in math.

4. **Keep copies of the exams you've taken.**
   * Make sure you understand all the solutions.
   * Don't be afraid to ask your instructor to explain something twice.
   * Review these exams for the final.

You could also have said that the ratio is $\dfrac{6 \text{ students}}{2 \text{ calculators}}$; note that $\dfrac{6}{2}$ reduces to $\dfrac{3}{1}$.

The ratios 6:2 and 3:1 are equal. When we write $\dfrac{6}{2} = \dfrac{3}{1}$ we have a *proportion*. A **proportion** expresses the equality of two ratios.

In the proportion $\dfrac{6}{2} = \dfrac{3}{1}$ multiply *crosswise*:

$$\dfrac{6}{2} \diagdown\!\!\!\!\diagup \dfrac{3}{1}$$

$$6 \times 1 \overset{?}{=} 3 \times 2$$

$$6 = 6 \quad \checkmark$$

Here, 6×1 and 3×2 are known as the **cross products**. Note that both cross products equal 6.

In the following proportions, determine the cross products:

$$\dfrac{3}{5} = \dfrac{21}{35}; \qquad \dfrac{18}{12} = \dfrac{3}{2}; \qquad \dfrac{10}{25} = \dfrac{20}{50}$$

If you calculated correctly, you found:

In a proportion, cross products are equal.

This property is useful in a proportion in which one term is unknown. For example:

$$\frac{3}{11} = \frac{12}{x}$$

Since this is a proportion, the cross products are equal.

$$3x = 12 \times 11$$

that is,

$$3x = 132$$

Solve the equation: _____

Check. $\quad \frac{3}{11} \stackrel{?}{=} \frac{12}{44}$

$$\frac{3}{11} = \frac{3}{11} \quad \sqrt{}$$

Proportions are important in problems such as the following:

If in a math lab the ratio of calculators to students is 1 to 3, how many calculators are needed for 27 students?

Let the number of calculators be represented by x.

How do you write the ratio of calculators to 27 students?

_____

Since $\frac{1}{3}$ and $\frac{x}{27}$ represent the ratio of calculators to students in the same math lab, both ratios should be equal.

$$\frac{\text{Calculators}}{\text{Students}} = \frac{\text{Calculators}}{\text{Students}}$$

$$\frac{1}{3} = \frac{x}{27}$$

Cross multiply: $\qquad\qquad\qquad 3x = 27$

Solve the equation: $\qquad\qquad\quad x = 9$

Check.

Is it true that 9 calculators for 27 students is in the ratio of 1:3?

_____

The order in which the ratios in a proportion are written is important. Consider the following quantities:

| 1 calculator | 3 students |
| x calculators | 27 students |

Write the ratio of calculators: _____

Write the ratio of students: _____

Express as a proportion: _____

Cross multiply: _____

Solve for x: _____

Do you still get 9? _____

We have found the following proportions:

$$\frac{\text{Calculators}}{\text{Students}} = \frac{\text{Calculators}}{\text{Students}}$$

and

$$\frac{\text{Calculators}}{\text{Calculators}} = \frac{\text{Students}}{\text{Students}}$$

Example 1   Write the ratio 6:10 in reduced form.

Solution.

$$\frac{6}{10} = \frac{3}{5}$$ ■

Example 2   Find x, if

$$\frac{5}{7} = \frac{x}{3}$$

Solution.

Cross multiply: $5(3) = 7x$

Divide by 7: $\frac{15}{7} = x$ ■

Example 3   The ratio of two numbers is 3:4. Find the larger number, if the smaller number is 12.

Solution.   Call the larger number x. Set up the ratios of smaller numbers to larger numbers:

$$\frac{3}{4} \text{ and } \frac{12}{x}$$

These ratios are equal, so we have the proportion

$$\frac{3}{4} = \frac{12}{x}$$

Cross multiply: $3x = 4(12)$

$3x = 48$

Divide by 3: $x = 16$

The larger number is 16.

Check.

Is it true that $12:16 = 3:4$? _____   ■

Strategy for solving proportion problems:

1. Write the numbers in columns representing the same units (for example, write "calculators" under "calculators").
2. Form ratios with numbers in each column.
3. State a proportion.
4. Solve the proportion.

MINI-PRACTICE
(Use this space for your work.)

(a) Write the ratio 25:35 in reduced form.   *Answer:* _____

(b) Find x if $\dfrac{4}{5} = \dfrac{36}{x}$.   *Answer:* _____

(c) There are 28 students in a class of which 16 are women. Find the ratio of men to women in this class.

Answer: _____

IN YOUR OWN WORDS   explain the relationship between a ratio and a proportion: _____

_____

_____

Answer:

A ratio is the comparison of two quantities through division. An equation that states the equality of two ratios is called a proportion. That is, a proportion is a statement such as

$$\frac{a}{b} = \frac{c}{d}$$

where b and d are not equal to 0.

Practice Exercises

In Exercises 1–10, there are 20 people at a party. Among them 8 are women; 15 are married; 7 are black; 6 are white; 4 are Hispanic; and 3 are Asian. What is the ratio of:

1. women to men?   *Answer:* _____
2. blacks to whites?   *Answer:* _____
3. married to single?   *Answer:* _____
4. whites to Asians?   *Answer:* _____
5. men to women?   *Answer:* _____
6. Asians to Hispanics?   *Answer:* _____
7. Hispanics to blacks?   *Answer:* _____
8. single to married?   *Answer:* _____
9. women to the total number of people at the party?   *Answer:* _____
10. married persons to the total number of people at the party?   *Answer:* _____

Answers to Mini-Practice:

(a) 5:7   (b) 45   (c) 12:16 or 3:4

In Exercises 11–16, write each ratio in reduced form.

11. 4:14   *Answer:* _____   **12.** 12:18   *Answer:* _____   **13.** 22:99   *Answer:* _____

14. 3:9   *Answer:* _____   **15.** 10:2   *Answer:* _____   **16.** 12:16   *Answer:* _____

In Exercises 17–30, determine the value of the unknown quantity in each proportion.

17. $\dfrac{x}{4} = \dfrac{45}{36}$ *Answer:* _____   **18.** $\dfrac{3}{7} = \dfrac{x}{56}$ *Answer:* _____   **19.** $\dfrac{9}{x} = \dfrac{63}{35}$ *Answer:* _____

20. $\dfrac{2}{13} = \dfrac{10}{x}$ *Answer:* _____   **21.** $\dfrac{a}{15} = \dfrac{22}{165}$ *Answer:* _____   **22.** $\dfrac{4}{9} = \dfrac{20}{y}$ *Answer:* _____

23. $\dfrac{2}{3} = \dfrac{x}{12}$ *Answer:* _____   **24.** $\dfrac{x}{27} = \dfrac{5}{9}$ *Answer:* _____   **25.** $\dfrac{7}{a} = \dfrac{21}{36}$ *Answer:* _____

26. $\dfrac{81}{93} = \dfrac{27}{b}$ *Answer:* _____   **27.** $\dfrac{4}{5} = \dfrac{c}{40}$ *Answer:* _____   **28.** $\dfrac{y}{13} = \dfrac{33}{143}$ *Answer:* _____

29. $\dfrac{10}{d} = \dfrac{110}{121}$ *Answer:* _____   **30.** $\dfrac{10}{18} = \dfrac{15}{x}$ *Answer:* _____

In Exercises 31–40, solve each of the following problems by setting up a proportion.

31. A typist makes 6 errors on 15 pages. At this rate, how many errors will she make on 45 pages?

     Answer: _____

32. In a parking lot there are 3 foreign cars for every 5 domestic ones. If there are 36 foreign cars, how many domestic cars are there?   *Answer:* _____

33. In a small town 2 out of 3 cars are domestic. If there are 240 cars in town, how many domestic cars are there?

     Answer: _____

34. A car travels 24 kilometers on 1 gallon of gas. How far can it travel on 15 gallons?

     Answer: _____

35. A baseball team won 10 out of 12 games. If it played 48 games in a season:

     (a) How many games did it win?   *Answer:* _____

     (b) How many games did it lose?   *Answer:* _____

36. A photograph 6 inches by 7 inches in size is to be enlarged so that its shorter side will be 12 inches. What will be the size of the larger side?   *Answer:* _____

37. A real-estate broker receives a $6 commission for every $100 of sales. At this rate, how much does she receive on a $96,000 sale?   *Answer:* _____

38. A recipe requires 3 cups of rice for 8 portions. How many cups are needed for 24 portions?

     Answer: _____

39. The scale of a map is $\dfrac{1}{2}$ inch for every 3 miles. How far apart are two towns that are $3\dfrac{1}{4}$ inches apart on the map?

     Answer: _____

40. In a village in Alaska there are 3 women for every 7 men. How many women are there in the town if the total adult population is 3750?   *Answer:* _____

13.2 Percent Problems

Percent Problems as Proportions

In the sentence

$$50\% \text{ of } 60 \text{ equals } 30$$

50% is the **rate**, 60 is the **whole**, and 30, the result, is the **part**. The rate 50% can be expressed as $\frac{50}{100}$. In general, for a rate of $r\%$, we have

$$r\% = \frac{r}{100}$$

Now the rate $\frac{r}{100}$ expresses the particular ratio of $\frac{Part}{Whole}$. Thus,

$$\text{Rate} = \text{Ratio} = \frac{Part}{Whole}$$

or, if we let R = ratio, P = part, and W = whole,

$$R = \frac{r}{100} = \frac{P}{W}$$

For now, we work with the proportion,

$$\frac{r}{100} = \frac{P}{W}$$

Use this proportion to solve:

$$\text{What } \% \text{ of } 40 \text{ is } 60?$$

Set up the proportion: _____

Cross multiply: _____

Solve for x: _____

Example 1 5% of what number is 15?

Solution.

Formula: $\dfrac{r}{100} = \dfrac{P}{W}$

Proportion: $\dfrac{5}{100} = \dfrac{15}{x}$

Cross multiply: $5x = 1500$

Divide by 5: $x = 300$

Check.

Does 5% *of* 300 equal 15? The word *of* indicates *multiplication*.

$$5\% \cdot 300 \overset{?}{=} 15$$
$$0.05 \times 300 \overset{?}{=} 15$$
$$15.00 = 15 \quad \sqrt{}$$

∎

Example 2   What % of 40 is 2?

Solution.

Formula: $$\frac{r}{100} = \frac{P}{W}$$

Proportion: $$\frac{r}{100} = \frac{2}{40}$$

Cross multiply: $$40r = 200$$

Solve: $$r = \frac{200}{40}$$
                  $$r = 5$$

Check. $5\% \cdot 40 \overset{?}{=} 2$

$0.05 \times 40 \overset{?}{=} 2$

$2.00 = 2$ $\checkmark$ ■

Percent Problems as Direct Translations

In the sentence

$$50\% \text{ of } 60 \text{ equals } 30$$

50% can be considered the *ratio*, 60 is the *whole,* and, 30, the result, is the *part.* Recall that

$$\text{Ratio} = \frac{\text{Part}}{\text{Whole}}$$

If the ratio is R, the whole is W, and the part is P, then

$$R = \frac{P}{W}$$

or, multiplying both sides by W,

$$R \cdot W = P$$

This last equation expresses:

> The ratio *of* the whole is equal to the part

Remember that "of" indicates multiplication.

Now, use your formula to write an equation for:

60% of what number is 24? _____

To solve the equation

$$60\% \cdot x = 24$$

you can change 60% (a) to a decimal or (b) to a common fraction. Solve the equation:

(a) _____

(b) _____

Did you get the answer 40 in both cases?

Solutions.

(a) $$0.60x = 24$$
$$6x = 240$$
$$x = 40$$

(b) $$\frac{60}{100}x = 24$$

Reduce the fraction: $$\frac{3}{5}x = 24$$

Multiply by $\frac{5}{3}$: $$\frac{5}{3} \cdot \frac{3}{5}x = 24 \cdot \frac{5}{3}$$
$$x = 40$$

Check. $60\% \cdot 40 \overset{?}{=} 24$
$.60 \times 40 \overset{?}{=} 24$
$24.00 = 24$ $\checkmark$

Translate: What percent of 40 is 24? _____

You probably wrote: $x\% \cdot 40 = 24$

Solve this equation by changing % to $\frac{1}{100}$: _____

Did you get 60?

Solution. $$\frac{x}{100} \cdot 40 = 24$$

Multiply by $\frac{100}{40}$: $$\frac{100}{40} \cdot \frac{x}{100} \cdot \frac{40}{1} = 24 \cdot \frac{100}{40}$$
$$x = \frac{2400}{40}$$
$$x = 60$$

Check. $60\% \cdot 40 \overset{?}{=} 24$
$0.60 \times 40 \overset{?}{=} 24$
$24 = 24$ $\checkmark$

Consider this problem: In a class of 30 students, 18 are women. What percent of the students are men? How do you translate:

"What percent"? _____

"the students" _____

"men" _____

Hopefully, you realized that we must calculate:

$$30 \text{ (students)} - 18 \text{ (women)} = 12 \text{ (men)}$$

Write the equation: _____

Solve the equation: _____

You probably wrote:

$$x\% \cdot 30 = 12$$

$$\frac{x}{100} \cdot 30 = 12$$

To isolate x, multiply both sides by 100 and divide by 30, that is, multiply by $\frac{100}{30}$.

$$\frac{x}{100} \cdot 30 \cdot \frac{100}{30} = 12 \cdot \frac{100}{30}$$

$$x = 40 \text{ (percent)}$$

$$x\% = 40\%$$

Check. $40\% \cdot 30 \overset{?}{=} 12$

$0.40 \times 30 \overset{?}{=} 12$

$12.00 = 12$ $\sqrt{}$

A baseball player gets 45 hits in 60 at bats. What percent of at bats are hits?

What does 45 represent? _____

What does 60 represent? _____

What is the unknown? _____

Write the equation: _____

Solve the equation: _____

Since $W = 60$ and $P = 45$, we see that

$$x\% \cdot 60 = 45$$

Thus, by solving the equation, we get $x = 75$, so that

$$x\% = 75\%$$

Example 3   A real-estate commission rate is 6%. If an agent's commission on the sale of an apartment is $13,500, how much did the apartment sell for?

Solution.   The rate is 6%. The whole is the unknown. The part is 13,500.

Equation: $6\% \cdot x = 13,500$

$0.06x = 13,500$

$6x = 1,350,000$

$x = 225,000$

Check. $6\% \cdot 225,000 \overset{?}{=} 13,500$

$0.06 \times 225,000 \overset{?}{=} 13,500$

$13,500 = 13,500$ $\sqrt{}$ ∎

Example 4   Write as an equation and solve the equation: The price of a dozen eggs rose from 90¢ to $1.08. What was the price increase in percent?

Solution. What $\% = x\%$. The whole $= 90$¢.

$$\text{The part} = \text{the increase} = \$1.08 - \$0.90 = \$0.18$$

Equation:

$$x\% \cdot 90 = 18$$

$$\frac{x}{100} \cdot 90 = 18$$

$$x \cdot 90 = 18 \cdot 100$$

$$x = 18 \cdot \frac{100}{90}$$

$$x = 20$$

$$x\% = 20\%$$

Check. $20\% \cdot 90 \stackrel{?}{=} 18$

$0.20 \times 90 \stackrel{?}{=} 18$

$18.00 = 18$ √   ∎

MINI-PRACTICE
(Use this space for your work.)

Write as equations and then solve these equations:

    (a) 3 is 10% of what number?   (b) What is 120% of 75?

    (c) 49 is what percent of 70?

    (d) The price of eggs decreased from $1 to 90¢. Find the percent of decrease.

Answers: (a) _____ (b) _____ (c) _____ (d) _____

IN YOUR OWN WORDS   explain how to solve percent problems by using direct translation: _____

_____

_____

Rule:

> **1. Determine the role played by each known quantity.**
>
> **2. Determine what is unknown.**
>
> **3. Translate the problem into an equation of the form:**
> $$R \cdot W = P$$
> **where R = ratio, P = part, and W = whole.**
>
> **4. Solve the equation.**

Answers to Mini-Practice:

(a) 30   (b) 90   (c) 70 (percent)   (d) 10 (percent)

Practice Exercises

In Exercises 1–16, translate the following problems into equations of the form

$$R \cdot W = P$$

and determine the value of the unknown quantity.

1. What is 36% of 144?   *Answer:* _____

2. 35 is what percent of 350?   *Answer:* _____

3. 26 is 25% of what number?   *Answer:* _____

4. 3.5% of 54 is what number?   *Answer:* _____

5. 100% of 11 is what number?   *Answer:* _____

6. 75 is what percent of 25?   *Answer:* _____

7. What percent of 60 is 80?   *Answer:* _____

8. 70% of what number is 35?   *Answer:* _____

9. What is $8\frac{1}{4}$ % of 28?   *Answer:* _____

10. 6% of what number is 24?   *Answer:* _____

11. What percent of 150 is 120?   *Answer:* _____

12. 12% of what number is 8.4?   *Answer:* _____

13. 6.12 is what percent of 51?   *Answer:* _____

14. $8\frac{1}{2}$ % of what number is 4.25?   *Answer:* _____

15. What is 19% of 3020?   *Answer:* _____

16. 125% of what number is 3.2?   *Answer:* _____

17. A family makes $1200 monthly. Leisure expenses take 4% of the monthly income. How much is spent for leisure?   *Answer:* _____

18. Don answered 85% of the 20 questions on a test correctly. How many questions did he answer correctly?
    Answer: _____

19. On a test consisting of 16 problems, Dot answered 12 of them correctly. What percent were correct?
    Answer: _____

20. An alloy contains 65% iron. How much iron is there in 300 tons of this alloy?   *Answer:* _____

21. A bank charges 18% annual interest on loans. How much interest per year will a customer pay on a $6500 loan?
    Answer: _____

22. A telephone-answering machine, which normally costs $139, was sold for 30% off the normal price. How much was the machine sold for?   *Answer:* _____

23. Three gallons of pure milk were mixed with 1.8 gallons of water. What percent of the mixture was milk?
    Answer: _____

24. A waitress got a 15% tip on a $58 check. How much did she get?   *Answer:* _____

25. 18% of the workers in a factory smoke. If there are 72 smokers there, what is the total number of workers in the factory?   *Answer:* _____

26. The population of a village increased from 540 to 675. What was the percent of increase?

    Answer: _____

27. A TV sold for $350 last year and for $420 this year. What was the percent of increase?

    Answer: _____

28. A woman on a diet goes from 140 pounds to 126 pounds. What is the percent of decrease in her weight?

    Answer: _____

29. John bought a new car for $13,500. If the value of the car decreased to $10,800 in the first year, what was the percent of decrease?   *Answer:* _____

30. The subway fare in a Western city went from $0.80 to $1.00. What was the percent of increase?

    Answer: _____

13.3 GEOMETRY PROBLEMS

What is this shape called? _____

You probably know that this is a rectangle.

Which sides of a rectangle are equal?_____

Opposite sides are equal, that is, the two longer sides are equal and the two shorter sides are equal.

The measure of each of the two longer sides is usually called the **length**. The measure of each of the two shorter sides is usually called the **width**. The **perimeter**, which measures the total distance around the rectangle, is the sum of the measures of the four sides. The **area** is length times width.

Suppose that in a rectangle the length is three meters longer than the width. The sum of the length and the width is twenty-seven meters. Find the length and the width.

Draw a rectangle in the space provided. Call the width x.

The length = _____

Write an expression for "The sum of the length and width":

_____

But this sum is also equal to 27.

Write the equation that states that "the sum of the length and the width is twenty-seven": _____

Solve the equation:

$x =$ _____

width = _____

length = _____

Repeat this problem, but call the length x.

The width = _____

The sum of the length and the width = _____

But this sum also = 27.

Write this as an equation: _____

Solve this equation.

$x =$ _____

length = _____

width = _____

Did you get the same answer no matter what you called x? _____

If not, check your calculations.

Example 1   The length of a rectangle is six meters longer than the width. The sum of twice the length and the width is fifty-seven meters. Find the length and the width.

Solution.   width = x;   length = $x + 6$

The sum of

twice the length and the width is fifty-seven (meters).

$$2(x + 6)   +   x   =   57$$

Equation: $2(x + 6) + x = 57$
$2x + 12 + x = 57$
$3x + 12 = 57$
$3x = 45$
$x = 15$
$x + 6 = 21$

$x = 15$

$x + 6 = 21$

Here is the problem redone by letting x represent the length.

$$\text{length} = x; \quad \text{width} = x - 6$$

The sum of

twice the length and the width is fifty-seven (meters).

$$2x \quad + \quad x - 6 \quad = \quad 57$$

Equation: $\quad 2x + x - 6 = 57$

$$3x - 6 = 57$$
$$3x = 63$$
$$x = 21$$

$$\text{length} = 21$$
$$\text{width} = 21 - 6 = 15$$ ∎

Example 2 The length of a rectangle is seven inches longer than the width. If each side is decreased by three inches, then the length is twice the width. Find the length and the width.

Solution. $\qquad \text{length} = x; \quad \text{width} = x - 7$

Decrease by three: $\qquad \text{new length} = x - 3$
$\qquad\qquad\qquad\qquad \text{new width} = (x - 7) - 3 = x - 10$

The length is twice the width: $\qquad x - 3 = 2(x - 10)$

Solve the equation: $\qquad\qquad x - 3 = 2x - 20$
$$-3 = x - 20$$
$$17 = x$$

$$\text{length} = 17 \text{ inches}; \quad \text{width} = 17 \text{ inches} - 7 \text{ inches} = 10 \text{ inches}$$

Check. $\qquad 17 - 3 \overset{?}{=} 14$
$$10 - 3 \overset{?}{=} 7$$
$$14 = 2(7) \quad \checkmark$$ ∎

Example 3 The width of a rectangle is five centimeters less than the length. If the perimeter is fifty-four centimeters, find the length and the width.

Solution.

Call the length x. Then the width is $x - 5$. Opposite sides of a rectangle are equal, so the perimeter equals

$$\text{length} + \text{width} + \text{length} + \text{width}$$

$$x \quad + x - 5 + \quad x \quad + x - 5$$

But the perimeter is fifty-four centimeters, so

$$x + (x - 5) + x + (x - 5) = 54$$

Solve the equation:

$$4x - 10 = 54$$
$$4x = 64$$
$$x = 16$$

The length is 16 cm. and the width is 11 cm.

Check. $16 + 11 + 16 + 11 = 54$ ■

MINI-PRACTICE
(Use this space for your work.)

Solve the following problems:

(a) The length of a rectangle is nine feet longer than the width. The sum of the length and the width is thirty-nine feet. Find the length and the width.

(b) The width of a rectangle is eleven meters shorter than the length. The perimeter of the rectangle is 70 meters. Find the length and the width.

(c) The length of a rectangle is eight meters longer than the width. If each side is decreased by five meters, then the length is twice the width. Find the length and the width of the original rectangle.

Answers: (a) _____   (b) _____   (c) _____

Practice Exercises

1. The length of a rectangle is eight meters longer than the width. Twice the width is four meters less than the length. Find the length and the width.   *Answer:* _____

2. The length of a rectangle is seventeen meters longer than the width. The sum of the length and the width is forty-seven meters. Find the length and the width.   *Answer:* _____

3. The width of a rectangle is eight centimeters shorter than the length. The difference between five times the length and the width is forty-eight centimeters. Find the length and the width.   *Answer:* _____

4. The width of a rectangle is four feet shorter than the length. If each side is shortened by one foot, then the length is three times the width. Find the length and the width of the original rectangle.
    Answer: _____

5. The length of a rectangle is nine feet longer than the width. The perimeter is seventy-eight feet. Find the length and the width.   *Answer:* _____

6. The width of a rectangle is eight centimeters shorter than the length. The perimeter is twenty-four centimeters. Find the length and the width.   *Answer:* _____

7. The width of a rectangle is eleven feet shorter than the length. If the perimeter is two hundred twenty-six feet, find the length and the width.   *Answer:* _____

8. The perimeter of a square is sixty-eight centimeters. Find the length of a side.   *Answer:* _____

Answers to Mini-Practice:
(a) $l = 24$ feet, $w = 15$ feet   (b) $l = 23$ meters, $w = 12$ meters   (c) $l = 21$ meters, $w = 13$ meters

9. The perimeter of a square is one hundred forty-eight yards. Find the length of a side.
    Answer: _____

10. What is the length of the side of a square whose perimeter is one hundred twelve meters?
    Answer: _____

11. The perimeter of a rectangle is 20 feet. If the length is 3 feet longer than the width, find the length and width.
    Answer: _____

12. If the sides of a square are increased by 2 inches each, the perimeter is then 20 inches. Find a side of the original square.   *Answer:* _____

13. The length of a rectangle is 10 centimeters longer than the width. If twice the length equals seven times the width, find the width.   *Answer:* _____

14. The width of a rectangle is $\frac{1}{4}$ of the length. Half the perimeter is 10 inches. Find the length.
    Answer: _____

15. The length of a rectangle is twice the width. If the length is shortened by 5 centimeters and the width is lengthened by the same amount, the rectangle becomes a square. Find the original length and width.
    Answer: _____

16. If two opposite sides of a square are increased by 5 inches each and the other two sides are increased by 4 inches each, a rectangle is formed. The perimeter of this rectangle is 6 times the perimeter of the square. Find a side of the square.   *Answer:* _____

13.4 AGE PROBLEMS

Age problems, though not terribly important in and of themselves, afford us useful practice in setting up and solving equations. A typical age problem is:

Jack is three years older than Susan. The sum of their ages is twenty-seven. Find their ages.

In order to solve this problem mathematically, let's analyze the problem:

Who is older? _____

How much older? _____

Who is younger? _____

Jack is three years older than Susan.

Who is being compared with whom? _____

Susan's age is being used as a base for comparison. Call the base *x*.

What do we compare? _____

We compare Jack's age with Susan's age.

Mathematically, how do we write Jack's age in terms of Susan's age?

_____

No doubt you wrote: Jack's age = $x + 3$.

Use this information to get an expression for "the sum of their ages":

_____

You probably wrote $x + (x + 3)$. We also have the information that the sum of their ages is 27. In other words,

$$x + (x + 3) = 27$$

Solve this equation.   _____

    _____

    _____

Therefore, $x =$ _____

Thus, Susan's age is 12.

How do we find Jack's age?   _____

Since $x = 12$, Jack's age is given by

$$x + 3 = 12 + 3 = 15$$

Check.

Is it true that 15 is 3 more than 12?   _____

Is it true that $15 + 12 = 27$?   _____

You were asked to call Susan's age x.

Would it have been possible to denote Jack's age by x?   _____

Of course, we can call Jack's age x. Jack's age is now the base.

Rewrite the problem so that you use the word *younger* instead of *older*.

_____

Who is older?   _____

Who is younger?   _____

How much younger is Susan?   _____

How do you write Susan's age?   _____

Jack's age = x.

Susan's age =   _____

Use this information to get an expression for:

"the sum of the ages" =   _____

But this sum also = 27. Write the equation stating that "the sum of their ages is twenty-seven."

$$\underline{\hspace{4cm}}$$

Solve the equation: $\underline{\hspace{3cm}}$

$$\underline{\hspace{3cm}}$$

$$\underline{\hspace{3cm}}$$

Therefore, $x = $ $\underline{\hspace{1.5cm}}$

Jack's age = $\underline{\hspace{3cm}}$

Susan's age = $\underline{\hspace{3cm}}$

Example 1   Jaime is six years older than Carla. The sum of twice Jaime's age and Carla's age is fifty-seven years. Find their ages.

Solution.   Let Carla's age = x; then Jaime's age = $x + 6$.

The sum of
twice Jaime's age and Carla's age is fifty-seven years.

$$2(x + 6) \quad + \quad x \quad = \quad 57$$

Equation: $2(x + 6) + x = 57$

Solve the equation:
$$2x + 12 + x = 57$$
$$3x + 12 = 57$$
$$3x = 45$$
$$x = 15$$

Carla is 15 years old. Jaime is 15 + 6 = 21 years old.

Here is the problem solved by letting Jaime's age be x.

$$\text{Jaime's age} = x; \quad \text{Carla's age} = x - 6$$

The sum of
twice Jaime's age and Carla's age is fifty-seven years.

$$2x \quad + \quad x - 6 \quad = \quad 57$$

Equation: $x - 6 + 2x = 57$

Solve the equation:
$$3x - 6 = 57$$
$$3x = 63$$
$$x = 21$$

Jaime's age = 21 (years); Carla's age = 21 − 6 = 15 (years)   ∎

Example 2   Chris is seven years younger than Mary. Chris' age subtracted from three times Mary's age is forty-three. Find their ages.

Solution.   Let Mary's age $= x$; then Chris' age $= x - 7$.

Recall that "subtracted from" reverses the order.

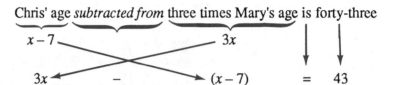

Chris' age *subtracted from* three times Mary's age is forty-three

$x - 7$ $3x$

$3x$ $-$ $(x - 7)$ $=$ 43

Equation: $3x - (x - 7) = 43$

Solve the equation: $3x - x + 7 = 43$

$2x + 7 = 43$

$2x = 36$

$x = 18$

Mary is 18 years old. Chris is 18 years $- 7$ years $= 11$ years old.

Check. $3(18) - 11 = 54 - 11 = 43$ $\checkmark$ ∎

MINI-PRACTICE   Solve the following problems:
(Use this space for your work.)

(a) Robert is nine years older than Anna. The sum of their ages is thirty-nine years. Find their ages.

(b) Henry is eleven years younger than Randy. The sum of three times Randy's age and Henry's age is two hundred thirty-seven years. Find their ages.

(c) If six is added to the sum of Glenda's and Olga's ages, the result is thirty-one years. Knowing that Olga is five years older than Glenda, fine their ages.

Answers: (a) _____   (b) _____   (c) _____

IN YOUR OWN WORDS   try to explain which name to call x in order to translate the problem smoothly.

_____

_____

_____

Answer:

Let x be the variable that is being "manipulated" or "compared with."

Answers to Mini-Practice:
(a) Robert: 24; Anna: 15   (b) Randy: 62; Henry: 51   (c) Glenda: 10; Olga: 15

Practice Exercises

Solve the following problems.

1. Deborah is eight years younger than Edward. Five times Edward's age minus Deborah's age is forty-four years. Find their ages.

   Answer: _____

2. Eileen is fifteen years older than Irene. Five times Irene's age minus twice Eileen's age is nine years. How old are they?

   Answer: _____

3. When Mark's age is subtracted from twice David's age, the result is thirty-six years. Given that David is fifteen years older than Mark, find their ages.

   Answer: _____

4. Pablo is eight years older than Neil. Twice Neil's age is four years less than Pablo's age. Find their ages.

   Answer: _____

5. Sarah is two years older than Martin. Three times Martin's age is fourteen years more than Sarah's age. Find their ages.

   Answer: _____

6. Juan is seven years younger than Alicia. Twice Juan's age is three years less than Alicia's age. Find their ages.

   Answer: _____

7. Ellen is two years older than Eva. The sum of their ages is 64. How old is Ellen?

   Answer: _____

8. Nancy is five years younger than Susan. Two years ago, Susan was twice Nancy's age. How old is Nancy now?

   Answer: _____

9. Ed is eight years older than Peter. Two times Ed's age equals three times Peter's age. Find Peter's age.

   Answer: _____

10. Elaine is eight years younger than Betty. The sum of four times Elaine's age and twice Betty's age is 226. Find their ages.

    Answer: _____

11. If fifteen is subtracted from the sum of Tony's and Anne's ages, the result is 90. Knowing that Tony is five years younger than Anne, find their ages.

    Answer: _____

12. Lee is four years older than Ken. If Ken's age is subtracted from Lee's age, the result is four years less than Lee's age. Find their ages.

    Answer: _____

> ## SUMMARY
>
> A ratio compares two quantities, usually in fractional form. A proportion expresses the equality of two ratios. To solve a proportion, use the fact that the cross products are equal.
>   Percent problems are often translated into an equation of the form
>
> $$R \cdot W = P$$
>
> where R = ratio, P = part, and W = whole.
>   The perimeter of a rectangle is the sum of the measures of the four sides. The area is length times width.
>   In age problems let one person's age be x, and express the other parts of the problem in terms of x.

Review Practice Exercises

In Exercises 1–4, determine the value of the unknown quantity in each proportion.

1. $\dfrac{3}{4} = \dfrac{x}{16}$ *Answer:* _____   2. $\dfrac{7}{15} = \dfrac{21}{a}$ *Answer:* _____

3. $\dfrac{p}{3} = \dfrac{16}{24}$ *Answer:* _____   4. $\dfrac{1}{6} = \dfrac{t}{36}$ *Answer:* _____

In Exercises 5 and 6, solve each problem by setting up a proportion.

5. If you exchange 12 Swedish kronor for $2, how many dollars do you get for 30 kronor?

    Answer: _____

6. If 0.5 pound of ground meat serves two people, how many pounds do you need for 25 people?

    Answer: _____

In Exercises 7–10, translate the following problems into equations of the form
$$R \cdot W = P$$
and determine the value of the unknown quantity.

7. 15 is what percent of 96?   *Answer:* _____

8. 3% of what number is 18?   *Answer:* _____

9. 4.2% of 18.75 is what number?   *Answer:* _____

10. 40% of 80 is what number?   *Answer:* _____

11. If two pounds of pecans are sold for $1.10, how much will five pounds cost?

    Answer: _____

12. If two oranges will yield 3/4 of a glass of juice, how many oranges will be needed for 21 glasses of juice?

    Answer: _____

13. What percent of 520 is 208?

    Answer: _____

14. A $180 bicycle was reduced by $45. What percent was the reduction?

    Answer: _____

15. If a basketball team wins sixty out of eighty games, what percent of its games does it win?

    Answer: _____

16. A person's savings account increased from $1500 to $1800. Find the percent of increase.

    Answer: _____

17. Seventy-two percent of all the dogs in town get along with cats. If four hundred thirty-two dogs get along with cats, how many dogs are in town?

    Answer: _____

18. The width of a rectangle is eleven feet shorter than the length. The sum of three times the length and the width is two hundred thirty-seven feet. Find the length and the width.

    Answer: _____

19. The length of a rectangle is nine meters longer than the width. If the length and the width are each shortened by three meters, the resulting perimeter is then thirty meters. Find the length and the width.

    Answer: _____

20. The perimeter of a square is seventy-six feet. Find the length of a side.

    Answer: _____

21. Joan is six years older than Tom. The sum of their ages is 30 years. How old is each of them?

    Answer: _____

22. Peter is six years younger than Tom. Three times Peter's age equals twice Tom's age. How old is each of them?

    Answer: _____

23. What is your percent score if you get 120 out of 150 problems correct?

    Answer: _____

24. A $250 coat was reduced to $150. What was the percent reduction?

    Answer: _____

25. The width of a rectangle is one-third of the length. If the perimeter is forty-eight feet, what is the length?

    Answer: _____

26. Agnes is four years younger than Joan. The sum of their ages is fifty-two years. How old is Agnes?

Answer: _____

27. Raymond is two years younger than Sigmund. Three times Sigmund's age subtracted from five times Raymond's age equals Sigmund's age. How old are the boys?

Answer: _____

If you have questions about this chapter, write them here: _____

_____

_____

_____

_____

_____

_____

PART VI

Graphing

14 Points and Ordered Pairs

Remember to

* *REVIEW THE CONCEPT OF A NUMBER LINE IN CHAPTER 1*
* *USE A RULER AND GRAPH PAPER*
* *GO STEP BY STEP*
* *LEARN WHAT ALL THE NEW VOCABULARY WORDS MEAN*

In Chapter 1 we constructed number lines. We selected a starting point on the line to represent the number 0. Then we decided how long one unit should be. We marked off units to the right of 0 to represent positive numbers, and units to the left of 0 to represent negative numbers. It might be useful to review the material on pages 21–25 at this time.

We now consider graphs in two "dimensions." How do we read and locate points on such graphs? The inventor of the so-called "rectangular coordinate system" was René Descartes. He started a new field of mathematics, known as analytic geometry. During the first half of the 17th century he began to draw mathematical shapes, such as lines, triangles, squares, and circles on grids.

14.1 ORDERED PAIRS

Look at a street map of New York or another big city where there are numbered streets and avenues:

Mark the intersection of 5th Avenue and 3rd Street.

Mark the intersection of 3rd Avenue and 5th Street.

Are these intersections the same? _____

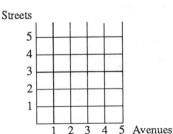

299

Reviewing

1. **Reviewing is a crucial part of learning math.**
   - Most people do not learn math all at once.
   - When you review your work, it helps it to sink in.
   - Repetition helps you to remember.

2. **Reread your notes and the text.**
   - Add to today's lesson by first reviewing yesterday's.

3. **Review frequently.**
   - Never let material for the next exam pile up.
   - Every few days review some past material.
   - By reviewing frequently, you gain a deeper understanding.

4. **Read the summaries at the end of each chapter.**
   - This will help to pinpoint topics for in-depth study.

If you marked your intersections correctly, you would find that they are not the same.

Look at the points labelled *A* and *B* on the following grid. Both points can be labeled by the numbers 2 and 3.

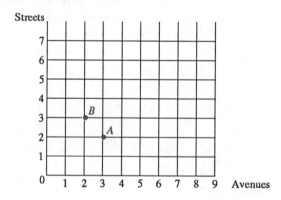

If this were a street map with avenues and streets, which avenue and street would intersect at *A*?

_____ Avenue and _____ Street

What about *B*? _____ Avenue and _____ Street

In mathematics, we would say that *A* corresponds to (3, 2) and *B* to (2, 3). Just as with avenues and streets, we look at the numbers of each pair. Since we use numbers only, we have to make a commitment about which to say first. It is generally agreed that we use the order "avenue/street." Since we write this "pair of numbers" in a definite order, the numbers form an **ordered pair**.

Example 1   Which ordered pair represents point *C*?

Solution.

Point *C* is *above* 6 and to the *right* of 5. The ordered pair is (6, 5).

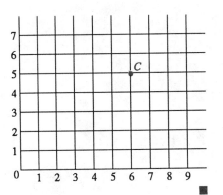

Example 2   Which ordered pair represents point *D*?

Solution.

Point *D* is *above* 8 and to the *right* of 1. The ordered pair is (8, 1).

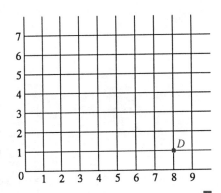

MINI-PRACTICE
(Use this space for your work.)

Read the points on the graph below and record them as ordered pairs.

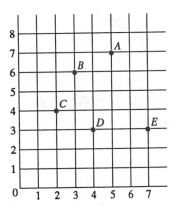

Answers: (a) *A* = (,   )   (b) *B* = (,   )   (c) *C* = (,   )

(d) *D* = (,   )   (e) *E* = (,   )

Answers to Mini-Practice:

(a) *A* = (5, 7)   (b) *B* = (3, 6)   (c) *C* = (2, 4)   (d) *D* = (4, 3)   (e) *E* = (7, 3)

IN YOUR OWN WORDS   explain what is meant by an "ordered pair." _____

_____

_____

Answer:

An ordered pair consists of two numbers written in a definite order. In the usual order, the first number refers to the horizontal axis and the second number refers to the vertical axis.

Practice Exercises

In Exercises 1–26, read the points on the graph below and record them as ordered pairs.

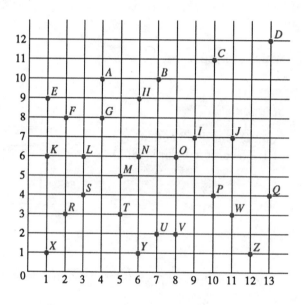

1. $A = ($, $)$ 2. $B = ($, $)$ 3. $C = ($, $)$ 4. $D = ($, $)$ 5. $E = ($, $)$

6. $F = ($, $)$ 7. $G = ($, $)$ 8. $H = ($, $)$ 9. $I = ($, $)$ 10. $J = ($, $)$

11. $K = ($, $)$ 12. $L = ($, $)$ 13. $M = ($, $)$ 14. $N = ($, $)$ 15. $O = ($, $)$

16. $P = ($, $)$ 17. $Q = ($, $)$ 18. $R = ($, $)$ 19. $S = ($, $)$ 20. $T = ($, $)$

21. $U = ($, $)$ 22. $V = ($, $)$ 23. $W = ($, $)$ 24. $X = ($, $)$ 25. $Y = ($, $)$

26. $Z = ($, $)$

14.2 RECTANGULAR COORDINATE SYSTEMS

Using the space provided, draw a horizontal number line and locate the integers from –5 to 5 on it. Now draw a vertical number line through the zero point of your horizontal line. Then, using the same distance unit, label this vertical line with positive numbers upward from zero and with negative numbers downward from zero. Label the horizontal number line with an *x*, and the vertical number line with a *y*.

Your two number lines should look something like this:

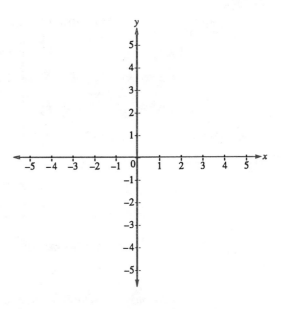

Which ordered pair represents the intersection of these two number lines? _____

These two number lines intersect at the zero point of each line. The ordered pair is (0, 0), and it is called the **origin**.

The system formed by these two number lines is called a **rectangular coordinate system**. The horizontal line is called the **x-axis** and the vertical line is called the **y-axis**. These two lines are also referred to as the **coordinate axes**. The two numbers in the ordered pair that refer to a specific location of a point are called **coordinates**; the first number is the **x-coordinate** and the second, the **y-coordinate**. Mathematicians have selected the letters *x* and *y*, but we could use any letters of the alphabet.

The coordinate axes divide the plane into four parts. They are called **quadrants**. The quadrants are numbered from I to IV, following a counter-clockwise motion.

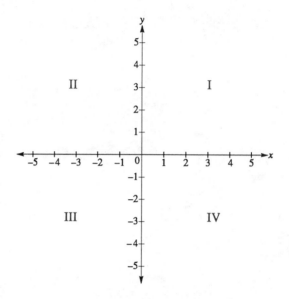

Determine the signs of the coordinates x and y in each quadrant.

Quadrant	Sign of x	Sign of y
I		
II		
III		
IV		

Answers: I: $+, +$;   II: $-, +$;   III: $-, -$;   IV: $+, -$

Example 1   Which ordered pair represents point A?

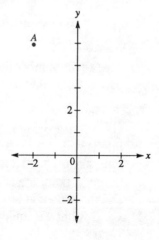

Solution.

Look at point *A*. Point *A* is in the second quadrant; it has *x*-coordinate –2 and *y*-coordinate 5. Point *A* is represented by the ordered pair (–2, 5).

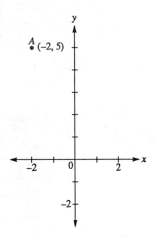

Example 2   Which ordered pair represents point *B*?

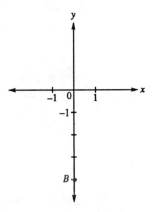

Solution.

Look at point *B*. Point *B* is on the negative part of the *y*-axis. Every point on the *y*-axis has *x*-coordinate equal to 0. The *y*-coordinate is –4. Point *B* is represented by the ordered pair (0, –4). Points that are on a coordinate axis are not considered to be in any quadrant.

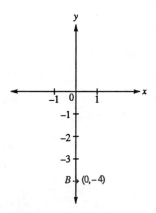

MINI-PRACTICE   Read the points on the graph below and record them as ordered pairs.
(Use this space for your work.)

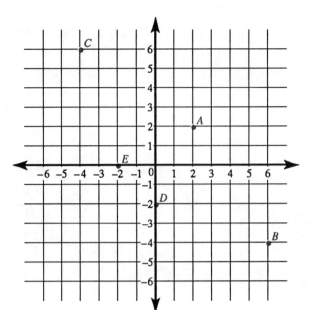

Answers: (a) $A = ($, $)$ (b) $B = ($, $)$ (c) $C = ($, $)$

(d) $D = ($, $)$ (e) $E = ($, $)$

IN YOUR OWN WORDS   explain how to determine the coordinates of points in a rectangular coordinate system. Pay special attention to points on the coordinate axes: _____

_____

_____

Answer:

If you pretend that you are standing at the point in question, look directly at the x-axis and read the number you see. Then look directly at the y-axis and read the number you see.

Practice Exercises

In Exercises 1–26, read the points on the graph at the top of page 307 and record them as ordered pairs.

1. $A = ($, $)$ 2. $B = ($, $)$ 3. $C = ($, $)$ 4. $D = ($, $)$ 5. $E = ($, $)$

6. $F = ($, $)$ 7. $G = ($, $)$ 8. $H = ($, $)$ 9. $I = ($, $)$ 10. $J = ($, $)$

11. $K = ($, $)$ 12. $L = ($, $)$ 13. $M = ($, $)$ 14. $N = ($, $)$ 15. $O = ($, $)$

16. $P = ($, $)$ 17. $Q = ($, $)$ 18. $R = ($, $)$ 19. $S = ($, $)$ 20. $T = ($, $)$

21. $U = ($, $)$ 22. $V = ($, $)$ 23. $W = ($, $)$ 24. $X = ($, $)$ 25. $Y = ($, $)$

26. $Z = ($, $)$

_____

Answers to Mini-Practice:

(a) $A = (2, 2)$ (b) $B = (6, -4)$ (c) $C = (-4, 6)$ (d) $D = (0, -2)$ (e) $E = (-2, 0)$

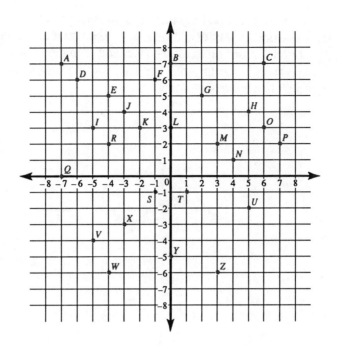

14.3 Plotting Points

In mathematics, locating a point on a graph is called **plotting** the point. Let's plot the point (2, 3).

What is the value of x? _____

What is the value of y? _____

We can plot the point represented by the ordered pair (2, 3) by drawing a vertical line through $x = 2$ and a horizontal line through $y = 3$.

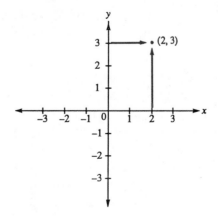

At which point do these two lines intersect? _____

You probably realized that this point is represented by the ordered pair (2, 3).

Another way of plotting (x, y) is to start at the origin, move to the right or left according to the x-value, and then move up or down according to the y-value. Thus, to plot (2, 3), start at the origin, move to the right 2 units, and then move up 3 units.

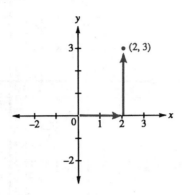

Example 1   Plot (3, 2).

Solution.

Start at (0, 0). Move 3 units to the right. Continue 2 units upward.

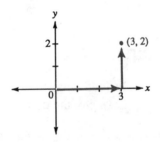

Example 2   Plot (–4, 0).

Solution.

Start at (0, 0). Move 4 units to the left. Stay there. The point (–4, 0) is on the negative portion of the x-axis.

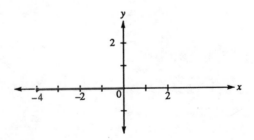

MINI-PRACTICE
(Use this space for your work.)

Plot $F(3, -4)$, $G(2, 1)$, $H(-2, -3)$, and $I(0, 2)$ on this coordinate system.

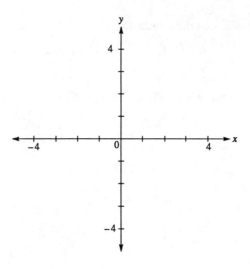

IN YOUR OWN WORDS   explain how to plot points on a rectangular coordinate system: _____

_____

_____

Answer:

To plot a point, start at the origin, and move left or right on the *x*-axis, according to the *x*-value. Then continue to move up or down parallel to the *y*-axis, according to the *y*-value.

Answer to Mini-Practice:

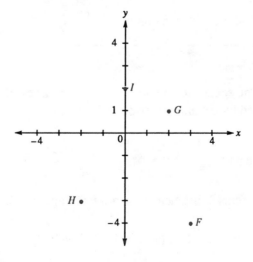

Practice Exercises

Plot the following points on the coordinate system at the right.

1. $A = (0, 1)$ 2. $B = (1, 1)$ 3. $C = (-1, 2)$

4. $D = (-3, 1)$ 5. $E = (4, 0)$ 6. $F = (0, 0)$

7. $G = (-2, 1)$ 8. $H = (2, -3)$ 9. $I = (3, 3)$

10. $J = (-3, 0)$

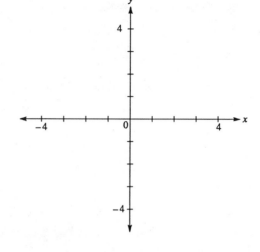

14.4 SCALES FOR NUMBER LINES

How can you represent 25 on the following number line?

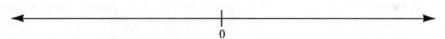

Maybe you took a small distance unit and then tried to apply it 25 times to the right of zero. This task involves a lot of work! It would be almost impossible to represent numbers like 100, 5000, or 25,000 by using a *small* distance unit just to represent *one* unit.

Find a better way to represent 25:

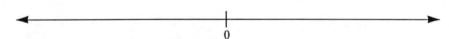

A solution to this problem is to let the unit

⊢——⊣

represent 5 distance units.

The **scale** for a number line is the series of markings laid down, determined by the unit you use.

Example 1   If the unit

⊢——⊣

represents 5 distance units, draw the number line.

Solution.   The following number line is obtained:

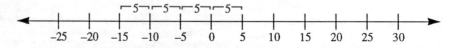

How can you represent 0.5 on the following number line?

0

If you have to work with decimals, you can choose the units to be large enough so that you can estimate the decimal. You can also estimate fractions in this way.

Example 2   If the unit

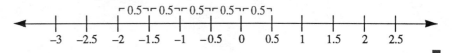

represents 0.5 distance unit, draw the number line.

Solution.   The following number line is obtained:

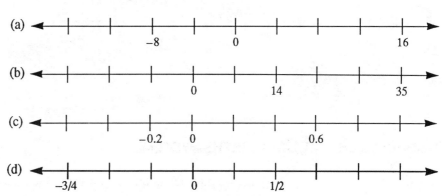

MINI-PRACTICE
(Use this space for your work.)

Complete the following number lines:

(a)

| | | | | | | | | |
| | | −8 | | 0 | | | | 16 |

(b)

(c)

(d)

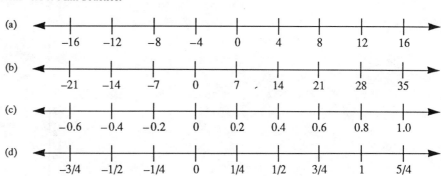

IN YOUR OWN WORDS   explain how to select a scale when you put numbers onto a number line. _____

_____

_____

_____

Answers to Mini-Practice:

(a) −16 −12 −8 −4 0 4 8 12 16

(b) −21 −14 −7 0 7 14 21 28 35

(c) −0.6 −0.4 −0.2 0 0.2 0.4 0.6 0.8 1.0

(d) −3/4 −1/2 −1/4 0 1/4 1/2 3/4 1 5/4

Answer:

To select a scale first consider the *range* (spread of numbers) you have to plot. You have a certain limited space on a piece of paper to represent all these numbers. Decide if you want small or large units.

Practice Exercises

Represent the indicated numbers on the number lines at the right.

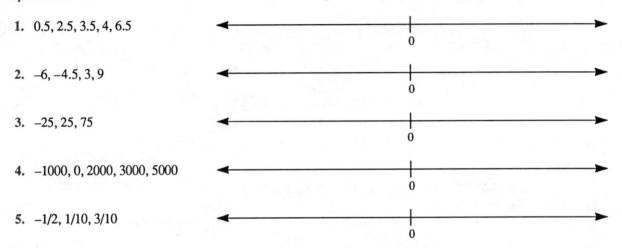

1. 0.5, 2.5, 3.5, 4, 6.5

2. –6, –4.5, 3, 9

3. –25, 25, 75

4. –1000, 0, 2000, 3000, 5000

5. –1/2, 1/10, 3/10

14.5 SCALES FOR COORDINATE SYSTEMS

When we plot points on a coordinate system, we modify the technique used on a number line. We start out with two number lines for the *x*-axis and the *y*-axis. The two axes can have the same scale or different ones; it depends on the problem you have.

Make one scale for the *x*-axis and another for the *y-axis so that* one unit on the *x-axis* has the same length as *two* units on the *y-axis*. Your grid should look something like this:

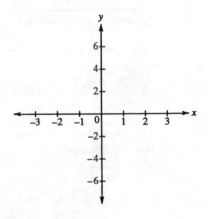

Make another grid so that *two* units on the *x-axis* is the same as *one* unit on the *y-axis*. Your grid should look something like the one on the right:

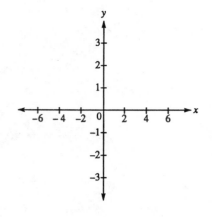

Example 1 Plot (–0.75, 1).

Solution.

Let each marker represent 0.25 unit on both the *x*-axis and the *y*-axis.

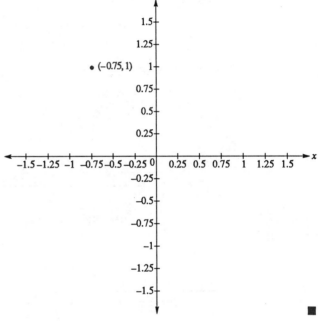

Example 2 Plot (3, 50).

Solution.

Let each marker represent 1 unit on the *x*-axis and 10 units on the *y*-axis.

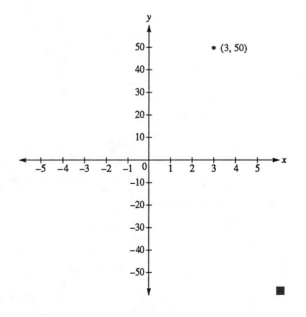

MINI-PRACTICE   Use suitable scales in each case and plot the following points.

(Use this space for your work.)

(a) $A = (5, 20)$ (b) $B = (-15, 2.5)$ (c) $C = (100, 200)$ (d) $D = (0.2, 50)$

Answers:

(a)

(b)
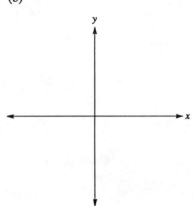

(c)

(d)

_____

Answers to Mini-Practice:

(a)
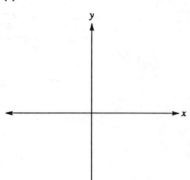

(b)
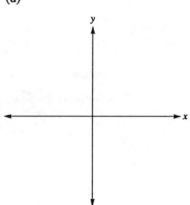

IN YOUR OWN WORDS   explain how we can change the scales on the coordinate axes when we graph either large numbers or small numbers: _____

_____

_____

Answer:

Decide on the range (spread of numbers) on the *x*-axis and on the *y*-axis. Divide each of these lines into units suitable for graphing.

Practice Exercises

1. *Plot the following points on the given coordinate system.*

$A = (-2.5, 0)$	$B = (0, 1.5)$
$C = (-1.5, 3.5)$	$D = (0.5, -2.5)$
$E = (-3.5, -4.5)$	$F = (2.5, 3.5)$
$G = (-3.5, 0)$	$H = (1.5, -2)$
$I = (4.5, -1.5)$	$J = (2.5, -1.5)$
$K = (0, 3.5)$	

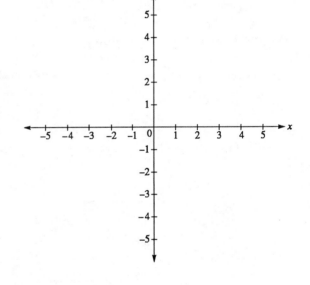

Answers to Mini-Practice:

(c)

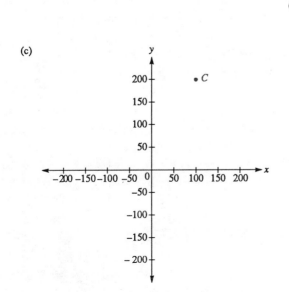

(d)

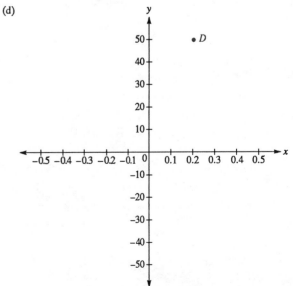

SUMMARY

A system formed by a horizontal and a vertical line is called a rectangular coordinate system. The horizontal line is the *x*-axis and the vertical line is the *y*-axis.

Ordered pairs of numbers can be plotted as points on a rectangular coordinate system. To plot a point, start at the origin, (0, 0), and move left or right on the *x*-axis according to the *x*-value. Then continue to move up or down parallel to the *y*-axis, according to the *y*-value.

Different scales can be used on the *x*- or *y*-axis to represent large or small numbers. Decide on the range on each axis. Divide each of these lines into units suitable for graphing.

Review Practice Exercises

Plot the following points on the given coordinate systems.

1. $A = (1, -10)$ $B = (-5, -50)$
    $C = (-3, -40)$ $D = (2, -20)$
    $E = (5, 40)$

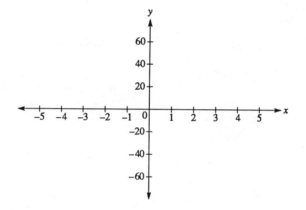

2. $F = (10, 1.5)$ $G = (20, -2)$
    $H = (-30, 3.5)$ $I = (-40, 0.5)$
    $J = (30, -4)$

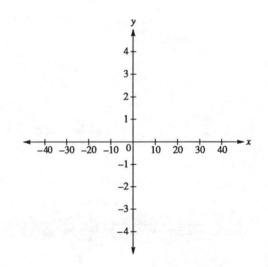

3. $V = (200, 400)$ $W = (-300, 500)$

     $X = (-500, 700)$ $Y = (400, -200)$

     $Z = (1000, -500)$

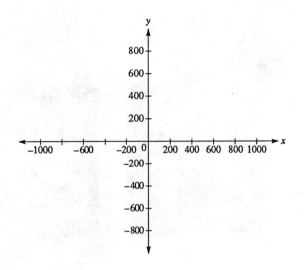

If you have questions about this chapter, write them here: _____

_____

_____

_____

_____

15 Graphing Equations

Remember to

* ✳ *REVIEW THE LITERAL EQUATIONS IN CHAPTER 11*
* ✳ *REVIEW THE FORMULAS IN CHAPTERS 7 AND 11*
* ✳ *COMPLETE ALL TABLES SO THAT YOU UNDERSTAND THE RELATIONSHIP BETWEEN THE TWO VARIABLES*
* ✳ *PLOT THE POINTS CAREFULLY AND CONNECT THE POINTS ACCORDING TO THE SHAPE YOU DETECT*

You are ready to learn about relations in mathematics and about solving equations in two variables. You will also learn how to graph equations in two variables when both variables are raised to the first power. The graphs of these equations are lines. You will learn about the points where the lines cross the coordinate axes as well as about the steepness of the lines. You will also study horizontal and vertical lines.

15.1 EQUATIONS IN TWO VARIABLES

If one pound of potatoes costs 30¢, how much do you pay for

3 pounds? _____

5 pounds? _____

x pounds? _____

For x pounds you pay $30x$ (cents).

We can say that

$$y = 30x$$

and thereby establish a *formula* for the price of potatoes. Remember that a formula is represented by an equation in more than one variable.

In mathematics we say that we have a **relation** with two variables, x and y. Since y (the price we pay) depends on x (the weight of the potatoes), we call y the **dependent variable** and x the **independent variable**.

Write an equation for: "One number is always two more than another number." _____

No doubt you wrote $y = x + 2$.

Write an equation for: "One number is the square of another number." _____

Write an equation for: "The sum of the squares of two numbers is four." _____

The equations

$$y = 30x$$
$$y = x + 2$$
$$y = x^2$$

and

$$x^2 + y^2 = 4$$

all show a relationship between two variables, x and y.

Write a formula for the relationship between the following ordered pairs: (2, 3), (3, 4), (5, 6) _____

Since y is always one more than x, we have the equation

$$y = x + 1$$

Example 1   Write an equation for the relationship:

A number minus another number is five

Solution.   Call the numbers x and y.

Direct translation: $x - y = 5$ ■

Example 2   Write a formula for the relationship between the following ordered pairs:

$$(3, 9), \quad (4, 16), \quad (-1, 1)$$

Solution.

$$(3, 9), \quad (4, 16), \quad (-1, 1)$$
$$\downarrow \qquad \quad \downarrow \qquad \qquad \downarrow$$
$$(3, 3^2), \quad (4, 4^2), \quad (-1, (-1)^2)$$

The second coordinate is always the square of the first coordinate.

Formula: $y = x^2$ ■

The Final Exam

1. Plan ahead.
- Allow enough time to review.
- This will help avoid finals-week jitters.

2. A month before:
- Begin to organize your review.
- Get copies of old finals.
- Many types of questions will be repeated each term.
- Ask your instructor and classmates for help.

3. Study the exams you've already taken.
- Similar questions may appear on the final.
- Be sure you understand why you made mistakes.
- Rework all of the problems.

4. At the exam:
- Do the problems you know first.
- Take short breaks and tell yourself you can do it.
- If you can't solve a whole problem, do as much of it as you can.
- Take the full time.
- Check your work carefully.

MINI-PRACTICE
(Use this space for your work.)

(a) Write a formula for the relationship: "The sum of twice a number and another number is six."

(b) Write an equation for the relationship: "The square of a number plus the square of another number is sixteen."

(c) Write a formula for the relationship between the following ordered pairs:

$$(1, 0), (2, 2), (-1, -4), (-2, -6)$$

Answers: (a) _____   (b) _____   (c) _____

Practice Exercises

In Exercises 1–12, write formulas for the following relationships.

1. A number minus another number is ten.   *Answer:* _____

2. The product of two numbers is ten.   *Answer:* _____

3. A number divided by another number is two.   *Answer:* _____

Answers to Mini-Practice:

(a) $2x + y = 6$ (b) $x^2 + y^2 = 16$ (c) $y = 2x - 2$

4. One number plus twice another number is four.   *Answer:* _____

5. The sum of the squares of two numbers is twenty-five.   *Answer:* _____

6. One number plus three times another is ten.   *Answer:* _____

7. The difference of twice a number and another number (in this order) is fifteen.

Answer: _____

8. Twice a number divided by three times another number is twenty-one.

Answer: _____

9. The square of a number added to another number is sixteen.   *Answer:* _____

10. One number minus the quotient of that number divided by another number is one hundred.

Answer: _____

11. The difference of the squares of two numbers is twelve.   *Answer:* _____

12. One number is two more than half of another number.   *Answer:* _____

In Exercises 13–22, write formulas for the relationships between the following ordered pairs.

13. (1, 1), (2, 4), (3, 9) (4, 16)   *Answer:* _____

14. (1, 3), (2, 4), (3, 5), (4, 6)   *Answer:* _____

15. (–1, –1), (2, 8), (–2, –8)   *Answer:* _____

16. (–1, 2), (1, –2), (–2, 4), (2, –4)   *Answer:* _____

17. (–1, –1), (1, 3), (–2, –3), (2, 5)   *Answer:* _____

18. (–2, –5), (–1, –3), (0, –1), (1, 1), (2, 3)   *Answer:* _____

19. (2, 4), (3, 6), (4, 8), (5, 10)   *Answer:* _____

20. (1, 1), (4, 2), (9, 3), (16, 4)   *Answer:* _____

21. (1, 0), (2, 3), (3, 8), (4, 15)   *Answer:* _____

22. (0, 0), (1, 1), (2, 8), (3, 27)   *Answer:* _____

15.2 SOLUTIONS OF EQUATIONS IN TWO VARIABLES

In the equation $x + y = 10$, if $x = 2$, what is y? _____

If $x = -5$, what is y? _____

If $x = 0.3$, what is y? _____

The ordered pair (2, 8) is called a **solution** to the equation $x + y = 10$ because when 2 replaces x and 8 replaces y, we obtain the equality $2 + 8 = 10$. Other solutions are (–5, 15) and (0.3, 9.7).

How many solutions can you find of this equation? _____

You probably said that you can find an *infinite* number of solutions of the equation $x + y = 10$.

Give 5 other solutions of the equation $x + y = 10$:

$x =$ _____ $y =$ _____ $(x, y) =$ _____

$x =$ _____ $y =$ _____ $(x, y) =$ _____

$x =$ _____ $y =$ _____ $(x, y) =$ _____

$x =$ _____ $y =$ _____ $(x, y) =$ _____

$x =$ _____ $y =$ _____ $(x, y) =$ _____

For (x, y), you might have said: $(1, 9)$, $(3, 7)$, $(-1, 11)$, $(0, 10)$ and $(1.5, 8.5)$, for example.

Example 1   Complete the table for the equation

$$y = 3x - 5$$

x	y	(x, y)
0		$(0,\ )$
	4	$(\ , 4)$
-1		
1		

Solution.

$x = 0 \ \rightarrow \ y = 3(0) - 5 = 0 - 5 = -5$

$y = 4 \ \rightarrow \ 4 = 3x - 5$

$\qquad\qquad 9 = 3x$

$\qquad\qquad x = 3$

$x = -1 \rightarrow \ y = 3(-1) - 5 = -3 - 5 = -8$

$x = 1 \ \rightarrow \ y = 3(1) - 5 = 3 - 5 = -2$

x	y	(x, y)
0	-5	$(0, -5)$
3	4	$(3, 4)$
-1	-8	$(-1, -8)$
1	-2	$(1, -2)$

■

Example 2   Complete the table for the equation

$$y = x^2$$

x	y	(x, y)
0		
-1		
1		
-2		
2		

Solution.

$x = 0 \rightarrow y = 0^2 = 0$

$x = -1 \rightarrow y = (-1)^2 = 1$

$x = 1 \rightarrow y = 1^2 = 1$

$x = -2 \rightarrow y = (-2)^2 = 4$

$x = 2 \rightarrow y = 2^2 = 4$

x	y	(x, y)
0	0	(0, 0)
–1	1	(–1, 1)
1	1	(1, 1)
–2	4	(–2, 4)
2	4	(2, 4)

■

MINI-PRACTICE

(Use this space for your work.)

Complete the tables for the following equations:

(a) $y = 2x - 5$

x	y	(x, y)
0		(0,)
	0	(, 0)
–1		
1		

(b) $y = 2x + 1$

x	y	(x, y)
0		
	9	
–1		
1		

(c) $x^2 + y^2 = 25$

x	y	(x, y)
3	4 and –4	(3, 4) and (3, –4)
0	and	(,) and (,)
4 and –4	–3	(,) and (,)
and	4	(,) and (,)
and	0	(,) and (,)

IN YOUR OWN WORDS

explain how to solve an equation in two variables: _____

_____

_____

Answer:

When you solve an equation in two variables, you give a value to one variable and substitute that in the equation. Then you solve the equation for the other variable.

Answers to Mini-Practice:

(a) (0, –5), $\left(\frac{5}{2}, 0\right)$, (–1, –7), (1, –3)   (b) (0, 1), (4, 9), (–1, –1), (1, 3)

(c) (0, 5) and (0, –5); (4, –3) and (–4, –3); (3, 4) and (–3, 4); (5, 0) and (–5, 0)

Practice Exercises

In Exercises 1–8, complete the tables.

1. $y = 8x + 1$

x	y	(x, y)
0		
	0	
−1		
	−15	
2		
−3		

2. $y = 6x + 6$

x	y	(x, y)
0		
	0	
1		
	−6	
	9	
2		

3. $y = -2x + 3$

x	y	(x, y)
0		
	0	
−1		
	1	
	−3	
2		

4. $y = 4x - 2$

x	y	(x, y)
0		
	0	
−1		
	1	
2		
3		
	−10	

5. $y = -3x - 1$

x	y	(x, y)
0		
	0	
−2		
	−7	
3		
−3		
	11	

6. $y = 3x + 2$

x	y	(x, y)
0		
	0	
−2		
	−1	
	8	
1		
3		

7. $y = -4x - 1$

x	y	(x, y)
0		
	0	
1		
	−9	
3		
−2		
	11	

8. $y = 5x - 4$

x	y	(x, y)
0		
	0	
1		
	6	
	−9	
3		
−3		

In Exercises 9–30, find three solutions of each of the following equations.

9. $y = 2x - 5$ Answer: _____   _____   _____

10. $y = -x + 3$ Answer: _____   _____   _____

11. $y = -2x - 5$ Answer: _____   _____   _____

12. $x + y = 5$ Answer: _____   _____   _____

13. $x - y = 3$ Answer: _____   _____   _____

14. $2x + 3y = 6$ Answer: _____   _____   _____

15. $2x - 3y = 6$ Answer: _____   _____   _____

16. $2x + 5y = 10$ Answer: _____   _____   _____

17. $2x - 5y = 10$ Answer: _____   _____   _____

18. $x - 3y = 15$ Answer: _____   _____   _____

19. $y = 2x + 1$ Answer: _____   _____   _____

20. $y = 3x - 2$ Answer: _____   _____   _____

21. $y = 2x$ Answer: _____   _____   _____

22. $3x + 4y = 12$ Answer: _____   _____   _____

23. $2x - 5y = 10$ Answer: _____   _____   _____

24. $x + 2y = 6$ Answer: _____   _____   _____

25. $y = 2x - 3$ Answer: _____   _____   _____

26. $y = -2x + 2$ Answer: _____   _____   _____

27. $3x - y = 3$ Answer: _____   _____   _____

28. $2x + y = 4$ Answer: _____   _____   _____

29. $x - 2y = 4$ Answer: _____   _____   _____

30. $3x + 2y = 4$ Answer: _____   _____   _____

15.3 GRAPHS OF EQUATIONS IN TWO VARIABLES

Recall the Mini-Practice on page 323 and plot the points you found by completing the tables for Exercises (a)–(c) on three separate coordinate systems, using a suitable scale for each:

(a) $y = 2x - 5$

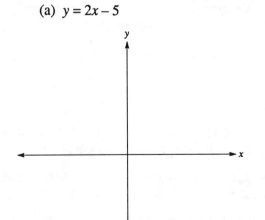

x	y	(x, y)
0	–5	(0, –5)
2.5	0	(2.5, 0)
–1	–7	(–1, –7)
1	–3	(1, –3)

(b) $y = 2x + 1$

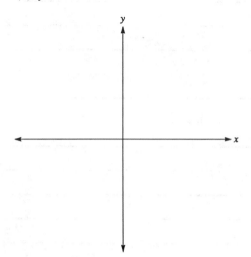

x	y	(x, y)
0	1	(0, 1)
4	9	(4, 9)
−1	−1	(−1, −1)
1	3	(1, 3)

(c) $x^2 + y^2 = 25$

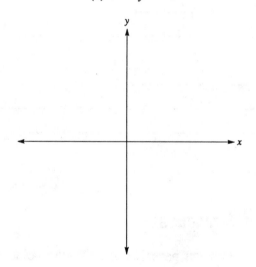

x	y	(x, y)
3	4 and −4	(3, 4) and (3, −4)
0	5 and −5	(0, 5) and (0, −5)
4 and −4	−3	(4, −3) and (−4, −3)
3 and −3	4	(3, 4) and (−3, 4)
5 and −5	0	(5, 0) and (−5, 0)

Look at your points to see if they seem to form straight lines or curves. Connect the points on each coordinate system, going from left to right, according to the shapes suggested. This is called **graphing**.

What shapes do you see?   (a) _____

(b) _____

(c) _____

If you plotted your points correctly, you found that graphs (a) and (b) are straight lines, whereas graph (c) suggests a circle.

Look again at the equation $y = x^2$. As you can see from the table in Example 2 on page 323, both $x = 1$ and $x = -1$ result in a y-value of 1.

Which x–values give a y-value of 4? _____

What shape does this graph have? _____

The graph $y = x^2$ is shaped somewhat like the letter U. In mathematics this graph is called a **parabola**.

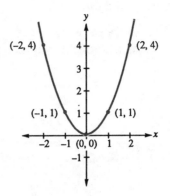

Example 1   Graph the equation $y = 2x$ by first plotting three points (x, y).

Solution.   Choose three values for x: (We choose 0, –1, and 1.)

x	y
0	
–1	
1	

Calculate the corresponding y-values:

$$y = 2(0) = 0$$
$$y = 2(-1) = -2$$
$$y = 2(1) = 2$$

Note that the points (x, y) can be read off from the two columns of the table, so that it is unnecessary to have an (x, y)-column. Thus, from the completed table

x	y
0	0
–1	–2
1	2

we obtain the points

$$(0, 0), \quad (-1, -2), \quad (1, 2)$$

Plot these three points on the coordinate system at the right. All three lie on a straight line. Draw this line through these points.

The graph of $y = 2x$ is a straight line.

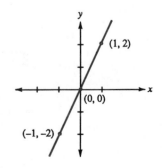

■

Example 2   Graph the equation $x = y^2$ by first plotting five points (x, y).

Solution.

Because y is squared, it is easier to choose values for y, and then find the corresponding values for x. Let y be 0, 1, –1, 2, and –2.

x	y
	0
	1
	–1
	2
	–2

Calculate x:

$x = 0^2 = 0$

$x = 1^2 = 1$

$x = (-1)^2 = 1$

$x = 2^2 = 4$

$x = (-2)^2 = 4$

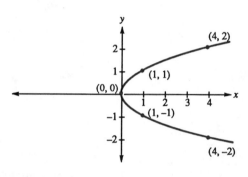

Plot the points (0, 0), (1, 1), (1, –1), (4, 2), and (4, –2).

This graph is a parabola that opens to the right.   ■

Example 3   Graph the equation $y = 2x^2$ by first plotting five points (x, y).

Solution.

Because x is squared, we choose values for x first, and then square them. But here, $y = 2x^2$. So, we will take an extra column to double the value of x^2 found. This will give the value of $y = 2x^2$. Let $x = 0, 1, -1, 2, -2$. The corresponding values of x^2 and $2x^2$ are shown in the table.

x	x^2	$y = 2x^2$
0	0	0
1	1	2
–1	1	2
2	4	8
–2	4	8

Plot the points (0, 0), (1, 2), (–1, 2), (2, 8), and (–2, 8).

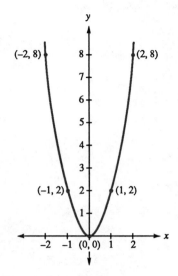

This graph is a parabola that opens upward.   ■

MINI-PRACTICE
(Use this space for your work.)

Graph the following equations. Begin by completing the tables.

(a) $y = 2x + 2$

x	y
0	
–1	
1	

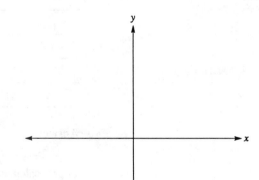

(b) $y = -2x + 2$

x	y
0	
1	
2	

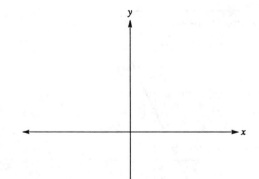

(c) $y = x^2 + 1$

x	x^2	$y = x^2 + 1$
0		
−1		
1		
−2		
2		

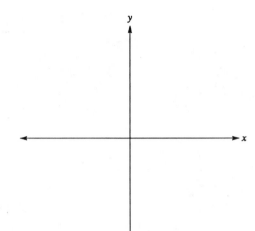

IN YOUR OWN WORDS   explain how to graph (simple) equations. Explain how you can tell from an equation if the graph will be a straight line: _____

_____

_____

_____

_____

_____

Answer:

In order to graph an equation, you select values for one of the variables (usually x) and calculate the corresponding values of the other variable (usually y). Then you plot the points and connect the points, usually from left to right, by following the shape you detect.

The graph of an equation in two variables is a line if each variable only occurs with exponent 1.

_____

Answers to Mini-Practice:

(a)

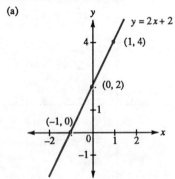

(b)

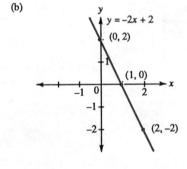

(c)

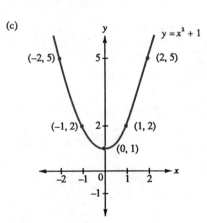

Practice Exercises

Go back to pages 324–325, Practice Exercises 9-18. Use your solutions and graph the equations on the coordinate systems provided.

9.

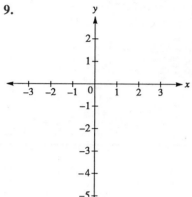

10.

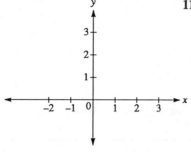

11.

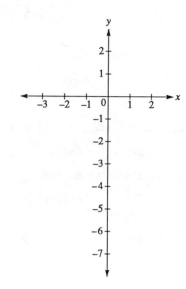

12.

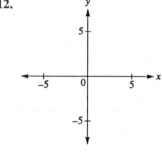

13.

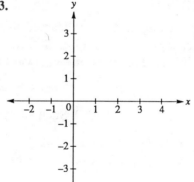

14.

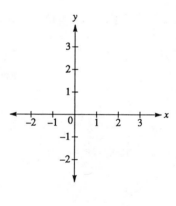

15.

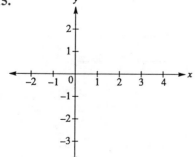

16.

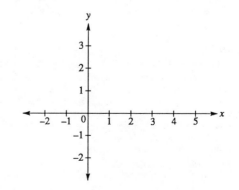

17.

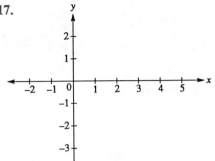

18.

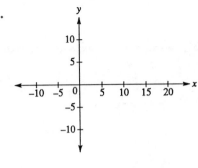

In Exercises 19–22, plot five points in order to graph each equation.

19. $y = x^2 - 1$ *Answer:* _____

20. $y = x^2 + 2$ *Answer:* _____

21. $y = 3x^2$ *Answer:* _____

22. $y = \dfrac{x^2}{2}$ *Answer:* _____

15.4 LINES AND THEIR INTERCEPTS

In the preceding section we learned that the graph of an equation in two variables, say x and y, is a line if each variable can only occur with exponent 1. This kind of equation can always be written in the form

$$ax + by = c$$

where a and b are not *both* 0. Such an equation is called a **linear equation**.

Graph the equation

$$x + y = 3$$

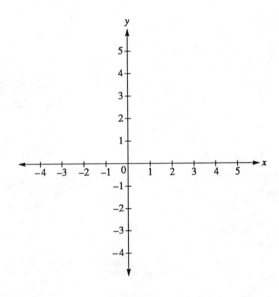

At which point does this line cross the *x*-axis? _____

The *x*-coordinate of this point (3, 0) is called the **x-intercept** of the graph. Thus, 3 is the *x*-intercept of this graph.

At which point does the line cross the *y*-axis? _____

The *y*-coordinate of this point (0, 3) is called the **y-intercept** of the graph. Thus, 3 is the *y*-intercept of this graph.

Graph the equation

$$y = 3x - 6$$

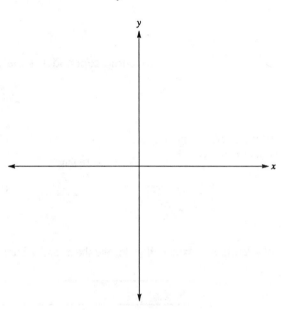

What is the *x*-intercept? _____

What is the *y*-intercept? _____

The line intersects the *x*-axis at (2, 0), so the *x*-intercept is 2. The line intersects the *y*-axis at (0, −6), so the *y*-intercept is −6.

Example 1   Find the *x*- and *y*-intercepts of the line $x - y = 3$.

Solution.

x	*y*
0	−3
3	0

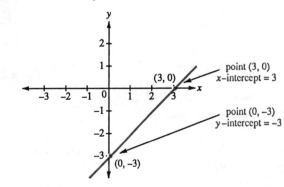

3 is the *x*-intercept and −3 is the *y*-intercept.   ∎

Example 2   Find the x- and y-intercepts of the line $y = 2x + 1$.

Solution.

x	y
0	1
$-\dfrac{1}{2}$	0

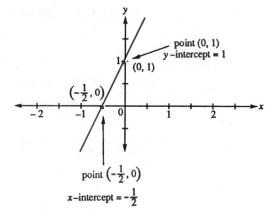

$-\dfrac{1}{2}$ is the x-intercept and 1 is the y-intercept.   ■

MINI-PRACTICE
(Use this space for your work.)

Find the x- and y-intercepts of the following lines:

(a) $y = 2x + 4$ (b) $x - y = 2$ (c) $2x + 3y = 6$

Answers: (a) _____   (b) _____   (c) _____

IN YOUR OWN WORDS   explain how to find the x- and y-intercepts of an equation of a line: _____

_____

_____

_____

Answer:

   See the following rule.

Rule:

1. **Replace y with 0. Solve for x.**

   This value of x is the x-intercept.

2. **Replace x with 0. Solve for y.**

   This value of y is the y-intercept.

Practice Exercises

Go back to pages 331–332, Practice Exercises 9–18. Find the x- and y-intercepts. See also Practice Exercises 9–18, pages 324–325.

Answers to Mini-Practice:

(a) x-intercept $= -2$, y-intercept $= 4$ (b) x-intercept $= 2$, y-intercept $= -2$
(c) x-intercept $= 3$, y-intercept $= 2$

15.5 SLOPE

Consider the equations: (a) $y = x + 2$

                                   (b) $y = 2x + 2$

                                   (c) $y = -2x + 2$

Find the y-intercepts.

(a) _____   (b) _____   (c) _____

Hopefully, you concluded that all three equations had the same y-intercept, namely 2.

Go back to the equations and find the coefficients of the x-terms.

(a) _____   (b) _____   (c) _____

Graph the three equations on one coordinate system and label the lines:

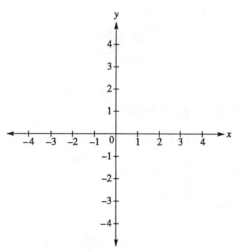

Compare the *steepness* of lines (a) and (b): _____

_____

The equation with the larger x-coefficient corresponds to the steeper line.

Compare the directions of lines (b) and (c): _____

_____

If your graphs are correct, you saw that a line with a negative coefficient of x leans backward,

and that a line with a positive x-coefficient leans forward,

This "leaning" together with "steepness" are encoded in the notion of the **slope** of a line.

Go back and look at the equations (a), (b), and (c) again. Each equation is written in the form:

$$y = \text{slope} \cdot x \ + \ y\text{-intercept}$$

As an equation we write

$$y = mx + b$$

where m is the slope and b the y-intercept. This is known as the **slope-intercept form** of the equation of a line.

Example 1   Find the slope and the y-intercept of the line $2y = 3x + 4$.

Solution.

Solve for y.

$$y = \frac{3}{2}x + 2$$

Slope $= \dfrac{3}{2}$

y-intercept $= 2$

■

Example 2   Find the slope and the y-intercept of the line $x - y = 5$.

Solution.

Solve for y.

$$-y = -x + 5$$
$$y = x - 5$$

Slope $= 1$
y-intercept $= -5$

■

MINI-PRACTICE
(Use this space for your work.)

Find the slopes of the lines given by the following equations:

    (a) $y = 5x - 3$ (b) $x + y = 4$ (c) $2x - 5y = 10$

Answers: (a) _____ (b) _____ (c) _____

IN YOUR OWN WORDS

explain how to find the slope of a given line: _____

_____

_____

_____

Answers to Mini-Practice:

(a) 5   (b) −1   (c) $\dfrac{2}{5}$

Answer:

To find the slope of a line we can first solve the equation of the line for *y*. The slope is then the coefficient of the *x*-term.

Practice Exercises

Find the slope and the y-intercept of the lines given by the following equations.

1. $y = x + 1$ Answer: _____

2. $y = 3x - 2$ Answer: _____

3. $y = \frac{x}{2} + 1$ Answer: _____

4. $y = x - 3$ Answer: _____

5. $y = 3x + 1$ Answer: _____

6. $y = -2x - 1$ Answer: _____

7. $y = \frac{2}{3}x - 2$ Answer: _____

8. $y = 1 - \frac{3}{4}x$ Answer: _____

9. $2y = x - 3$ Answer: _____

10. $2y = 3x - 5$ Answer: _____

11. $4y = 3 - 5x$ Answer: _____

12. $3x + y = 4$ Answer: _____

13. $4x - y = 6$ Answer: _____

14. $3x + 2y = 6$ Answer: _____

15. $2x + y = 3$ Answer: _____

16. $3x - y = 1$ Answer: _____

17. $x - 2y = 4$ Answer: _____

18. $x + 3y = 6$ Answer: _____

19. $y = -x + 1$ Answer: _____

20. $3x - 4y = 12$ Answer: _____

21. $5x - 2y = 10$ Answer: _____

22. $y = -4x + 2$ Answer: _____

23. $5y = 20 - 4x$ Answer: _____

24. $-x - 3 + y = 0$ Answer: _____

15.6 HORIZONTAL LINES AND VERTICAL LINES

Graph the equation $y = 3$ by first completing the following table:

x	*y*
0	
1	
−1	

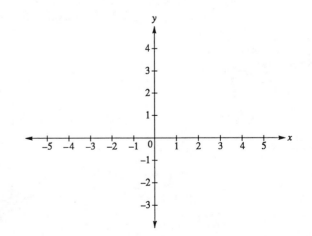

The equation specifies that *y* is always 3, no matter what *x* is. Therefore, the line passes through *all* points with a *y*-coordinate of 3. The line *y* = 3 is horizontal.

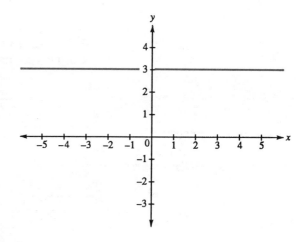

Write the equation *y* = 3 in slope-intercept form: _____

What is the coefficient of *x*? _____

You should have written $y = 0x + 3$ in slope-intercept form. Thus, the coefficient of *x* is 0. In general, the slope of any horizontal line is 0.

Graph the equation *x* = 2 by first completing the following table:

x	*y*
	0
	1
	−1

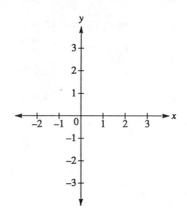

The equation specifies that *x* is always 2, no matter what *y* is. Therefore, the line passes through *all* points with the *x*-coordinate equal to 2. The line *x* = 2 is vertical.

The slope of a vertical line is undefined. The equation *x* = 2 cannot be written in slope-intercept form.

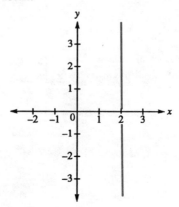

Example 1   Graph the equation $y = -1$.

Solution.

x	y
0	−1
1	−1
−1	−1

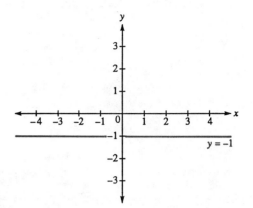

y is always equal to −1.

Example 2   Graph the equation $x = -3$.

Solution.

x	y
−3	−1
−3	0
−3	1

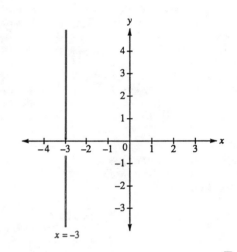

x is always equal to −3.

MINI-PRACTICE
(Use this space for your work.)

Graph the following equations on the coordinate system provided:

(a) $x = 1$ (b) $y = 2$ (c) $x = -1$ (d) $y = 0$

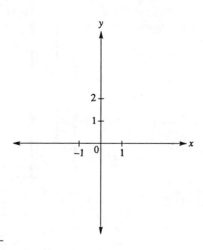

Answer to Mini-Practice
is on page 340.

IN YOUR OWN WORDS   describe the equations of horizontal and vertical lines: _____

_____

_____

Answer:

See the following rule.

Rule:

1. **A horizontal line has an equation of the form**

   $$y = a$$

   where a is a constant.

2. **A vertical line has an equation of the form**

   $$x = a$$

   where a is a constant.

Practice Exercises

Graph the following equations.

1. $x = 5$
2. $x = -2$
3. $y = 1$
4. $y = -2$
5. $x = 4$
6. $x = -5$
7. $y = 4$
8. $y = -5$
9. $x = 20$
10. $y = -50$

Answers to Mini-Practice:

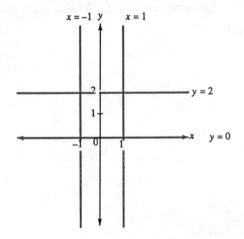

SUMMARY

1. A linear equation in two variables x and y is an equation of the form

$$ax + by = c$$

where not both a and b are zero.

2. A linear equation whose graph is not a vertical line can be written in slope-intercept form

$$y = mx + b$$

where m is the slope and b the y-intercept of the line.

3. To graph a linear equation: select two or three values for x or y. Then substitute these values in the equation and calculate the corresponding values of the other variable. Plot these points on a coordinate system and connect the points with a straight line.

4. To graph a parabola, it is best to graph five points. Select values for x or y. If more than squaring is involved, you may want to use another column to find the value of the other variable. Plot these points on a coordinate system and connect the points by means of a parabola.

Review Practice Exercises

In Exercises 1–4, write formulas for the following relationships.

1. Twice a number minus three times another number is six.   *Answer:* _____

2. One number plus four times another number equals eight.   *Answer:* _____

3. One number is three more than the opposite of another number.   *Answer:* _____

4. When one-half of a number is added to one-quarter of another number, the result is twelve.

   Answer: _____

In Exercises 5 and 6, write formulas for the relationships between the following ordered pairs.

5. $(4, 3)$, $(5, 4)$, $(6, 5)$, $(7, 6)$, $(8, 7)$

   Answer: _____

6. $(-4, -5)$, $(-3, -4)$, $(-2, -3)$, $(4, 3)$, $(5, 4)$

   Answer: _____

In Exercises 7 and 8, complete the tables.

7. $y = 2x$

x	y	(x, y)
0		
1		
2		
	6	
	10	

8. $y = 3x - 2$

x	y	(x, y)
0		
	1	
	-5	
2		
3		

In Exercises 9–15, find the x- and y-intercepts of the following lines. Also, find the slope of each line.

9. $y = 2x + 2$ Answer: _____

10. $y = 3x + 12$ Answer: _____

11. $3x - 2y = 6$ Answer: _____

12. $6x - 4y = 12$ Answer: _____

13. $y = \frac{1}{2}x + 1$ Answer: _____

14. $3x - y = 1$ Answer: _____

15. $2x + 3y = 2$ Answer: _____

In Exercises 16–44, graph the following equations.

16. $x + 3y = 15$ 17. $5x - 2y = 10$ 18. $5x + 2y = 10$ 19. $3x + 7y = 21$

20. $3x - 7y = 21$ 21. $y = -1 - 2x$ 22. $y = -2x + 2$ 23. $y = -2x - 2$

24. $x - 2y = 1$ 25. $2x - 3y = 6$ 26. $2x + y = 1$ 27. $2x = y + 4$

28. $3x - 2y = -3$ 29. $x - y + 5 = 0$ 30. $5 - y + 5x = 0$ 31. $x - y = 0$

32. $5 + 2y - x = 0$ 33. $2x - \frac{y}{3} = 10$ 34. $8 - 4x = \frac{y}{2}$ 35. $\frac{x}{3} + y = \frac{1}{2}$

36. $x = -4$ 37. $x = 6$ 38. $y = -3$ 39. $y = -6$

40. $y = x^2$ 41. $x = y^2$ 42. $y = x^2 + 3$ 43. $y = x^2 - 3$

44. $y = 2x^2$

If you have questions about this chapter, write them here: _____

_____   _____

_____

_____

_____

_____

APPENDIX

Review of Arithmetic

APPENDIX

Review of Arithmetic

A.1 ADDITION AND SUBTRACTION

The decimal system is built on place values. By looking at any whole number, such as 123, we know that the "3" has a value of three *units* (or *ones*), the "2" a value of two *tens*, or twenty, and the "1" is worth one *hundred*. By putting a period at the end of the number, we state that this is the end of the whole number. The places after the decimal point also have values and names:

thousands I hundreds I tens I ones I . I tenths I hundredths I thousandths

1 thousand = 10 hundreds

1 hundred = 10 tens

1 ten = 10 ones

1 one = 10 tenths

1 tenth = 10 hundredths

1 hundredth = 10 thousandths

Both whole numbers and decimals are added by lining up the numbers with the same place values in the same column. For example, add 23 and 140:

$$
\begin{array}{r}
23 \\
+\ 140 \\
\hline
163
\end{array}
$$

Example 1   Add: 2.3 + 14.0

Solution.
$$
\begin{array}{r}
2.3 \\
+\ 14.0 \\
\hline
16.3
\end{array}
$$
∎

Example 2   Add: 0.12 + 1.87

Solution.   0.12   (The zero is not necessary.)
                  + 1.87
                  ———
                    1.99   ■

Example 3   Add: 2 + 0.3 + 1.45

Solution.   2.   (The decimal point is after the 2.)
                    0.3
                  + 1.45
                  ———
                    3.75   ■

Example 4   Subtract: 12.56 – 3.42

Solution.   12.56
                  – 3.42
                  ———
                    9.14   ■

Example 5   Subtract: 5 – 1.7

Solution.   5.0   (5 equals 5.0.)
                  – 1.7
                  ———
                    3.3   ■

Example 6   Subtract: 11.4 – 5

Solution.   11.4
                  – 5.0
                  ———
                    6.4   ■

Practice Exercises

In Exercises 1–10, add.

1. 1.5 + 0.24   Answer: _____   2. 16.83 + 3.25   Answer: _____

3. 11.28 + 0.72   Answer: _____   4. 6.76 + 4   Answer: _____

5. 0.2 + 2 + 2.22 Answer: _____   6. 3.2 + 1.56 + 4 Answer: _____

7. 5.3 + 0.78 + 1 Answer: _____   8. 6.13 + 2.5 + 0.01 Answer: _____

9. 8.43 + 0.2 + 5 Answer: _____   10. 9 + 0.25 + 3.7 Answer: _____

In Exercises 11–20, subtract.

11. 11.55 – 2.8   Answer: _____   12. 3 – 1.7   Answer: _____

13. 4.2 – 2   Answer: _____   14. 5.4 – 0.7   Answer: _____

15. 0.011 – 0.009 Answer: _____   16. 3.5 – 2.54   Answer: _____

17. $5.9 - 3.2$ *Answer:* _____

18. $9.32 - 0.12$ *Answer:* _____

19. $8.7 - 5.9$ *Answer:* _____

20. $3.79 - 0.124$ *Answer:* _____

A.2 MULTIPLICATION AND DIVISION

Multiplication   Two whole numbers can be multiplied in column form.

Example 1   Multiply: 24×5

    Solution.

$$\begin{array}{r} 24 \\ \times 5 \\ \hline 120 \end{array}$$
■

Example 2   Multiply: 45×32

    Solution.

$$\begin{array}{r} 45 \\ \times 32 \\ \hline 90 \\ + 135 \\ \hline 1440 \end{array}$$
■

Example 3   Multiply: 103×67

    Solution.

$$\begin{array}{r} 103 \\ \times 67 \\ \hline 721 \\ + 618 \\ \hline 6901 \end{array}$$
■

Every time we multiply, we put the result in the correct place. For example: $7 \times 3 = 21$, but $60 \times 3 = 180$; we shift one step to the left. In Example 3, the multiplication could have been carried out in the reverse order:

$$\begin{array}{r} 67 \\ \times 103 \\ \hline 201 \\ 000 \\ + 67 \\ \hline 6901 \end{array}$$

    000   Move one step to the left.
    + 67   Move another step to the left.

The step involving adding 000 could have been skipped, but then we would have to move two steps to the left.

Decimals are multiplied as whole numbers. We first ignore the decimal points.

Example 4   Multiply: (a) 0.5×24;   (b) 0.5×2.4;   (c) 0.5×0.24

Solution.   First consider:

$$5 \times 24 = 120$$

(a) When multiplying 0.5×24, we have one decimal place among the factors and need one decimal in the product:

$$0.5 \times 24 = 12.0 = 12$$

(b) Next consider

$$0.5 \times 2.4$$

First we multiply:

$$5 \times 24 = 120$$

We have two decimal places among the factors and need two decimal places in the resulting product.

$$0.5 \times 2.4 = 1.20$$

The final decimal place 0 can be eliminated, so

$$0.5 \times 2.4 = 1.2$$

(c) Similarly,

$$0.5 \times 0.24 = 0.120 = 0.12$$

3 decimal places   3 decimal places   ∎

Example 5   Multiply: 1.56×2.95

Solution. $156 \times 295 = 46,020$

We have $2 + 2$ or 4 decimal places among the factors; thus,

$$1.56 \times 2.95 = 4.6020 = 4.602$$ ∎

Example 6   Multiply: 0.039×3.61

Solution. $39 \times 361 = 14,079$

$$0.039 \times 3.61 = 0.14079$$

5 decimal places   5 decimal places   ∎

Division   When dividing by a one-placed integer, the following format can be used.

Example 7   Divide: $8451 \div 9$

Solution.

$$9 \overline{\smash{\big)}\, 8451} \quad \overset{939}{}$$

Here,

$$84 \div 9 = 9; \text{ carry the remainder } 3$$
$$35 \div 9 = 3; \text{ carry the remainder } 8$$
$$81 \div 9 = 9$$

■

Example 8   Divide: $6.18 \div 0.6$

Solution.   To obtain an integer divisor, multiply both numbers by 10. Thus, move both decimal points one place to the right.

$$6.18 \div 0.6 = 61.8 \div 6$$

$$6 \overline{\smash{\big)}\, 61.8} \quad \overset{10.3}{}$$

■

Example 9   Divide: (a) $5 \div 0.2$ (b) $0.5 \div 2$

Solution.

(a) $5 \div 0.2 = 50 \div 2 = 25$

(b) $2 \overline{\smash{\big)}\, 0.50} \quad \overset{0.25}{}$

Here we inserted a 0 to the right of 5. Note that

$$0.5 = 0.50$$

■

Long Division   Long division is performed in the following way:

Example 10   Divide: $7868 \div 14$

Solution.

$$
\begin{array}{r}
562 \\
14 \overline{\smash{\big)}\, 7868} \\
-70 \\
\hline
86 \\
-84 \\
\hline
28 \\
-28 \\
\hline
0
\end{array}
$$

14 goes into 78 five times.
$5 \times 14 = 70$. Subtract; bring down the 6.
14 goes into 86 six times.

14 goes into 28 two times.

■

With decimals we do the same thing:

Example 11   Divide: (a) $78.68 \div 14$ (b) $7.868 \div 1.4$

Solution.

(a) Mark the decimal point and divide as in Example 10.

$$\frac{5.62}{14\,\overline{)\,78.68}}$$

(b) To obtain an integer divisor, multiply both numbers by 10. Thus, move decimal points one place to the right:

$$78.68 \div 14 = 5.62$$

as before.   ∎

Practice Exercises

In Exercises 1–10, multiply.

1. 3.67×8.42 Answer: _____

2. 0.051×16.3 Answer: _____

3. 0.02×0.04 Answer: _____

4. 1000×0.0045 Answer: _____

5. $52,000 \times 0.0025$ Answer: _____

6. 4.5×2.1 Answer: _____

7. 0.12×3 Answer: _____

8. 6.1×0.001 Answer: _____

9. 7.2×10.2 Answer: _____

10. 8.3×0.34 Answer: _____

In Exercises 11–20, divide.

11. $1.32 \div 0.03$ Answer: _____

12. $5 \div 0.0005$ Answer: _____

13. $7.8 \div 0.65$ Answer: _____

14. $10.5 \div 2.1$ Answer: _____

15. $0.39 \div 13$ Answer: _____

16. $6 \div 0.2$ Answer: _____

17. $0.55 \div 11$ Answer: _____

18. $8.6 \div 0.01$ Answer: _____

19. $9.3 \div 3$ Answer: _____

20. $10.5 \div 0.5$ Answer: _____

A.3 ROUNDING

Whole numbers and decimals often have to be rounded to a specified place. We use the most common way of rounding numbers:

1. Identify the given place value.

2. Look at the number to the right of the given place value:
   (a) if that number is less than 5, drop it;
   (b) if the number is 5 or more, add 1 to the given place value.

Whole Numbers   Round 2586 to the nearest hundred:

> 5 is in the hundred's place. The next number is 8 (more than 5). Add 1 to 5 and show the place values with zeroes.

$$2586$$
$$+ \ 1$$
$$2600$$

Thus 2586 rounds to 2600 (to the nearest hundred).

Example 1   Round 9526 to the nearest (a) ten, (b) hundred, (c) thousand.

Solution.

(a) ten: 2 is in the ten's place.
       6 is more than 5.   *Answer:* 9530

(b) hundred: 5 is in the hundred's place.
         2 is less than 5.   *Answer:* 9500

(c) thousand: 9 is in the thousand's place.
          5 is equal to 5.   *Answer:* 10,000   ■

Decimals   Decimals are rounded as are whole numbers, but to show place values, zeros are not usually required.

Example 2   Round 26.4395 to the nearest (a) tenth, (b) hundredth, (c) thousandth.

Solution.

(a) tenth: 4 is in the tenth's place.
        3 is less than 5.   *Answer:* 26.4

(b) hundredth: 3 is in the hundredth's place.
           9 is more than 5.   *Answer:* 26.44

(c) thousandth: 9 is in the thousandth's place.
            5 is equal to 5.   *Answer:* 26.440
     (9 + 1 = 10; here, we need the zero since we were required to round to the thousandth's place.)   ■

Practice Exercises

In Exercises 1–10, round each whole number to the indicated place.

1. 3784; thousand   *Answer:* _____   2. 10,369; ten thousand   *Answer:* _____

3. 45; ten   *Answer:* _____   4. 397,974,238; million   *Answer:* _____

5. 43,956; hundred   *Answer:* _____   6. 1250; ten   *Answer:* _____

7. 35,825; hundred   *Answer:* _____

8. 135,025; thousand   *Answer:* _____

9. 2,456,278; ten thousand *Answer:* _____

10. 3,406,003; hundred thousand *Answer:* _____

In Exercises 11–20, round each decimal to the indicated place.

11. 42.864; hundredth   *Answer:* _____

12. 0.967; tenth   *Answer:* _____

13. 1.529; hundredth   *Answer:* _____

14. 3.958; tenth   *Answer:* _____

15. 15.63; whole number   *Answer:* _____

16. 6.007; whole number   *Answer:* _____

17. 35.8906; hundredth   *Answer:* _____

18. 0.396539; thousandth   *Answer:* _____

19. 5937.050798; ten thousandth   *Answer:* _____

20. 0.05723075; hundred thousandth   *Answer:* _____

A.4 COMMON FRACTIONS

Vocabulary   A common fraction contains a fraction bar. The number above the fraction bar is the *numerator* and the number below the bar is the *denominator.* For example, in $\frac{3}{4}$, 3 is the numerator and 4 is the denominator.

Instead of the fraction bar one may use a slash, as in 3/4. Both the bar and the slash indicate division. For example, the fraction $\frac{6}{3}$, or 6/3, equals 2.

A fraction in which the numerator is smaller than the denominator is called *proper* and a fraction where the numerator is greater than or equal to the denominator is *improper.*

Examples of proper fractions:

$$\frac{1}{2}, \frac{2}{3}, \frac{5}{8}, \frac{21}{25}, \frac{100}{101}$$

Examples of improper fractions:

$$\frac{2}{1}, \frac{5}{2}, \frac{11}{9}, \frac{17}{3}, \frac{4}{4}, \frac{101}{100}$$

An improper fraction can be changed into an integer or into a *mixed number*, that is, a whole number plus a fraction. In the improper fraction $\frac{17}{3}$, we see that 3 goes into 17 five times. Since $3 \times 5 = 15$, we have 2 left over. In other words: $\frac{17}{3} = 5\frac{2}{3}$.(Read five *and* two thirds.)

Examples of improper fractions changed into mixed numbers:

$$\frac{11}{4} = 2\frac{3}{4}; \qquad \frac{34}{11} = 3\frac{1}{11}; \qquad \frac{82}{9} = 9\frac{1}{9}$$

A mixed number can be changed into an improper fraction in the following way. Multiply the whole number by the denominator, then add the numerator and keep the same denominator. For example, for $3\frac{5}{6}$:

$$3 \times 6 = 18$$
$$18 + 5 = 23$$
$$3\frac{5}{6} = \frac{23}{6}$$

Example 1   Change $1\frac{1}{2}$ to an improper fraction.

Solution.

$$1\frac{1}{2} = \frac{(2 \times 1) + 1}{2} = \frac{2 + 1}{2} = \frac{3}{2}$$ ∎

Example 2   Change $4\frac{3}{5}$ to an improper fraction.

Solution.

$$4\frac{3}{5} = \frac{(5 \times 4) + 3}{5} = \frac{20 + 3}{5} = \frac{23}{5}$$ ∎

Example 3   Change $11\frac{3}{4}$ to an improper fraction.

Solution.

$$11\frac{3}{4} = \frac{(4 \times 11) + 3}{4} = \frac{44 + 3}{4} = \frac{47}{4}$$ ∎

Equivalent Fractions   When we cut a cake into two equal pieces, each piece is half a cake. Each half can be cut into two quarters and each quarter into two eighths. $\frac{1}{2}$ is the same as $\frac{2}{4}$ and $\frac{1}{4}$ is the same as $\frac{2}{8}$. We say that

$$\frac{1}{2} \text{ is } \textit{equivalent} \text{ to } \frac{2}{4}$$

$$\frac{1}{4} \text{ is } \textit{equivalent} \text{ to } \frac{2}{8}$$

We can change any fraction to an equivalent fraction by either multiplying or dividing both the numerator and the denominator by the same nonzero number. For example,

$$\frac{1}{4} = \frac{2}{8} = \frac{3}{12} = \frac{4}{16} = \frac{5}{20}$$

$$\frac{12}{16} = \frac{24}{32} = \frac{36}{48} = \frac{6}{8} = \frac{3}{4}$$

A whole number, n, can be written with n as the numerator and 1 as the denominator. For example,

$$4 = \frac{4}{1}$$

When the numerator and denominator of a fraction are divided by the same number, we say that we *reduce* the fraction. For example, $\frac{16}{24}$ can be reduced as follows:

$$\frac{16}{24} = \frac{8}{12} = \frac{4}{6} = \frac{2}{3}$$

When it is impossible to reduce further, as in $\frac{2}{3}$, the fraction is said to be *reduced to lowest terms.*

Examples of fractions reduced to lowest terms:

$$\frac{6}{9} = \frac{2}{3}; \qquad \frac{12}{16} = \frac{3}{4}; \qquad \frac{-12}{18} = \frac{-2}{3}; \qquad \frac{24}{40} = \frac{3}{5}$$

It is often necessary in mathematics to change a fraction into an equivalent fraction with a given denominator. To do this, we have to find what number the old denominator has to be multiplied by to change it into the new denominator.

For example, to change $\frac{2}{3}$ into twenty-fourths, we realize that 3 goes into 24 eight times and

$$\frac{2}{3} = \frac{16}{24} \quad \text{because} \quad \frac{2(8)}{3(8)} = \frac{16}{24}$$

Example 4   Change $\frac{4}{5}$ to an equivalent fraction with denominator 15.

Solution. $$\frac{4}{5} = \frac{?}{15}$$

$$5 \times 3 = 15; \qquad 4 \times 3 = 12$$

$$\frac{4}{5} = \frac{12}{15}$$ ∎

Practice Exercises

In Exercises 1–5, change into mixed numbers.

1. $\frac{20}{3}$ Answer: _____   2. $\frac{18}{11}$ Answer: _____   3. $\frac{13}{4}$ Answer: _____

4. $\frac{21}{10}$ Answer: _____   5. $\frac{37}{7}$ Answer: _____

In Exercises 6–10, change into improper fractions.

6. $1\frac{4}{5}$ Answer: _____   7. $2\frac{3}{4}$ Answer: _____   8. $4\frac{1}{2}$ Answer: _____

9. $1\frac{9}{11}$ *Answer:* _____   10. $6\frac{2}{3}$ *Answer:* _____

In Exercises 11–15, change to equivalent fractions with the indicated denominators.

11. $\frac{3}{4} = \frac{?}{12}$ *Answer:* _____   12. $\frac{5}{11} = \frac{?}{33}$ *Answer:* _____   13. $\frac{7}{9} = \frac{?}{72}$ *Answer:* _____

14. $\frac{3}{16} = \frac{?}{48}$ *Answer:* _____   15. $\frac{11}{13} = \frac{?}{65}$ *Answer:* _____

In Exercises 16–20, reduce to lowest terms.

16. $\frac{15}{25}$ *Answer:* _____   17. $\frac{24}{36}$ *Answer:* _____   18. $\frac{39}{65}$ *Answer:* _____

19. $\frac{11}{143}$ *Answer:* _____   20. $\frac{49}{91}$ *Answer:* _____

A.5 DECIMALS AND FRACTIONS

Conversions from Fractions to Decimals

The fraction bar in a common fraction indicates division; to change a common fraction into a decimal, we work out that division. For example,

$$\frac{1}{2} = 1 \div 2$$

Thus, divide 1 by 2:

$$2\overline{\smash{\big)}\,1.0}^{.5}$$

Therefore, $\frac{1}{2} = 0.5$.

Example 1   Convert $\frac{5}{6}$ to a decimal.

Solution.

$$\frac{5}{6} = 5 \div 6 = 0.83333333\ldots$$

The 3's repeat indefinitely, so we round to a suitable place. For example, to the hundredth's place:

$$\frac{5}{6} = 0.83$$ ∎

Example 2   Convert $1\frac{4}{5}$ to a decimal.

Solution. $1\frac{4}{5} = 1.8$

The whole number does not change and $4 \div 5 = 0.8$. ∎

Conversions from Decimals to Fractions

A decimal is converted into a fraction in the following way: 0.12 is read "twelve hundredths," which is $\frac{12}{100}$. This can be reduced by dividing numerator and denominator by 4:

$$0.12 = \frac{12}{100} = \frac{3}{25}$$

1.125 is read "one and one hundred twenty-five thousandths."

$$1.125 = 1\frac{125}{1000} = 1\frac{5}{40} = 1\frac{1}{8} = \frac{9}{8}$$

Example 3   Convert 0.7 to a fraction.

Solution. $0.7 = \frac{7}{10}$ ∎

Example 4   Convert 6.4 to a fraction in lowest terms.

Solution. $6.4 = 6\frac{4}{10} = 6\frac{2}{5} = \frac{32}{5}$ ∎

Example 5   Convert 2.04 to a fraction in lowest terms.

Solution. $2.04 = 2\frac{4}{100} = 2\frac{1}{25} = \frac{51}{25}$ ∎

Practice Exercises

In Exercises 1–10, convert each fraction to a decimal. (Round to the nearest hundredth, if the division does not come out evenly.)

1. $\frac{2}{5}$ Answer: _____

2. $\frac{3}{4}$ Answer: _____

3. $1\frac{2}{5}$ Answer: _____

4. $\frac{4}{7}$ Answer: _____

5. $\frac{13}{11}$ Answer: _____

6. $\frac{3}{5}$ Answer: _____

7. $3\frac{1}{100}$ Answer: _____

8. $\frac{3}{8}$ Answer: _____

9. $2\frac{9}{10}$ Answer: _____

10. $\frac{3}{20}$ Answer: _____

In Exercises 11–20, convert each decimal to a fraction. (Reduce to lowest terms.)

11. 0.45   Answer: _____

12. 1.5   Answer: _____

13. 2.03   Answer: _____

14. 0.625   Answer: _____

15. 0.006   Answer: _____

16. 0.25   Answer: _____

17. 1.75   Answer: _____

18. 0.125   Answer: _____

19. 0.4   Answer: _____

20. 2.8   Answer: _____

A.6 ADDITION AND SUBTRACTION OF FRACTIONS

Addition: Same Denominator

When two of a kind is added to five of the same kind, we get seven of that kind. (For instance, 2 apples added to 5 apples give you 7 apples.)

In case "the kind" is *ninths*, for example, we have

$$\frac{2}{9} + \frac{5}{9} = \frac{7}{9}$$

With common fractions, as long as the denominators are the same, we add the numerators.

Here are examples of addition of fractions in which the denominators are the same:

$$\frac{2}{3} + \frac{1}{3} = \frac{3}{3} = 1$$

$$\frac{4}{11} + \frac{5}{11} = \frac{9}{11}$$

$$\frac{7}{25} + \frac{28}{25} = \frac{35}{25} = \frac{7}{5}$$

Addition of Mixed Numbers

Consider mixed numbers in which the fractional parts have the same denominators. We add the whole number parts separately and the proper fractions separately. For example:

$$3\frac{2}{5} + 4\frac{3}{5} = 3 + 4 + \frac{2+3}{5} = 7\frac{5}{5} = 7 + 1 = 8$$

Example 1   Add $1\frac{1}{2} + 3\frac{1}{2}$.

Solution. $1\frac{1}{2} + 3\frac{1}{2} = 4\frac{1+1}{2} = 4\frac{2}{2} = 4 + 1 = 5$ ■

Example 2   Add $5\frac{4}{5} + 11\frac{3}{5}$.

Solution. $5\frac{4}{5} + 11\frac{3}{5} = 16\frac{7}{5} = 16 + 1 + \frac{2}{5} = 17\frac{2}{5}$ ■

Addition: Different Denominators

When two fractions with different denominators are to be added, we must first find equivalent fractions with the same denominator. For example, consider

$$\frac{2}{5} + \frac{3}{4}$$

The denominators of the equivalent fractions could be 20, 40, 60, etc.; 20 is the smallest "common denominator" and therefore the easiest number to use:

$$\frac{2}{5} = \frac{8}{20}$$

$$\frac{3}{4} = \frac{15}{20}$$

Add: $\frac{8}{20} + \frac{15}{20} = \frac{23}{20}$

Example 3   Add $\frac{1}{2} + \frac{1}{3}$.

Solution. $\frac{1}{2} + \frac{1}{3}$: $\frac{1}{2} = \frac{3}{6}$

$$\frac{1}{3} = \frac{2}{6}$$

Add: $\frac{3}{6} + \frac{2}{6} = \frac{5}{6}$ ∎

Example 4   Add $\frac{3}{4} + \frac{5}{8}$.

Solution. $\frac{3}{4} + \frac{5}{8}$: $\frac{3}{4} = \frac{6}{8}$

$$\frac{5}{8} = \frac{5}{8}$$

Add: $\frac{6}{8} + \frac{5}{8} = \frac{11}{8}$ ∎

Example 5   Add $5\frac{3}{4} + 4\frac{1}{6}$.

Solution. $5\frac{3}{4} + 4\frac{1}{6}$: $5\frac{3}{4} = 5\frac{9}{12}$

$$4\frac{1}{6} = 4\frac{2}{12}$$

$$9\frac{11}{12}$$ ∎

Subtraction without Borrowing   Subtraction is done in a similar way to addition: we subtract the second numerator from the first. If the denominators are not alike, we must first find equivalent fractions with common denominators. For example,

$$\frac{3}{4} - \frac{2}{4} = \frac{1}{4}$$

Example 6   Subtract $\frac{1}{2} - \frac{3}{8}$.

Solution.

$$\frac{1}{2} - \frac{3}{8}: \qquad \frac{1}{2} = \frac{4}{8}$$

$$-\frac{3}{8} = -\frac{3}{8}$$

Subtract: $\dfrac{4}{8} - \dfrac{3}{8} = \dfrac{1}{8}$ ∎

Example 7   Subtract $\frac{8}{15} - \frac{11}{25}$.

Solution.

$$\frac{8}{15} - \frac{11}{25}: \qquad \frac{8}{15} = \frac{40}{75}$$

$$-\frac{11}{25} = -\frac{33}{75}$$

Subtract: $\dfrac{40}{75} - \dfrac{33}{75} = \dfrac{7}{75}$ ∎

Subtraction with Borrowing

When at least one of the numbers is a mixed number, we cannot always subtract immediately. For example, suppose both fractional parts have the same denominator, but the second numerator is larger than the first, as in the case of

$$4\frac{2}{7} - \frac{5}{7}$$

Here are two choices:

The first method is to make each mixed number into an improper fraction and then subtract. For example:

$$4\frac{2}{7} = \frac{30}{7}$$

$$-\frac{5}{7}$$

$$\frac{30}{7} - \frac{5}{7} = \frac{25}{7} \text{ or } 3\frac{4}{7}$$

In the second method change $4\frac{2}{7}$ to $3\frac{9}{7}$. Here we "borrow" 1 from 4, changing the 4 to a 3; $1 = \frac{7}{7}$ and this is added to $\frac{2}{7}$ to obtain $\frac{9}{7}$:

$$3\frac{9}{7}$$

$$-\frac{5}{7}$$

$$\overline{\quad 3\frac{4}{7}\quad}$$

Example 8   Subtract $1\frac{1}{4} - \frac{3}{4}$.

Solution.   We use Method 1.

$$1\frac{1}{4} = \frac{5}{4}$$
$$-\frac{3}{4} = -\frac{3}{4}$$

Subtract: $\frac{5}{4} - \frac{3}{4} = \frac{2}{4} = \frac{1}{2}$ ∎

Example 9   Subtract $2\frac{3}{5} - \frac{4}{5}$.

Solution.

Method 1:

$$2\frac{3}{5} = \frac{13}{5}$$
$$-\frac{4}{5} = -\frac{4}{5}$$

Subtract: $\frac{13}{5} - \frac{4}{5} = \frac{9}{5}$ or $1\frac{4}{5}$

Method 2:

$$2\frac{3}{5} = 1\frac{8}{5}$$
$$-\frac{4}{5} = -\frac{4}{5}$$
$$1\frac{4}{5}$$ ∎

Example 10   Subtract $7\frac{1}{15} - 4\frac{7}{10}$.

Solution.   We use Method 2.

$$7\frac{1}{15} = 7\frac{2}{30} = 6\frac{32}{30}$$
$$-4\frac{7}{10} = \qquad -4\frac{21}{30}$$
$$2\frac{11}{30}$$ ∎

Example 11   Subtract $3 - \frac{4}{7}$.

Solution.

Method 1:

$$3 = \frac{21}{7}$$
$$-\frac{4}{7}$$

Subtract: $\frac{21}{7} - \frac{4}{7} = \frac{17}{7}$ or $2\frac{3}{7}$

Method 2:

$$3 = 2\frac{7}{7}$$
$$-\frac{4}{7}$$
$$2\frac{3}{7}$$ ∎

Practice Exercises

In Exercises 1–5, add.

1. $\frac{3}{7} + \frac{4}{7}$ *Answer:* _____

2. $\frac{2}{9} + \frac{1}{9}$ *Answer:* _____

3. $\frac{7}{8} + \frac{5}{8}$ *Answer:* _____

4. $2\frac{3}{5} + 3\frac{1}{5}$ *Answer:* _____

5. $4\frac{5}{6} + 2\frac{1}{6}$ *Answer:* _____

In Exercises 6–10, add.

6. $\frac{2}{5} + \frac{2}{7}$ *Answer:* _____   **7.** $\frac{5}{12} + \frac{7}{16}$ *Answer:* _____   **8.** $\frac{4}{9} + \frac{11}{12}$ *Answer:* _____

9. $8\frac{2}{3} + 1\frac{7}{12}$ *Answer:* _____   **10.** $4\frac{7}{10} + 12\frac{11}{15}$ *Answer:* _____

In Exercises 11–15, subtract.

11. $\frac{21}{23} - \frac{10}{23}$ *Answer:* _____   **12.** $\frac{11}{15} - \frac{7}{15}$ *Answer:* _____   **13.** $2\frac{3}{4} - 1\frac{1}{4}$ *Answer:* _____

14. $\frac{5}{6} - \frac{2}{3}$ *Answer:* _____   **15.** $4\frac{5}{7} - 3\frac{3}{14}$ *Answer:* _____

In Exercises 16–20, subtract.

16. $1\frac{1}{4} - \frac{3}{4}$ *Answer:* _____   **17.** $3\frac{2}{5} - 1\frac{4}{5}$ *Answer:* _____

18. $12\frac{9}{32} - 10\frac{13}{32}$ *Answer:* _____   **19.** $4\frac{2}{15} - 3\frac{5}{12}$ *Answer:* _____

20. $3 - 1\frac{3}{5}$ *Answer:* _____

A.7 MULTIPLICATION AND DIVISION OF FRACTIONS

Multiplication of Proper Fractions

When two fractions are multiplied, the numerators are multiplied separately and the denominators are multiplied separately:

$$\frac{1}{2} \times \frac{1}{2} = \frac{1 \times 1}{2 \times 2} = \frac{1}{4}$$

$$\frac{5}{7} \times \frac{3}{10} = \frac{5 \times 3}{7 \times 10} = \frac{15}{70} = \frac{3}{14}$$

It is best to reduce to lowest terms *before* multiplying:

$$\frac{5}{7} \times \frac{3}{10} = \frac{\overset{1}{\cancel{5}} \times 3}{7 \times \cancel{10}} = \frac{3}{7 \times 2} = \frac{3}{14}$$

Example 1   Multiply $\frac{3}{4} \times \frac{8}{15}$.

Solution. $\frac{3}{4} \times \frac{8}{15} = \frac{\overset{1}{\cancel{3}} \times \overset{2}{\cancel{8}}}{\underset{1}{\cancel{4}} \times \underset{5}{\cancel{15}}} = \frac{2}{5}$ ■

Example 2   Multiply $\frac{10}{11} \times \frac{22}{25}$.

Solution. $\frac{10}{11} \times \frac{22}{25} = \frac{\overset{2}{\cancel{10}} \times \overset{2}{\cancel{22}}}{\underset{1}{\cancel{11}} \times \underset{5}{\cancel{25}}} = \frac{4}{5}$ ■

Example 3   Multiply $\frac{4}{9} \times \frac{3}{10} \times \frac{5}{8}$.

Solution. $\frac{4}{9} \times \frac{3}{10} \times \frac{5}{8} = \frac{\overset{1}{4} \times \overset{1}{3} \times \overset{1}{5}}{\underset{3}{9} \times \underset{2}{10} \times \underset{2}{8}} = \frac{1}{12}$ ∎

Multiplication of Mixed Numbers

All mixed numbers should be changed into improper fractions before they are multiplied.

$$1\frac{3}{4} \times 2\frac{2}{7} = \frac{7}{4} \times \frac{16}{7} = \frac{\overset{1}{7} \times \overset{4}{16}}{\underset{1}{4} \times \underset{1}{7}} = 4$$

Example 4   Multiply $1\frac{1}{2} \times 2\frac{1}{4}$.

Solution. $1\frac{1}{2} \times 2\frac{1}{4} = \frac{3}{2} \times \frac{9}{4} = \frac{3 \times 9}{2 \times 4} = \frac{27}{8} = 3\frac{3}{8}$ ∎

Example 5   Multiply $4 \times \frac{3}{4}$.

Solution. $4 \times \frac{3}{4} = \frac{4}{1} \times \frac{3}{4} = \frac{\overset{1}{4} \times 3}{1 \times \underset{1}{4}} = \frac{3}{1} = 3$ ∎

Example 6   Multiply $1\frac{1}{3} \times \frac{3}{8}$.

Solution. $1\frac{1}{3} \times \frac{3}{8} = \frac{4}{3} \times \frac{3}{8} = \frac{\overset{1}{4} \times \overset{1}{3}}{\underset{1}{3} \times \underset{2}{8}} = \frac{1}{2}$ ∎

Remember to write the 1's in the numerator!

Division of Fractions

To divide is to multiply by the reciprocal:

$$6 \div 2 = 3 \quad \text{and} \quad 6 \times \frac{1}{2} = \frac{6}{2} = 3$$

Division of fractions is changed to multiplication by inverting the second fraction.

$$\frac{4}{5} \div \frac{3}{10} = \frac{4}{\underset{1}{5}} \times \frac{\overset{2}{10}}{3} = \frac{4 \times 2}{1 \times 3} = \frac{8}{3}$$

Example 7   Divide $\dfrac{4}{5} \div \dfrac{8}{15}$.

Solution. $\dfrac{4}{5} \div \dfrac{8}{15} = \dfrac{4}{5} \times \dfrac{15}{8} = \dfrac{\overset{1}{\cancel{4}} \times \overset{3}{\cancel{15}}}{\underset{1}{\cancel{5}} \times \underset{2}{\cancel{8}}} = \dfrac{3}{2}$ ■

Example 8   Divide $\dfrac{1}{2} \div \dfrac{1}{4}$.

Solution. $\dfrac{1}{2} \div \dfrac{1}{4} = \dfrac{1}{2} \times \dfrac{4}{1} = \dfrac{1 \times \overset{2}{\cancel{4}}}{\underset{1}{\cancel{2}} \times 1} = \dfrac{2}{1} = 2$ ■

Division of Mixed Numbers

As in multiplication, mixed numbers must be changed to improper fractions:

$$2\dfrac{4}{5} \div 1\dfrac{2}{5} = \dfrac{14}{5} \div \dfrac{7}{5} = \dfrac{\overset{2}{\cancel{14}}}{\underset{1}{\cancel{5}}} \times \dfrac{\overset{1}{\cancel{5}}}{\underset{1}{\cancel{7}}} = \dfrac{2}{1} = 2$$

Example 9   Divide $1\dfrac{4}{5} \div 2\dfrac{1}{2}$.

Solution. $1\dfrac{4}{5} \div 2\dfrac{1}{2} = \dfrac{9}{5} \div \dfrac{5}{2} = \dfrac{9}{5} \times \dfrac{2}{5} = \dfrac{18}{25}$ ■

Example 10   Divide $2 \div \dfrac{1}{2}$.

Solution. $2 \div \dfrac{1}{2} = \dfrac{2}{1} \div \dfrac{1}{2} = \dfrac{2}{1} \times \dfrac{2}{1} = \dfrac{4}{1} = 4$ ■

Example 11   Divide $\dfrac{1}{2} \div 2$.

Solution. $\dfrac{1}{2} \div 2 = \dfrac{1}{2} \div \dfrac{2}{1} = \dfrac{1}{2} \times \dfrac{1}{2} = \dfrac{1}{4}$ ■

Practice Exercises

In Exercises 1–5, multiply.

1. $\dfrac{2}{3} \times \dfrac{3}{4}$ *Answer:* _____

2. $\dfrac{5}{11} \times \dfrac{22}{10}$ *Answer:* _____

3. $\dfrac{4}{15} \times \dfrac{3}{10}$ *Answer:* _____

4. $\dfrac{8}{9} \times \dfrac{3}{2}$ *Answer:* _____

5. $\dfrac{4}{7} \times \dfrac{3}{8} \times \dfrac{14}{15}$ *Answer:* _____

In Exercises 6–10, multiply.

6. $1\dfrac{1}{2} \times 1\dfrac{1}{3}$ *Answer:* _____

7. $4 \times \dfrac{3}{4}$ *Answer:* _____

8. $2\dfrac{1}{5} \times 2\dfrac{3}{11}$ *Answer:* _____

9. $5\dfrac{2}{3} \times 1\dfrac{1}{17}$ *Answer:* _____

10. $1\dfrac{1}{7} \times \dfrac{3}{8} \times 1\dfrac{1}{6}$ *Answer:* _____

In Exercises 11–15, divide.

11. $\dfrac{4}{5} \div \dfrac{1}{10}$ *Answer:* _____

12. $\dfrac{5}{8} \div \dfrac{2}{5}$ *Answer:* _____

13. $\dfrac{3}{14} \div \dfrac{6}{7}$ *Answer:* _____

14. $\dfrac{7}{18} \div \dfrac{91}{9}$ *Answer:* _____

15. $\dfrac{16}{15} \div \dfrac{8}{45}$ *Answer:* _____

In Exercises 16–20, divide.

16. $3 \div \dfrac{1}{4}$ *Answer:* _____

17. $\dfrac{1}{5} \div 5$ *Answer:* _____

18. $2\dfrac{1}{4} \div 1\dfrac{1}{2}$ *Answer:* _____

19. $3\dfrac{1}{5} \div 1\dfrac{1}{15}$ *Answer:* _____

20. $11\dfrac{2}{3} \div 2\dfrac{1}{3}$ *Answer:* _____

A.8 PERCENT

Definition   The word percent means "divided by one hundred." This can also be expressed in terms of "hundredths." In other words, 50% equals 50 ÷ 100 or 0.50. The whole is equal to 100%.

Conversions   To change a number expressed as a percent, we remove the percent symbol and divide by 100. For example, 45% is the same as $\dfrac{45}{100}$, which equals .45 or $\dfrac{9}{20}$.

Example 1   Convert 15% to (a) a fraction, (b) a decimal.

Solution.

(a) $$15\% = \dfrac{15}{100} = \dfrac{3}{20}$$

(b) Remove the percent symbol and move the decimal point two steps to the left.

$$15\% = .15$$ ∎

Example 2   Convert $4\frac{1}{5}$ % to (a) a fraction, (b) a decimal.

Solution.

(a)
$$4\frac{1}{5}\% = \frac{21}{5} \div 100 = \frac{21}{5} \times \frac{1}{100} = \frac{21}{500}$$

(b)
$$4\frac{1}{5}\% = 4.2\% = 0.042 \qquad\blacksquare$$

To change a fraction or a decimal to a percent, we go the opposite way and multiply by 100%. For example,

$$\frac{1}{2} = \frac{1}{2} \times 100\% = \frac{100}{2}\% = 50\%$$

Alternatively,

$$\frac{1}{2} = 0.5 = 0.50 = 50\%$$

Example 3   Convert $\frac{4}{5}$ to a percent.

Solution.
$$\frac{4}{5} = \frac{4}{5} \times 100\% = \frac{400}{5}\% = 80\% \qquad\blacksquare$$

Example 4   Convert $\frac{1}{3}$ to a percent.

Solution.
$$\frac{1}{3} = \frac{1}{3} \times 100\% = \frac{100}{3}\% = 33\frac{1}{3}\% \qquad\blacksquare$$

Example 5   Convert $\frac{7}{20}$ to a percent.

Solution.
$$\frac{7}{20} = \frac{7}{20} \times 100\% = \frac{700}{20}\% = 35\% \qquad\blacksquare$$

Example 6   Convert 0.8 to a percent.

Solution.
$$0.8 = 0.80 = 80\% \qquad\blacksquare$$

Example 7   Convert 1.4 to a percent.

Solution.
$$1.4 = 1.40 = 140\% \qquad\blacksquare$$

Example 8   Convert 0.05 to a percent.

Solution. $0.05 = 5\%$ ■

Practice Exercises

In Exercises 1–5, convert each percent to a common fraction.

1. 15%   *Answer:* _____   2. 70%   *Answer:* _____   3. 150%   *Answer:* _____

4. $\frac{1}{2}\%$ *Answer:* _____   5. $2\frac{1}{4}\%$ *Answer:* _____

In Exercises 6–10, convert each percent to a decimal.

6. 200%   *Answer:* _____   7. 30%   *Answer:* _____   8. 1.4%   *Answer:* _____

9. 83.6%   *Answer:* _____   10. 0.05%   *Answer:* _____

In Exercises 11–15, convert each fraction to a percent.

11. $\frac{5}{8}$ *Answer:* _____   12. $\frac{7}{100}$ *Answer:* _____   13. $\frac{2}{5}$ *Answer:* _____

14. $\frac{12}{25}$ *Answer:* _____   15. $\frac{1}{500}$ *Answer:* _____

In Exercises 16–20, convert each decimal to a percent.

16. 0.02   *Answer:* _____   17. 1.7   *Answer:* _____   18. 0.546   *Answer:* _____

19. 0.045   *Answer:* _____   20. 3   *Answer:* _____

Answers

Chapter 1

Practice Exercises for Section 1.1, page 20

1. 15   **2.** –3   **3.** 85   **4.** 1000   **5.** 0.02   **6.** –1.2   **7.** $2\frac{1}{5}$ **8.** –0.25   **9.** 0.75   **10.** –1.5   **11.** 3.25

12. –2.75   **13.** $\frac{1}{2}$ **14.** $-\frac{3}{4}$ **15.** $\frac{1}{8}$ **16.** $-3\frac{1}{4}$ **17.** $2\frac{1}{2}$ **18.** 5   **19.** 2.3   **20.** 11   **21.** 11   **22.** 2

23. 0.5   **24.** 0.75   **25.** 2.1   **26.** 4.07   **27.** 5.33   **28.** $\frac{1}{2}$ **29.** $\frac{1}{4}$ **30.** $\frac{3}{4}$ **31.** $\frac{7}{8}$ **32.** $\frac{1}{8}$

33. 6   **34.** 6   **35.** 1   **36.** 0   **37.** 16   **38.** 11   **39.** 24   **40.** 3   **41.** 1.2   **42.** 11   **43.** 1.2

44. 1.1   **45.** 1.3   **46.** 1.4   **47.** 0.92   **48.** $\frac{1}{2}$ **49.** $\frac{5}{8}$ **50.** $\frac{9}{10}$ **51.** $\frac{1}{6}$ **52.** $\frac{2}{5}$ **53.** $\frac{4}{5}$ **54.** $\frac{1}{2}$

55. $\frac{2}{5}$ **56.** $\frac{3}{28}$ **57.** absolute value   **58.** opposite

Practice Exercises for Section 1.2, page 24

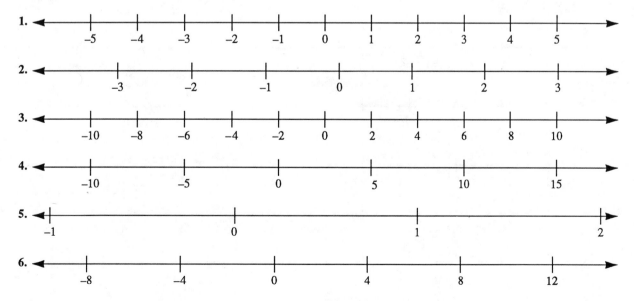

367

368   ANSWERS

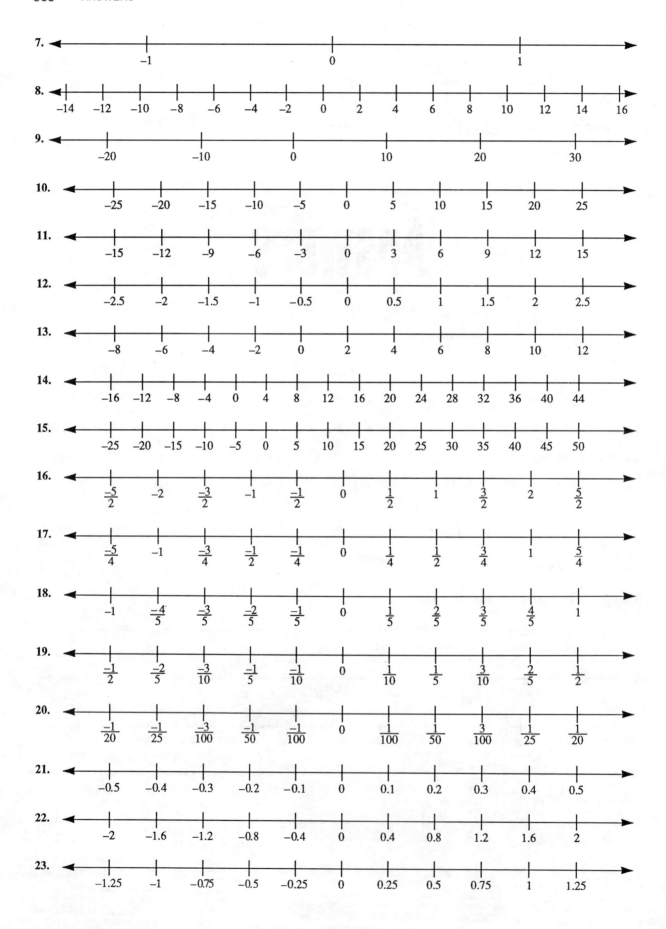

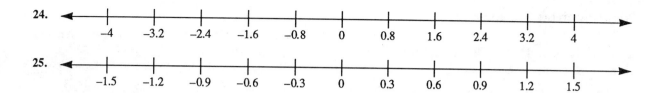

24.

$$-4 \quad -3.2 \quad -2.4 \quad -1.6 \quad -0.8 \quad 0 \quad 0.8 \quad 1.6 \quad 2.4 \quad 3.2 \quad 4$$

25.

$$-1.5 \quad -1.2 \quad -0.9 \quad -0.6 \quad -0.3 \quad 0 \quad 0.3 \quad 0.6 \quad 0.9 \quad 1.2 \quad 1.5$$

Practice Exercises for Section 1.3, page 27

1. < **2.** < **3.** > **4.** < **5.** > **6.** > **7.** < **8.** < **9.** > **10.** < **11.** > **12.** < **13.** <
14. > **15.** < **16.** > **17.** < **18.** < **19.** > **20.** > **21.** > **22.** < **23.** > **24.** > **25.** <
26. > **27.** > **28.** > **29.** > **30.** > **31.** > **32.** < **33.** < **34.** < **35.** > **36.** < **37.** <
38. >

Practice Exercises for Section 1.4, page 31

1. −22 **2.** −4 **3.** −8 **4.** −27 **5.** −9 **6.** −24 **7.** −13 **8.** −10 **9.** −39 **10.** −19 **11.** −30
12. −87 **13.** −23 **14.** −45 **15.** −60 **16.** −24 **17.** −30 **18.** −4 **19.** −62 **20.** −40 **21.** −64
22. −17 **23.** −48 **24.** −181 **25.** −28 **26.** −2 **27.** −0.33 **28.** −31.66 **29.** −13.34 **30.** −5.8
31. −18.9 **32.** −43.1 **33.** −5.16 **34.** −1 **35.** $\frac{3}{5}$ **36.** 1 **37.** −1 **38.** $\frac{-1}{2}$ **39.** $\frac{-40}{33}$ or $-1\frac{7}{33}$
40. $\frac{-19}{18}$ or $-1\frac{1}{18}$ **41.** $\frac{-15}{14}$ or $-1\frac{1}{14}$ **42.** $\frac{-7}{6}$ or $-1\frac{1}{6}$ **43.** $\frac{-23}{10}$ or $-2\frac{3}{10}$ **44.** $\frac{-21}{26}$

Practice Exercises for Section 1.5, page 35

1. 4 **2.** −51 **3.** −2 **4.** −2 **5.** −3 **6.** −5 **7.** 0 **8.** 4 **9.** 6 **10.** −5 **11.** −4 **12.** −9
13. 2 **14.** −1 **15.** −36 **16.** 3 **17.** −14 **18.** −5 **19.** −3 **20.** 0 **21.** 1 **22.** 37 **23.** −13
24. −1 **25.** −12 **26.** −11 **27.** −5 **28.** 16 **29.** −5 **30.** 49 **31.** 9 **32.** −18 **33.** −16 **34.** 11
35. −25 **36.** 4 **37.** 1 **38.** −40 **39.** −29 **40.** 11 **41.** −58 **42.** −4 **43.** 5.1 **44.** −3.15
45. −3.49 **46.** 2.7 **47.** 6.22 **48.** −5.58 **49.** 4.79 **50.** −10.81 **51.** 6.35 **52.** 8.6 **53.** $\frac{2}{5}$
54. $\frac{4}{5}$ **55.** $\frac{2}{3}$ **56.** $\frac{-1}{6}$ **57.** $\frac{3}{4}$ **58.** $\frac{-7}{12}$ **59.** $\frac{-3}{100}$ **60.** $\frac{4}{3}$ **61.** $\frac{-1}{13}$ **62.** $\frac{7}{8}$ **63.** $\frac{-1}{2}$ **64.** $\frac{5}{24}$

Practice Exercises for Section 1.6, page 38

1. −12 **2.** −22 **3.** 5 **4.** −41 **5.** 8 **6.** −3 **7.** −7 **8.** −5 **9.** −8 **10.** −9 **11.** −5 **12.** −1
13. −13 **14.** −6 **15.** −12 **16.** −8 **17.** −12 **18.** −10 **19.** −20 **20.** −1100 **21.** −4 **22.** −68
23. 1.8 **24.** −5.4 **25.** −20.75 **26.** −15.85 **27.** −1.8 **28.** 3.01 **29.** −7.35 **30.** −27.51 **31.** −117.5
32. −2.96 **33.** $\frac{-1}{5}$ **34.** −2 **35.** 1 **36.** $\frac{-1}{9}$ **37.** $\frac{-24}{23}$ **38.** $\frac{-1}{9}$ **39.** $\frac{-3}{4}$ **40.** $\frac{3}{100}$ **41.** $\frac{-5}{6}$ **42.** $\frac{1}{2}$

Practice Exercises for Section 1.7, page 40

1. −13 **2.** −50 **3.** −27 **4.** −48 **5.** −20 **6.** −69 **7.** −22 **8.** −10 **9.** −28 **10.** −70
11. −100 **12.** −12 **13.** −11 **14.** −131 **15.** −160 **16.** −700 **17.** −1659 **18.** −749 **19.** −1
20. −1.7 **21.** −9.4 **22.** −0.42 **23.** −2.27 **24.** −3.41 **25.** −64 **26.** −114.48 **27.** −0.56 **28.** −1.22
29. −68.85 **30.** −12.91 **31.** $\frac{-5}{3}$ **32.** $\frac{-2}{5}$ **33.** $\frac{-5}{3}$ **34.** $\frac{-5}{2}$ **35.** −3 **36.** $\frac{-20}{29}$ **37.** $\frac{-1}{7}$ **38.** $\frac{-17}{16}$
39. $\frac{-7}{12}$ **40.** $\frac{-9}{28}$ **41.** $\frac{-11}{12}$ **42.** $\frac{-39}{40}$

Practice Exercises for Section 1.8, page 42

1. −1 **2.** 19 **3.** −4 **4.** 2 **5.** 35 **6.** −7 **7.** −9 **8.** 20 **9.** 124 **10.** 72 **11.** −54 **12.** −57
13. 18 **14.** 7 **15.** 2 **16.** −8 **17.** 12 **18.** 12 **19.** −3 **20.** 75 **21.** 12 **22.** 159 **23.** −6
24. 83 **25.** −28 **26.** −49 **27.** −42 **28.** 8 **29.** 4.7 **30.** −1.45 **31.** 3.5 **32.** 5.8 **33.** −2.95
34. 11.3 **35.** −4.02 **36.** 2.75 **37.** 63 **38.** −26.84 **39.** 1 **40.** $\frac{6}{13}$ **41.** 0 **42.** $\frac{3}{5}$ **43.** $\frac{3}{4}$
44. 1 **45.** $\frac{5}{8}$ **46.** 0 **47.** $\frac{16}{11}$ **48.** $\frac{-1}{8}$

Practice Exercises for Section 1.9, page 44

1. −24 **2.** −28 **3.** −60 **4.** −8 **5.** −65 **6.** −85 **7.** −408 **8.** −74 **9.** −168 **10.** −576
11. −1363 **12.** −1932 **13.** −672 **14.** −5612 **15.** −60 **16.** −462 **17.** −56 **18.** 0 **19.** −15
20. −504 **21.** −390 **22.** −240 **23.** 0 **24.** 0 **25.** −24 **26.** −175 **27.** −0.6 **28.** −13.02
29. −0.1275 **30.** −1.02 **31.** −1.08 **32.** −0.0025 **33.** −2.03 **34.** −4.84 **35.** −41.76 **36.** −3.63
37. $\frac{-3}{8}$ **38.** −1 **39.** $\frac{-3}{8}$ **40.** $\frac{-1}{3}$ **41.** $\frac{-3}{1000}$ **42.** $\frac{-5}{51}$ **43.** $\frac{-1}{10}$ **44.** $\frac{-1}{2500}$

Practice Exercises for Section 1.10, page 47

1. 36 **2.** 8 **3.** 24 **4.** 35 **5.** 10 **6.** 0 **7.** −35 **8.** 0 **9.** 160 **10.** −735 **11.** −8
12. 30 **13.** −30 **14.** 20 **15.** −20 **16.** −20 **17.** −1200 **18.** 625 **19.** −12 **20.** −168 **21.** 168
22. −168 **23.** 168 **24.** 8 **25.** −8 **26.** −8 **27.** −15.96 **28.** 0.025 **29.** 1.02 **30.** 15.3
31. 11.6 **32.** −21.75 **33.** −35.131 **34.** 0.038 **35.** $\frac{7}{100}$ **36.** $\frac{3}{32}$ **37.** $\frac{3}{2}$ **38.** $\frac{3}{10}$ **39.** 2
40. $\frac{1}{6}$ **41.** $\frac{-5}{9}$ **42.** −6 **43.** 4 **44.** $\frac{-2}{3}$

Practice Exercises for Section 1.11, page 49

1. −7 **2.** −7 **3.** 7 **4.** 13 **5.** 7 **6.** −8 **7.** 7 **8.** −7 **9.** −3 **10.** 2 **11.** 6 **12.** −7
13. −3 **14.** 5 **15.** −4 **16.** −0.3 **17.** $\frac{-1}{2}$ **18.** −0.2 **19.** 0.04 **20.** −0.03 **21.** −2 **22.** −3.4
23. 3 **24.** −0.06 **25.** −16.7 **26.** 0 **27.** −0.1 **28.** −70 **29.** 2 **30.** −20,000 **31.** 50 **32.** −5

Review Practice Exercises, page 50

1. −8 **2.** 2 **3.** 3 **4.** −8 **5.** −8 **6.** 8 **7.** −6 **8.** −2 **9.** −5 **10.** −5 **11.** −2 **12.** −2
13. 2 **14.** 7 **15.** −48 **16.** 60 **17.** −1100 **18.** −15 **19.** 1 **20.** −8 **21.** 120 **22.** −5 **23.** 6
24. −4 **25.** 0 **26.** undefined **27.** 6 **28.** $\frac{1}{6}$ **29.** −4 **30.** 5 **31.** −4 **32.** −30 **33.** 211
34. −12 **35.** 1 **36.** −1 **37.** 42 **38.** 6 **39.** −8 **40.** 0 **41.** $\frac{2}{5}$ **42.** $\frac{5}{9}$ **43.** $\frac{1}{2}$ **44.** $\frac{1}{2}$
45. $\frac{1}{4}$ **46.** $\frac{1}{2}$ **47.** $\frac{-7}{24}$ **48.** $\frac{11}{10}$ **49.** $\frac{3}{8}$ **50.** $\frac{-2}{7}$ **51.** $\frac{-1}{8}$ **52.** $\frac{3}{2}$ **53.** 0.5 **54.** 0.9
55. 2.27 **56.** −10.7 **57.** −0.024 **58.** 0.124 **59.** 0.061 **60.** −0.0964

Chapter 2

Practice Exercises for Section 2.1, page 55

1. 9^4 **2.** 4^9 **3.** 8^5 **4.** 9^{44} **5.** 1^{100} **6.** 7^3 **7.** 10^6 **8.** 5^7 **9.** 2^{15} **10.** 3^{11} **11.** 1^{19} **12.** 1^{22}
13. 8 **14.** 7 **15.** 3 **16.** 7 **17.** 2 **18.** 4 **19.** 1 **20.** 2 **21.** 2 **22.** 1 **23.** 2 **24.** 2
25. 3 **26.** 1 **27.** 1 **28.** (a) 9 (b) 3 **29.** (a) 4 (b) 2 **30.** (a) 3 (b) 9 **31.** (a) 7 (b) 49
32. (a) 4 (b) 8 (c) 2 **33.** (a) 5 (b) 125 **34.** (a) 10 (b) 100 **35.** (a) 1000 (b) 100 (c) 10

Practice Exercises for Section 2.2, page 58

1. base = 5, exponent = 3   **2.** base = 6, exponent = 2   **3.** base = –3, exponent = 4   **4.** base = –10, exponent = 2
5. 1^2 **6.** 2^6 **7.** 4^7 **8.** 15^4 **9.** $(-3)^5$ **10.** $(0.1)^3$ **11.** $(2.5)^2$ **12.** $(-3.8)^3$ **13.** $(7.6)^4$ **14.** $(0.25)^2$
15. $\left(\frac{1}{2}\right)^2$ **16.** $\left(\frac{1}{4}\right)^2$ **17.** $\left(\frac{-1}{2}\right)^2$ **18.** $\left(\frac{-1}{4}\right)^2$ **19.** 729   **20.** 1   **21.** 32   **22.** 64   **23.** 100,000   **24.** 125
25. 81   **26.** 49   **27.** 512   **28.** 20   **29.** 0   **30.** 0   **31.** 0.25   **32.** 0.000001   **33.** 10.24   **34.** 0.0016
35. 0.01   **36.** 0.09   **37.** 8   **38.** 0.16   **39.** 0.25   **40.** 0.001   **41.** 6^3 **42.** 3^4 **43.** 5^3 **44.** $(5.1)^3$
45. $(3.5)^4$ **46.** $\left(\frac{1}{2}\right)^5$ **47.** $\left(\frac{1}{8}\right)^2$

Practice Exercises for Section 2.3, page 60

1. 16   **2.** 8   **3.** 9   **4.** 6   **5.** 216   **6.** 18   **7.** 1   **8.** 3   **9.** 512   **10.** 24   **11.** 25   **12.** 10
13. 343   **14.** 21   **15.** 6561   **16.** 36   **17.** 1024   **18.** 20   **19.** 121   **20.** 22   **21.** 1000   **22.** 30
23. 10.2   **24.** 26.01   **25.** 0.000008   **26.** 0.06   **27.** 0.001   **28.** 0.3   **29.** 0.25   **30.** 1   **31.** 0.09
32. 0.6   **33.** $\frac{1}{4}$ **34.** 1   **35.** $\frac{1}{9}$ **36.** $\frac{2}{3}$ **37.** $\frac{27}{8}$ **38.** $\frac{9}{2}$

Practice Exercises for Section 2.4, page 63

1. 64   **2.** –243   **3.** 25   **4.** 144   **5.** –216   **6.** –343   **7.** –125   **8.** 1   **9.** 81   **10.** –64   **11.** 64
12. 100   **13.** –100   **14.** 1   **15.** 121   **16.** –12   **17.** 0.01   **18.** –0.125   **19.** –0.25   **20.** –1.5
21. 0.81   **22.** 9.61   **23.** –166.375   **24.** 0.09   **25.** 0.0016   **26.** 50.41   **27.** $\frac{1}{4}$ **28.** $\frac{-1}{8}$ **29.** $\frac{1}{25}$
30. $\frac{-1}{64}$ **31.** $\frac{1}{100}$ **32.** $\frac{-8}{125}$ **33.** $\frac{-3}{2}$ **34.** $\frac{16}{25}$ **35.** $\frac{4}{9}$ **36.** $\frac{-729}{1000}$

Practice Exercises for Section 2.5, page 64

1. –4   **2.** 4   **3.** 16   **4.** –16   **5.** –81   **6.** 81   **7.** –216   **8.** –216   **9.** 8   **10.** –16   **11.** –64
12. –64   **13.** –49   **14.** 49   **15.** –81   **16.** –81   **17.** –125   **18.** 125   **19.** –100   **20.** –100
21. –0.001   **22.** –0.001   **23.** –0.25   **24.** 0.25   **25.** –0.064   **26.** –0.064   **27.** 0.0016   **28.** –0.0016
29. –0.0625   **30.** –0.0625   **31.** $\frac{-1}{8}$ **32.** $\frac{1}{16}$ **33.** $\frac{1}{16}$ **34.** $\frac{-1}{16}$ **35.** $\frac{-1}{125}$ **36.** $\frac{-1}{125}$

Practice Exercises for Section 2.6, page 66

1. base = 5, exponent = 1   **2.** base = –7, exponent = 1   **3.** 10   **4.** 1000   **5.** –35   **6.** 0   **7.** –9   **8.** –4
9. 6   **10.** 12   **11.** 24   **12.** 36   **13.** 0.5   **14.** –0.1   **15.** $\frac{2}{3}$ **16.** $\frac{-4}{5}$

Practice Exercises for Section 2.7, page 68

1. 226   **2.** 17   **3.** 61   **4.** 20   **5.** 80   **6.** 20   **7.** –54   **8.** –12   **9.** 20   **10.** 20   **11.** 9000
12. 101   **13.** 12   **14.** 57   **15.** –18   **16.** 100   **17.** –875   **18.** 89   **19.** –76   **20.** –89   **21.** 0.14
22. –24.7   **23.** 0.125   **24.** –0.009   **25.** 31.11   **26.** 1.043   **27.** 0   **28.** –40.432   **29.** 3.5625
30. 0.84   **31.** $\frac{1}{8}$ **32.** $\frac{101}{1000}$ **33.** $\frac{819}{10,000}$ **34.** $\frac{-11}{9}$ **35.** $\frac{-7}{36}$ **36.** $\frac{9}{25}$ **37.** $\frac{-65}{36}$ **38.** $\frac{-561}{100}$

Practice Exercises for Section 2.8, page 70

1. 3^5 **2.** 2^6 **3.** 5^4 **4.** 7^9 **5.** 3^4 **6.** 4^5 **7.** 9^9 **8.** 8^9 **9.** 2^3 **10.** 2^{10} **11.** 3^8 **12.** $(-10)^7$
13. $(-2)^9$ **14.** $(-4)^6$ **15.** $(-9)^{12}$ **16.** 5^{10} **17.** $(-6)^{10}$ **18.** $(-3)^{14}$ **19.** 7^{11} **20.** $(-8)^{15}$ **21.** 11^{16}
22. 100^{15} **23.** $(0.1)^8$ **24.** $(2.5)^6$ **25.** $(-7.1)^4$ **26.** $(5.8)^7$ **27.** $(-9.2)^7$ **28.** $(4.6)^{10}$ **29.** $\left(\frac{1}{2}\right)^5$
30. $\left(\frac{2}{5}\right)^4$ **31.** $\left(\frac{3}{10}\right)^5$ **32.** $\left(\frac{-1}{5}\right)^{10}$ **33.** $\left(\frac{2}{3}\right)^{10}$ **34.** $\left(\frac{-3}{4}\right)^{19}$

Practice Exercises for Section 2.9, page 71

1. 144   **2.** 216   **3.** 2000   **4.** 81   **5.** −72   **6.** 0.00125   **7.** 2.601   **8.** 124.59008   **9.** −0.0000624

10. 6548.76608   **11.** $\dfrac{1}{36}$ **12.** $\dfrac{27}{256}$ **13.** $\dfrac{1}{256,000}$ **14.** $3^6 \cdot 4^8$ **15.** $7^7 \cdot 9^{11}$ **16.** $(-3)^6 \cdot 5^8$

17. $2^3 \cdot 5^4 \cdot 7^5$ **18.** $5^4 \cdot 7^4$ **19.** $2^3 \cdot 3^5 \cdot 5^3$ **20.** $(-6)^5(9)^9$ **21.** $10^5 \cdot 11^5 \cdot 13^6$ **22.** $5^{11} \cdot 6^6 \cdot 7^6$ **23.** 0.018

24. 0.000001   **25.** 0.002   **26.** $(0.1)^8(0.5)^7$ **27.** $(-3.1)^3(0.1)^3$ **28.** $(5.2)(0.2)^8(0.3)^2$ **29.** $(9.2)^{10}(0.25)^5$

30. $(5.7)^5(5.3)^4$ **31.** $(4.1)^8(1.4)^8$ **32.** $\dfrac{1}{32}$ **33.** $\dfrac{4}{81}$ **34.** $\dfrac{9}{10,000}$

Practice Exercises for Section 2.10, page 73

1. 6^{36} **2.** 2^{35} **3.** 10^{40} **4.** 25^8 **5.** $(0.5)^{18}$ **6.** $(1.2)^6$ **7.** $(9.5)^{32}$ **8.** $(0.25)^{21}$ **9.** 3^8 **10.** 3^8

11. 10^{15} **12.** 5^{25} **13.** 11^{12} **14.** 7^{24} **15.** $\left(\dfrac{1}{2}\right)^6$ **16.** $\left(\dfrac{2}{3}\right)^{12}$ **17.** $\left(\dfrac{4}{5}\right)^8$ **18.** $\left(\dfrac{7}{9}\right)^{15}$

Practice Exercises for Section 2.11, page 75

1. $2^4 \cdot 3^4 \cdot 5^4$ **2.** $5^4 \cdot 6^8 \cdot 8^4$ **3.** $3^{18} \cdot 7^{30}$ **4.** $4^{14} \cdot 5^{28} \cdot 6^{35}$ **5.** $3^5 \cdot 4^{15} \cdot 5^{30} \cdot 7^{10}$ **6.** $2^2 \cdot 3^4 \cdot 5^6$

7. $5^9 \cdot 6^6 \cdot 7^{12}$ **8.** $3^{33} \cdot 5^{18} \cdot 9^6$ **9.** $2^{15} \cdot 8^6 \cdot 11^9$ **10.** $7^8 \cdot 12^{10} \cdot 13^6$ **11.** $2^5 \cdot 5^{20} \cdot 7^{10} \cdot 9^{15}$ **12.** $(0.1)^{10}(0.25)^6$

13. $(3.2)^{10}(4.5)^{15}$ **14.** $(-0.2)^{16}(0.3)^8$ **15.** $(2.9)^{15}(3.5)^{10}$ **16.** $(5.7)^{12}(-7.5)^{18}(0.1)^{30}$ **17.** $(0.1)^{14}(0.01)^{21}(0.2)^7$

18. $(3.6)^6(4.7)^4(5.9)^8$ **19.** $\left(\dfrac{1}{4}\right)^4\left(\dfrac{1}{2}\right)^6$ **20.** $\left(\dfrac{1}{2}\right)^{21}\left(\dfrac{2}{3}\right)^{28}$ or $\dfrac{2^7}{3^{28}}$ **21.** $\left(\dfrac{4}{7}\right)^8\left(\dfrac{1}{5}\right)^{12}$ **22.** $\left(\dfrac{1}{8}\right)^6\left(\dfrac{2}{9}\right)^{12}\left(\dfrac{3}{11}\right)^{21}$

23. $\left(\dfrac{3}{5}\right)^8\left(\dfrac{-2}{3}\right)^6\left(\dfrac{1}{8}\right)^{10}$ **24.** $\left(\dfrac{3}{5}\right)^{36}\left(\dfrac{1}{12}\right)^{18}\left(\dfrac{2}{7}\right)^{54}$

Practice Exercises for Section 2.12, page 77

1. 5^2 **2.** 7^3 **3.** 8   **4.** 2^2 **5.** 3   **6.** $(-4)^2$ **7.** $(-6)^2$ **8.** 9^3 **9.** 11^3 **10.** 10^7 **11.** $(-12)^2$

12. 153   **13.** $(-45)^2$ **14.** -9^3 **15.** $(0.7)^3$ **16.** $(0.1)^2$ **17.** 0.3   **18.** $(10.5)^4$ **19.** $(-5.2)^3$ **20.** $(9.6)^3$

21. $(0.02)^3$ **22.** $(-7.1)^3$ **23.** $(2 \cdot 5)^3$ **24.** $5 \cdot 9$ **25.** $\dfrac{1}{2}$ **26.** $\left(-\dfrac{2}{3}\right)^3$ **27.** $\left(\dfrac{4}{11}\right)^3$ **28.** $\left(\dfrac{5}{7}\right)^4$ **29.** $\left(\dfrac{-9}{8}\right)^2$

30. $\left(\dfrac{6}{7}\right)^3$

Practice Exercises for Section 2.13, page 80

1. $\dfrac{27}{1000}$ **2.** $\dfrac{9}{49}$ **3.** $\dfrac{8}{125}$ **4.** $\dfrac{1}{32}$ **5.** $\dfrac{125}{729}$ **6.** $\dfrac{8}{27}$ **7.** $\dfrac{1}{25}$ **8.** $\dfrac{4}{121}$ **9.** $\dfrac{1}{16}$ **10.** $\dfrac{27}{125}$

11. $\dfrac{1}{243}$ **12.** $\dfrac{27}{64}$ **13.** $\dfrac{7^4}{3^4}$ **14.** $\dfrac{5^7}{2^7}$ **15.** $\dfrac{9^3}{7^3}$ **16.** $\dfrac{4^5}{9^5}$ **17.** $\dfrac{6^8}{7^8}$ **18.** $\dfrac{11^5}{12^5}$ **19.** $\dfrac{4^7}{9^7}$ **20.** $\dfrac{3^{11}}{13^{11}}$

Practice Exercises for Section 2.14, page 82

1. 1   **2.** 1   **3.** 1   **4.** 1   **5.** 1   **6.** 1   **7.** 1   **8.** 1   **9.** 1   **10.** −1   **11.** 4   **12.** 3   **13.** 10

14. 5   **15.** 64   **16.** 21   **17.** 40   **18.** 729   **19.** 6.6   **20.** 2.1   **21.** 3.2   **22.** 0.1   **23.** 10.8

24. 12.3   **25.** 7.5   **26.** 5   **27.** $\dfrac{7}{4}$ **28.** $\dfrac{3}{8}$ **29.** $\dfrac{8}{9}$ **30.** $\dfrac{7}{6}$ **31.** 2   **32.** $\dfrac{19}{9}$

Practice Exercises for Section 2.15, page 84

1. $\dfrac{1}{3^3}$ **2.** $\dfrac{1}{4^5}$ **3.** $\dfrac{1}{6}$ **4.** $\dfrac{1}{7^3}$ **5.** $\dfrac{1}{11^4}$ **6.** $\dfrac{1}{16^6}$ **7.** $\dfrac{1}{20^9}$ **8.** 4^3 **9.** 15   **10.** 18   **11.** 9   **12.** 3

13. $\dfrac{1}{16}$ **14.** $\dfrac{1}{16}$ **15.** $\dfrac{-1}{8}$ **16.** $\dfrac{1}{25}$ **17.** $\dfrac{1}{27}$ **18.** $\dfrac{1}{7}$ **19.** $\dfrac{1}{12}$ **20.** $\dfrac{1}{216}$ **21.** $\dfrac{-1}{125}$ **22.** $\dfrac{1}{9}$

23. $\dfrac{-1}{64}$ **24.** $\dfrac{1}{81}$ **25.** $\dfrac{-1}{32}$ **26.** $\dfrac{-1}{64}$ **27.** $\dfrac{1}{100}$ **28.** −125   **29.** 81   **30.** 25   **31.** 81   **32.** −64

33. 49   **34.** 49   **35.** 216   **36.** 1000   **37.** 625   **38.** 6.25   **39.** 4   **40.** $\frac{1}{81}$ **41.** 16   **42.** $\frac{1}{27}$

43. $\frac{1}{625}$ **44.** $\frac{1}{64}$ **45.** $\frac{1}{81}$ **46.** 0.32   **47.** –10,000,000   **48.** 1   **49.** $\frac{16}{9}$ **50.** $\frac{512}{125}$ **51.** $\frac{1}{12,800}$

52. $\frac{16}{7}$ **53.** 36   **54.** –125   **55.** $\frac{1}{16}$ **56.** $\frac{1}{216}$ **57.** $\frac{1}{1000}$ **58.** 512

Practice Exercises for Section 2.16, page 87

1. 9   **2.** 12   **3.** 100   **4.** 13   **5.** 150   **6.** 16   **7.** 17   **8.** 18   **9.** 14   **10.** 19   **11.** 20   **12.** 25

13. 22   **14.** 200   **15.** 1000   **16.** 0.3   **17.** 0.04   **18.** 1.3   **19.** $\frac{4}{7}$ **20.** $\frac{8}{13}$

Practice Exercises for Section 2.17, page 88

1. 8   **2.** 9   **3.** 30   **4.** 0.1   **5.** 0.01   **6.** 0.8   **7.** 0.2   **8.** $\frac{1}{3}$ **9.** $\frac{2}{5}$ **10.** $\frac{3}{10}$ **11.** 100

12. 200   **13.** $\frac{2}{3}$ **14.** $\frac{3}{100}$ **15.** $\frac{5}{8}$

Practice Exercises for Section 2.18, page 89

1. $123^{1/2}$ **2.** $56^{1/3}$ **3.** $216^{1/6}$ **4.** $895^{1/8}$ **5.** $10{,}000^{1/10}$ **6.** $45^{1/2}$ **7.** $39^{1/3}$ **8.** $64^{1/4}$ **9.** $75^{1/5}$

10. $113^{1/7}$ **11.** $81^{1/9}$ **12.** $94^{1/2}$ **13.** $\sqrt{3}$ **14.** $\sqrt[3]{7}$ **15.** $\sqrt[3]{8}$ **16.** $\sqrt[4]{10}$ **17.** $\sqrt[6]{64}$ **18.** $\sqrt{25}$

19. $\sqrt[3]{93}$ **20.** $\sqrt[4]{11}$ **21.** $\sqrt[4]{15}$ **22.** $\sqrt[5]{100}$ **23.** $\sqrt[10]{1000}$ **24.** $\sqrt[8]{863}$ **25.** 4   **26.** 5   **27.** 10

28. 4   **29.** 3   **30.** $9^{1/2}=3$

Practice Exercises for Section 2.19, page 91

1. 5.9 (to the nearest tenth)   **2.** 13.7   **3.** 59.9   **4.** 0.2   **5.** 2   **6.** 11.2   **7.** 19.7   **8.** 93   **9.** 27

10. 7.7   **11.** 3.2   **12.** 15   **13.** 4.0   **14.** 5.9   **15.** 4.1   **16.** 4.7   **17.** 3   **18.** 2.7   **19.** 2   **20.** 2.9

21. 6.7   **22.** 11   **23.** 6   **24.** 2   **25.** 0.2   **26.** 39   **27.** 2.2   **28.** 4   **29.** 15.8   **30.** 5   **31.** 5.9

32. 4   **33.** 25   **34.** 10.0   **35.** 3   **36.** 10   **37.** 11   **38.** 8.2   **39.** 3.7   **40.** 2.4

Review Practice Exercises, page 93

1. base = 5, exponent = 2   **2.** base = –3, exponent = 4   **3.** base = 2, exponent = 3   **4.** base = 5.2, exponent = 3

5. base = 3.7, exponent = 3   **6.** base = –3.7, exponent = 3   **7.** base = 3, exponent = 2   **8.** base = $\frac{3}{4}$, exponent = 5

9. base = $-\frac{5}{8}$, exponent = 3   **10.** base = –5, exponent = 3   **11.** 3^5 **12.** 4^3 **13.** 5^7 **14.** $(5.1)^3$ **15.** $(0.05)^4$

16. $\left(\frac{3}{8}\right)^3$ **17.** $\left(-\frac{1}{2}\right)^5$ **18.** 64   **19.** 7   **20.** 125   **21.** 36   **22.** $\frac{9}{16}$ **23.** $\frac{8}{27}$ **24.** 90   **25.** 192

26. 80   **27.** 2   **28.** 72   **29.** 0.0001   **30.** 6.25   **31.** $\frac{27}{64}$ **32.** $\frac{25}{64}$ **33.** $\frac{3}{8}$ **34.** 4^5 **35.** 2^5

36. 5^7 **37.** 3^6 **38.** 4^6 **39.** $(3.1)^5$ **40.** $(0.05)^2$ **41.** $\left(\frac{3}{8}\right)^7$ **42.** $\left(\frac{6}{7}\right)^6$ **43.** 3^8 **44.** 5^{12} **45.** 4^6

46. 2^{15} **47.** 6^6 **48.** $2^2 \cdot 7^2$ **49.** $5^3 \cdot 2^3$ **50.** $5^0 \cdot 3^0$, or 1   **51.** $3^2 \cdot 5^4 \cdot 7^8$ **52.** $2^3 \cdot 3^9 \cdot 5^{18}$ **53.** 10,000

54. –16   **55.** –27   **56.** 125   **57.** 324   **58.** 12.25   **59.** –25.1001   **60.** –0.000001   **61.** $-\frac{27}{512}$

62. $-\frac{25}{49}$ **63.** $\frac{9}{5}$ **64.** 2^3 **65.** 8^6 **66.** 8^{-6} **67.** 5^6 **68.** 5^0 **69.** $(5.3)^4$ **70.** 6.8^{-7} **71.** $\left(\frac{7}{5}\right)^2$

72. $\left(\frac{9}{4}\right)^3$ **73.** 125   **74.** 1   **75.** 64   **76.** 36   **77.** 0.053   **78.** 0.0625   **79.** $\frac{4}{25}$ **80.** $\frac{1}{64}$ **81.** 21

82. 15   **83.** 0.2   **84.** $\frac{5}{6}$ **85.** 5   **86.** 0.3   **87.** 0.2   **88.** 13

Chapter 3

Practice Exercises for Section 3.1, page 98

1. 7.　**2.** 6　**3.** 20　**4.** 8　**5.** 7　**6.** 19　**7.** 1　**8.** 8　**9.** 25　**10.** 11　**11.** –5　**12.** 3　**13.** 1
14. 126　**15.** 60　**16.** –10　**17.** 56　**18.** –57　**19.** –18　**20.** –37　**21.** 0.6　**22.** 14.08　**23.** 5.4
24. 5.2　**25.** 264.34　**26.** 6.1　**27.** $-\dfrac{6}{5}$　**28.** $\dfrac{5}{3}$　**29.** $\dfrac{7}{6}$　**30.** $\dfrac{125}{48}$

Practice Exercises for Section 3.2, page 102

1. –8　**2.** –32　**3.** 8　**4.** 32　**5.** –23　**6.** –15　**7.** 19　**8.** 0　**9.** 26　**10.** 20　**11.** 20　**12.** 8
13. –1　**14.** –9　**15.** 19　**16.** 48　**17.** 1　**18.** 9　**19.** 1　**20.** –3　**21.** –48　**22.** $\dfrac{-2}{5}$　**23.** –26
24. –29　**25.** 19　**26.** 105　**27.** –0.4　**28.** 1.58　**29.** –18.4　**30.** –4.4　**31.** 4　**32.** $-\dfrac{5}{2}$　**33.** $\dfrac{4}{7}$
34. $\dfrac{-27}{100}$　**35.** 36　**36.** $-\dfrac{71}{15}$

Practice Exercises for Section 3.3, page 104

1. 3　**2.** 3　**3.** 2　**4.** 7　**5.** 7　**6.** 35　**7.** 9　**8.** 9　**9.** 2　**10.** –2　**11.** –2　**12.** –3　**13.** 7
14. 0　**15.** 2　**16.** 4　**17.** $\dfrac{30}{7}$　**18.** $\dfrac{-2}{3}$　**19.** 8　**20.** 5　**21.** –1　**22.** –4.6　**23.** 1.6　**24.** 52
25. 90　**26.** –22

Practice Exercises for Section 3.4, page 106

1. –2　**2.** 3　**3.** 45　**4.** 22　**5.** 19　**6.** 14　**7.** 23　**8.** –75　**9.** –36　**10.** 31　**11.** –8　**12.** 5
13. –1　**14.** 5　**15.** 5　**16.** 9　**17.** 11　**18.** 16　**19.** 0　**20.** –6　**21.** 4.482　**22.** –2.76　**23.** –117
24. –2.5　**25.** 0.052　**26.** 0.05　**27.** $-\dfrac{7}{2}$　**28.** $-\dfrac{5}{8}$　**29.** $\dfrac{35}{29}$　**30.** $\dfrac{23}{9}$　**31.** 42　**32.** 256　**33.** 340
34. 42

Review Practice Exercises, page 107

1. 7　**2.** 9　**3.** 1　**4.** 1　**5.** 19　**6.** 25　**7.** 33　**8.** 8　**9.** 1　**10.** 50　**11.** 45　**12.** 48
13. 23　**14.** 14　**15.** 23　**16.** 7　**17.** 1　**18.** 25　**19.** 8　**20.** 2　**21.** –1　**22.** –1　**23.** 142
24. 44　**25.** 0　**26.** –201　**27.** –38　**28.** 6　**29.** –74　**30.** –680　**31.** –52　**32.** 382　**33.** 112
34. 8　**35.** 4　**36.** $-\dfrac{5}{6}$　**37.** 8　**38.** –8　**39.** 4　**40.** –4　**41.** $\dfrac{3}{4}$　**42.** 4　**43.** 2　**44.** –16
45. –13　**46.** –51　**47.** 32　**48.** 78　**49.** 0　**50.** –55　**51.** 2　**52.** –6.2　**53.** 17.2　**54.** $-\dfrac{17}{4}$
55. $-\dfrac{7}{20}$　**56.** –1　**57.** –1　**58.** 7　**59.** $\dfrac{-8}{3}$　**60.** 1　**61.** –1　**62.** –1

Chapter 4

Practice Exercises for Section 4.1, page 115

1. $3+2$　**2.** $14-5$　**3.** $6(3)$　**4.** $4+5$　**5.** $4+5$　**6.** $5+4$　**7.** $5+4$　**8.** $10-4$　**9.** $17-9$
10. $8-6$　**11.** $7-4$　**12.** $12-9$　**13.** $15-10$　**14.** $10<15$　**15.** $7(4)$　**16.** $8(3)$　**17.** $10+5$　**18.** $8+4$
19. $10+5$　**20.** $8+4$　**21.** $5+10$　**22.** $4+8$　**23.** $5+10$　**24.** $4+8$　**25.** $17-9$　**26.** $30+6$　**27.** $6<7$
28. $7-6$

Practice Exercises for Section 4.2, page 117

1. $54+(12-3)$　**2.** $(15-11)5$　**3.** $(6+4)\div 2$　**4.** $4(13)-10$　**5.** $6(9)-7$　**6.** $3(4)-10$　**7.** $\dfrac{1}{4}(28)+10$
8. $6(3)+5$　**9.** $2(5)-2$　**10.** $3^2-2(3)$　**11.** $16-4$　**12.** $7-3$　**13.** $6(9-7)$　**14.** $10-(2\cdot 6)$　**15.** $\dfrac{8}{8-6}$
16. $\dfrac{9-6}{2\cdot 9}$　**17.** $(5\cdot 2)-19$　**18.** $\dfrac{12+9}{12}$　**19.** $\dfrac{4}{5}(8+17)$　**20.** $10+\dfrac{4}{8}$

Practice Exercises for Section 4.3, page 119

1. Five added to eight; the sum of eight and five; five more than eight   **2.** Seven subtracted from eighteen; seven less than eighteen; the difference between eighteen and seven (in this order)   **3.** Nine times three or the product of nine and three   **4.** Four multiplied by two   **5.** The product of six and five   **6.** Eight divided by four   **7.** Four divided by eight   **8.** Four is less than eight.   **9.** Four more than eight; eight increased by four   **10.** Ten decreased by seven; seven less than ten.   **11.** The difference of eight and ten (in this order); subtract ten from eight.   **12.** The quotient of twelve divided by four; twelve divided by four   **13.** The square of three; the second power of three   **14.** Two times three; the product of two and three   **15.** Eight minus six; the difference of eight and six (in this order)   **16.** Four-fifths of ten; four-fifths times ten

Review Practice Exercises, page 121

1. $5+6$ **2.** $3+6$ **3.** $4(3)-5$ **4.** $3(3+5)$ **5.** $4(27)-12$ **6.** $5(24)+3(24)$ **7.** $9-5+12$ **8.** $5(6+2)$ **9.** $5(3)+2$ **10.** $16-[2(4)+11]$ **11.** $6(5)+4(5)$ **12.** $6+(8-4)$ **13.** $2(7)+(7+3)$ **14.** $2(6)+15$ **15.** $(9+3)+2$ **16.** $2(10)-3(10)$ **17.** $(8-3)2+4$ **18.** 4^2-6 **19.** $3(9)+2(9)^2$ **20.** $4[3(11)+11^3]$ **21.** The difference between the square of ten and eight (in this order)   **22.** Six times the sum of four and five   **23.** One-half the sum of seven and nine   **24.** The quotient of eight divided by the difference of five and three (in this order)   **25.** Eight increased by the quotient of six and three   **26.** The sum of seven and five divided by the sum of six and three   **27.** The difference between ten and eight (in this order); ten decreased by eight; ten minus eight; subtract eight from ten.   **28.** The sum of nine and seven; nine increased by seven; seven more than nine; add seven to nine.

Chapter 5

Practice Exercises for Section 5.1, page 127

1. ⓐ   **2.** ②ⓐ   **3.** ⓐ + ②ⓑ   **4.** ⓐⓑ + ⓐ   **5.** ⓪⑦   **6.** ⓐ + ⓑ + ⓒ   **7.** ⓐⓑⓒ   **8.** ⓐ − ⓐⓑⓒⓓ   **9.** $\left(\dfrac{a}{2}\right) + \dfrac{2}{a}$ **10.** ⓐⓑ + ⓑⓒ + ⓓⓔ + ⓔⓕ   **11.** 2   **12.** 1   **13.** 3   **14.** 2   **15.** 1   **16.** 1   **17.** 4   **18.** 5   **19.** 1   **20.** 2

Practice Exercises for Section 5.2, page 130

1. y^3 **2.** $a^2b^3c^6$ **3.** $4x^2y^4z^2$ **4.** $-7u^2v^4w^3$ **5.** x^8 **6.** x^7 **7.** x^9 **8.** a^{11} **9.** y^9 **10.** a^5 **11.** p^2 **12.** b^9 **13.** c^{18} **14.** x^{1110} **15.** c^4 **16.** x^8 **17.** b^5 **18.** x^3 **19.** a^8 **20.** 1   **21.** $16a^4$ **22.** $25x^2$ **23.** $27a^3$ **24.** $16u^2$ **25.** $36y^2$ **26.** $8p^3q^3$ **27.** $16a^4b^4$ **28.** $-64x^3y^3$ **29.** $100{,}000c^5d^5$ **30.** $1{,}000{,}000y^6z^6$ **31.** $\dfrac{9}{x^2}$ **32.** $\dfrac{a^2}{25}$ **33.** $\dfrac{a^2}{100b^2}$ **34.** $\dfrac{x^3}{27}$ **35.** $4a^4$ **36.** a^5b^{10} **37.** x^4y^6 **38.** $64x^6y^9$ **39.** $10{,}000a^4b^8c^{12}$ **40.** $x^{20}y^5z^{15}$ **41.** $\dfrac{9b^2}{a^4}$ **42.** $\dfrac{1000x^6}{27y^9}$

Practice Exercises for Section 5.3, page 132

1. 13   **2.** 5   **3.** 2   **4.** 1   **5.** 1   **6.** −1   **7.** −1   **8.** $\dfrac{1}{2}$ **9.** $\dfrac{-2}{3}$ **10.** 0.04

Practice Exercises for Section 5.4, page 135

1. $-5xy$ **2.** $-11xy^3$ **3.** $18x$ **4.** $2p^2q$ **5.** $-7ab$ **6.** $-12x^3y$ **7.** $-6st$ **8.** $-10p$ **9.** $-88y$ **10.** $-39a$ **11.** $4P$ **12.** $2M$ **13.** $6M$ **14.** $7y^2$ **15.** y^2 **16.** t^3 **17.** $16a^2b^3$ **18.** $2a$ **19.** $-7xy^2$ **20.** $4x^2$ **21.** 0   **22.** $-18st$ **23.** $-9st$ **24.** $4a^2b$ **25.** $-1.5xy$ **26.** $-2.5xy^3$ **27.** $2.9a^2b^3$ **28.** $-2.3t^3$ **29.** $-3xy^2$ **30.** $-2a^3$ **31.** $\dfrac{1}{4}xy$ **32.** $\dfrac{5}{4}y^4$ **33.** $-\dfrac{1}{10}a^3c^2$ **34.** $\dfrac{1}{6}x^2y$ **35.** $10a$ **36.** $18x$ **37.** $4y$ **38.** $12z$ **39.** $7xy$ **40.** $4x^2y$

Practice Exercises for Section 5.5, page 138

1. monomial   **2.** binomial   **3.** trinomial   **4.** none of these   **5.** monomial   **6.** none of these   **7.** monomial   **8.** monomial   **9.** binomial   **10.** binomial

Practice Exercises for Section 5.6, page 139

1. $3p^6$ **2.** $-2y^3$ **3.** $15r^4$ **4.** $10a^3b^3$ **5.** $6x^5y^7$ **6.** $12p^5r^2$ **7.** $-x^8y^3$ **8.** $24a^3b^5c^3$ **9.** $-24x^6y^{13}$
10. $140x^5y^{11}$ **11.** $-\dfrac{1}{2}x^2y^2$ **12.** $-\dfrac{1}{6}p^5q^3$ **13.** $-\dfrac{15}{8}c^2d^3$ **14.** $-x^2y^4$

Practice Exercises for Section 5.7, page 141

1. $4x^2 + 12y$ **2.** $6x^2 + 9x^3$ **3.** $6x^5 + 8x^7$ **4.** $5x^5 + 15x^4 + 20x^3$ **5.** $7x^{-5} + 7x^{-6}$ **6.** $-12 + 8x + 20x^2$
7. $-2y^2 + 8y - 14$ **8.** $-18p^2r^3 + 24p^2r^4 - 42pr^5 + 12p^3r^5$ **9.** $-12x^4 - 8x^4y + 16x^3y^2$ **10.** $-\dfrac{3}{4}x^2y + \dfrac{3}{4}xy^2$
11. $-2a^2bc + 4ab^2c - 6abc^2$ **12.** $-4x^3y + x^2y^2 - 3x^2yz$ **13.** $10p^2q - 10pqr$ **14.** $abcx - 2abcy$
15. $4a^3bc^3 + 16a^3b^2c^3 - 12a^2b^3c^4$ **16.** $10x^3yz^4 + 50x^3y^2z^4 - 100x^3y^2z^6$

Practice Exercises for Section 5.8, page 143

1. $4a$ **2.** $-2b^4$ **3.** $2s$ **4.** $3m$ **5.** $\dfrac{x^5y^2}{6}$ **6.** $\dfrac{-x^2y^2}{2}$ **7.** $\dfrac{c^3d^3}{2}$ **8.** $\dfrac{p^2q}{2r}$ **9.** $\dfrac{n^4}{3}$ **10.** $\dfrac{2}{xy^4}$
11. $\dfrac{3a}{4b^2c^3}$ **12.** $\dfrac{-2e^2}{3d^3}$ **13.** $5x^2$ **14.** $6xy$ **15.** $2x^2y^2$ **16.** $-4x^3$ **17.** $-3y$ **18.** $3a^2b^2c$ **19.** $3x$
20. $12x^2$ **21.** $-36a^4b$ **22.** $8x^2y$ **23.** $-14ab$ **24.** $2y$ **25.** $-x$ **26.** $4x^2$ **27.** $-2y$ **28.** -4
29. $\dfrac{4x}{3y^2}$ **30.** $-\dfrac{9y}{4}$ **31.** $-\dfrac{x}{3}$ **32.** ab^4

Practice Exercises for Section 5.9, page 146

1. $x + 1$ **2.** $y + 1$ **3.** $2a - 5$ **4.** $2b - 5$ **5.** $-2y + 3$ **6.** $2a + 4$ **7.** $-3t^2 - 1$ **8.** $3a + 2$ **9.** $\dfrac{3y^2 + 2y}{3}$
10. $4b^2 - 3$ **11.** $12x - 7$ **12.** $x - 2$ **13.** $z + 2$ **14.** $7k - 6$ **15.** $2 - z$ **16.** $-x + 2$ **17.** $x^2 - 3x$
18. $a - 5$ **19.** $4p - 1$ **20.** $3 + 2s^2$ **21.** $-5 + 4a$ **22.** $-2x^2 + 3$ **23.** $-3y^3 + 5$ **24.** $4y^2 - 3y + 1$
25. $x^2 + 3x + 2$ **26.** $\dfrac{m^2 + 2m + 3}{m}$ **27.** $-4x^2 + 3x - 2$ **28.** $xy + 2$ **29.** $-xy + 3$ **30.** $5q + 3p - 2pq$
31. $-3z^2 - 2z + 1$ **32.** $\dfrac{-x^2 + 4x - 2}{x}$ **33.** $x + 2y - 3xy$ **34.** $4a - 5 + 6b$ **35.** $2a + 1 + 3b$ **36.** $3x + 2 - y$
37. $a + 3 - 6b$ **38.** $-2y^2 - 3y + 4$ **39.** $4ab^2 - 5a^2bc^3 + 6a^3c^2$ **40.** $x^2y + 2xy^2 - 3yz$ **41.** $6m^2np^2 + 5p + 4p^4$
42. $\dfrac{4a}{b} + 4b$ **43.** $3y^2z^3 + 3z$ **44.** $b + 10a$ **45.** $x - 4y$ **46.** $6xy^2$

Practice Exercises for Section 5.10, page 148

1. 8   **2.** 4   **3.** 5   **4.** 2   **5.** 6   **6.** 2   **7.** 9   **8.** 5   **9.** 1   **10.** 25   **11.** 16   **12.** 4   **13.** 6
14. 2   **15.** 2   **16.** 8   **17.** 4   **18.** 48   **19.** 4   **20.** 4

Practice Exercises for Section 5.11, page 151

1. $2xy^2$ **2.** xz^2 **3.** $10a^3b^2$ **4.** $8x^2y$ **5.** $2p$ **6.** $5mn$ **7.** $5x$ **8.** $25y^2$ **9.** $15xy$ **10.** $2a^2b^2$
11. x^3 **12.** y^6 **13.** xy^4 **14.** $6x$ **15.** $8a$ **16.** $3x^2$ **17.** x^2 **18.** $7a^3$ **19.** $4x^2y$ **20.** $3a^2b^2$
21. $3ab^2$ **22.** $4ab^2$ **23.** $5ab^2c^3$ **24.** $2x^3yz$ **25.** $3ab^2c$ **26.** $8xy^4z^2$ **27.** ab^2c **28.** $4x$ **29.** $10a$
30. $5xy^2$

Practice Exercises for Section 5.12, page 153

1. xz^2 **2.** $10a^2b$ **3.** $8x^2y$ **4.** $2ab$ **5.** $2pq$ **6.** $5mn$ **7.** $2x^5y^5$ **8.** $6ab$ **9.** $3xy$ **10.** $20xy^4z^5$

Practice Exercises for Section 5.13, page 155

1. $5(a + b)$ **2.** $7(b - c)$ **3.** $4x(2x + 1)$ **4.** $5y(2 - 5y)$ **5.** $8(2b - a^2)$ **6.** $12(x - y)$ **7.** $4(2x + 3y)$
8. $3y(4y - 5)$ **9.** $4(x^2 - 2)$ **10.** $a(4a + 5)$ **11.** $3a^2(1 + 2a^3)$ **12.** $x(9 - 5x)$ **13.** $y(14y - 15)$

14. $2b^2(3b-5)$ **15.** $2x^2(x^2+4)$ **16.** $6(5a-1)$ **17.** $5(4b+1)$ **18.** $8(2a-3)$ **19.** $3y(y^3-3)$
20. $2x^2(5x^2-6)$ **21.** $4a^2(3a^3-8)$ **22.** $4a^5(2a^3-1)$ **23.** $8y^4(2+y^3)$ **24.** $5(2z^3-x^2)$ **25.** $b(6b+7)$
26. $3(2+y)$ **27.** $5t(3t^2-1)$ **28.** $9xy(x-4)$ **29.** $3ab(5+ab)$ **30.** $12n^2(1+2n)$ **31.** $3ab(2a-b)$
32. $7r(s^3-2r)$ **33.** $xy(xy+1)$ **34.** $3xy(xy^3-2)$ **35.** $3ab(4ab^4+3)$ **36.** cannot be factored   **37.** $6b^2(a^2b-2)$
38. $4x^2(2y^3-1)$ **39.** $ab(ab+1)$ **40.** $3(p^2+p-1)$ **41.** $ab(7a+14b-2)$ **42.** $3x(x^3+x-2)$
43. $2x(x^4-2x^2-4)$ **44.** $5y(y^2-2+3y)$ **45.** $15ab(a+2ab^2-3)$ **46.** $4x(y-3xy^2+4)$

Review Practice Exercises, page 156

1. a^4 **2.** x^9 **3.** $8x^3$ **4.** $y^8 \cdot z^{12}$ **5.** x^{-3} **6.** 1 **7.** x^6 **8.** x^8 **9.** 1 **10.** y **11.** $9x^2$
12. x^{-2} **13.** $9x^4$ **14.** $27x^{12}$ **15.** $16a^4$ **16.** $27x^3$ **17.** $16x^6$ **18.** $8a^3b^6$ **19.** $9x^4y^2$ **20.** 0.00032
21. $2.25x^2$ **22.** 1 **23.** $\dfrac{9x^2y^6}{16}$ **24.** $\dfrac{a^5}{b^5}$ **25.** $\dfrac{9x^4}{4}$ **26.** $\dfrac{64a^6}{b^9c^{15}}$ **27.** $-48a^5b^6$ **28.** $100x^6y^3z^4$
29. $4a^3b^6c^3d$ **30.** $2x$ **31.** $-3a^3b^2$ **32.** $\dfrac{-2}{m^3n^2}$ **33.** $\dfrac{2}{x^5y^3}$ **34.** $3x+y$ **35.** $4a+5b$ **36.** $5x+6y-8xy$
37. $ab-2a^3b^2+3$ **38.** x **39.** cannot be done (unlike terms)   **40.** x^3 **41.** cannot be done   **42.** cannot be done
43. cannot be done   **44.** $4Y$ **45.** cannot be done   **46.** $3b$ **47.** y^5 **48.** cannot be done   **49.** cannot be done
50. $11Q^2$ **51.** $-1.9x$ **52.** $-1.5x^3$ **53.** $-2a^4b^5$ **54.** $-\dfrac{1}{8}xy^2$ **55.** cd^3 **56.** $\dfrac{1}{7}pqr$ **57.** $4(x+1)$
58. $6(y+3)$ **59.** $3a(3a+4)$ **60.** $5(4x+5)$ **61.** $6(2x-3)$ **62.** $3(x+2)$ **63.** $b(3b+7)$ **64.** $3y(4y^2+5)$
65. $2ab(2a+5b)$ **66.** $3x(3x+2)$ **67.** $5pq(3q-5p)$ **68.** $2x(3x+2)$ **69.** $4a(4ab-5)$ **70.** $7xy(2xy+3)$
71. $a(a+7)$ **72.** $6a(a-1)$ **73.** $3(b^2+3)$ **74.** $4x(x+2)$ **75.** $5(2a^2-5)$ **76.** $10x(2x^2+3)$ **77.** $9y(y^2+3)$
78. $2x(x+2y)$ **79.** $3ab(ab+3)$ **80.** $6pq(2p-3q)$ **81.** $4xy(3x^2+5y)$ **82.** $4x^2(2xy-3)$ **83.** $8x(3x^2+2x+1)$
84. $13x^2y(3x^2y-2xy+1)$ **85.** $5ab(3a+2b-1)$ **86.** $2a^2(7a^3+6a^2-4a+2)$ **87.** $4xyz(4x-3y)$
88. $y(9x^2y^3+6x^2y^2-3xy-2)$ **89.** $5a^2bc(5a^2b^2c-3ab+2c)$ **90.** $4x^2y^2z^3(4y^2-2x^2+3xyz)$

Chapter 6

Practice Exercises for Section 6.1, page 164

1. $4+n$ **2.** $n+4$ **3.** $n+4$ **4.** $4+n$ **5.** $4+n$ **6.** $3-n$ **7.** $n-3$ **8.** $n-4$ **9.** $n-4$ **10.** $n-4$
11. $4-n$ **12.** $n-4$ **13.** $4-n$ **14.** $n-4$ **15.** $2n$ **16.** $2n$ **17.** $\dfrac{n}{3}$ **18.** $\dfrac{3}{n}$ **19.** $\dfrac{n}{3}$ **20.** $\dfrac{3-x}{9}$

21. A number subtracted from seven   **22.** Eight times a number   **23.** A number increased by seven   **24.** The difference between a number and fifteen (in this order)   **25.** Eight divided by a number   **26.** The product of a number and nineteen   **27.** A number added to twenty-three   **28.** A number subtracted from twenty-one   **29.** Thirty-three subtracted from a number   **30.** A number times ninety-six

Practice Exercises for Section 6.2, page 167

1. $2n+6$ **2.** $n+6n$ **3.** $8n-5$ **4.** $n(n+5)$ **5.** $\dfrac{n}{3}-12$ **6.** $\dfrac{5}{6}n+\dfrac{3}{8}n$ **7.** $(n-5)+12$ **8.** $5(x+3)$

9. $5x+2$ **10.** $16-(2x+11)$ **11.** $6n+4n$ **12.** $8-(6+x)$ **13.** $2x+(x+3)$ **14.** $2n+15$ **15.** $\dfrac{x+3}{2}$

16. $2n-3n$ **17.** $\dfrac{2(x-3)}{4}$ **18.** n^2-6 **19.** $3n+2n^2$ **20.** $4(3n+n^3)$ **21.** $3n+n$ **22.** $4(x+9)$

23. $12x+2x$ **24.** $4+\dfrac{6}{x}$ **25.** $3x+4$ **26.** $2x-7$ **27.** $10x-7$ **28.** $\dfrac{x}{2}+6$ **29.** x^2-2x **30.** $\dfrac{x}{x^2}$

Review Practice Exercises, page 169

1. $n+10$ **2.** $n+5$ **3.** $n-5$ **4.** $n-6$ **5.** $n-12$ **6.** $12-n$ **7.** $5n$ **8.** $\dfrac{n}{5}$ **9.** $\dfrac{5}{n}$ **10.** $3n+8$

11. $5n - 6$ **12.** $n^2 - 1$ **13.** $2n + 1$ **14.** $\dfrac{n}{4} - 10$ **15.** $n^2 + 3$ **16.** $\dfrac{n-5}{3}$ **17.** $5n - 10$ **18.** $\dfrac{3x}{10}$
19. $\dfrac{2}{3}x + 2x$ **20.** $17 + \dfrac{8}{x}$ **21.** Ten times a number   **22.** A number divided by four   **23.** Four divided by a
number   **24.** A number subtracted from nine   **25.** A number added to nine   **26.** The product of a number and twenty-three   **27.** Fourteen more than a number   **28.** Fourteen divided by a number   **29.** A number divided by fourteen
30. Twenty times a number

Chapter 7

Practice Exercises for Section 7.1, page 176

1. one   **2.** two   **3.** one   **4.** two   **5.** three   **6.** two   **7.** two   **8.** one   **9.** three   **10.** two

Practice Exercises for Section 7.2, page 179

1. 2   **2.** 6859   **3.** 2366   **4.** 59   **5.** 14   **6.** –288   **7.** 75,625   **8.** 43   **9.** 336   **10.** 18   **11.** 200
12. 28   **13.** –30   **14.** 1   **15.** 10   **16.** 1464   **17.** –1   **18.** 249   **19.** 0   **20.** 24   **21.** 3   **22.** 6
23. 6   **24.** 201   **25.** 2   **26.** 1.85   **27.** 5.37   **28.** 7.6225   **29.** — **32.** depend on your values.   **33.** 18
34. 4   **35.** –2   **36.** 6   **37.** 5   **38.** –9   **39.** –6   **40.** –5   **41.** –11   **42.** 28

Practice Exercises for Section 7.3, page 184

1. 15 in.^{2} **2.** 3.22 in.^{2} **3.** 288 cm.^{2} **4.** 55.62 cm.^{2} **5.** $\dfrac{3}{8}$ in.^{2} **6.** $\dfrac{15}{4}$ cm.^{2} **7.** 0.046 km.^{2} **8.** 1 ft.^{2}
9. 7.5 in.^{2} **10.** 4 cm.^{2} **11.** 4.75 in.^{2} **12.** 11.13 in.^{2} **13.** $\dfrac{3}{16}$ in.^{2} **14.** $\dfrac{15}{8}$ cm.^{2} **15.** 7.248 ft.^{2}
16. $\dfrac{1}{2}$ km.^{2} **17.** 15.035 in.^{2} **18.** 3.705 in.^{2} **19.** 5.8 cm.^{2} **20.** $\dfrac{25}{16}$ in.^{2} **21.** $\dfrac{3}{5}$ cm.^{2} **22.** 15.95 ft.^{2}
23. 7.4 in.   **24.** 72 cm.   **25.** 31.4 cm.   **26.** $\dfrac{5}{2}$ in.   **27.** $\dfrac{11}{5}$ cm.   **28.** 49.84 km.   **29.** 18.5 in.
30. 69.2 cm.   **31.** $\dfrac{5}{2}$ in.   **32.** $\dfrac{23}{10}$ cm.   **33.** 90.25 km.   **34.** $\dfrac{15}{16}$ cm.

Practice Exercises for Section 7.4, page 188

1. $13,350   **2.** $18,510   **3.** $62,674.23   **4.** $51,849   **5.** $104,468.84   **6.** $2,081,218   **7.** $14,719.75
8. $19.50   **9.** $1490.97   **10.** $1062   **11.** –$1928.60   **12.** –$315.35   **13.** –$15.38   **14.** –$7295.48
15. $415.39   **16.** $142.12   **17.** $220.20   **18.** $1177.37   **19.** $1753.15   **20.** $5149.58   **21.** $3759.03
22. $915.21   **23.** $225.07   **24.** $207.05   **25.** $464.30   **26.** $1489.30   **27.** $687.05   **28.** $156
29. $13.93   **30.** $1161.52

Practice Exercises for Section 7.5, page 192

1. 35°C   **2.** 15°C   **3.** 60°C   **4.** –15°C   **5.** 5°C   **6.** –5°C   **7.** –20°C   **8.** 110°C   **9.** 55°C
10. 40°C   **11.** 59°F   **12.** 122°F   **13.** 212°F   **14.** –4°F   **15.** 14°F   **16.** 68°F   **17.** 77°F
18. 98.6°F   **19.** 95°F   **20.** –103°F   **21.** 208 g.   **22.** 40 kg.   **23.** 12 g.   **24.** 480 oz.   **25.** 2057 oz.
26. 2343 oz.   **27.** 611 g.   **28.** 435 g.   **29.** 50 m.p.h.   **30.** 50 m.p.h.   **31.** 71 m.p.h.   **32.** 862 m.p.h.
33. 2416 km./h.   **34.** 7 miles/sec.   **35.** 107 miles/min.   **36.** 235 km./h.

Review Practice Exercises, page 194

1. –12   **2.** 3   **3.** 26   **4.** 16   **5.** 35   **6.** 980   **7.** 14   **8.** 34   **9.** 43   **10.** 36   **11.** 80   **12.** 1
13. 77   **14.** –140   **15.** 17   **16.** 46   **17.** 25   **18.** 5   **19.** 36   **20.** 37   **21.** 0.25   **22.** 106
23. 35   **24.** 175   **25.** 11,000   **26.** 5   **27.** 12   **28.** 164   **29.** 97   **30.** 8   **31.** 1.6   **32.** 3.6
33. 2.1   **34.** 8.75   **35.** 6.56   **36.** (a) $858.75 (b) $685.57   **37.** (a) $499.99 (b) $2060.66   **38.** (a) 40°C
(b) 25°C (c) 131°F (d) 59°F   **39.** (a) 75 g. (b) 38 g.   **40.** (a) 21 m.p.h. (b) 52 cm. per sec.

Chapter 8

Practice Exercises for Section 8.1, page 198

1. $15a + 56b$ **2.** $-2a - 24b$ **3.** $9x - 8y$ **4.** $21a + 26b$ **5.** $13a - 9b$ **6.** $5a - 4b^2$ **7.** $5a^2 + 2b^2$
8. $-x + 11y$ **9.** $-9xy + 9x$ **10.** $4x + 7$ **11.** $5y^3 + 2y$ **12.** $9x + 10y$ **13.** $3x - y$ **14.** $30a - 13b$
15. $2xy - 9x - 5y$ **16.** $-x^2 + 13xy - 1$ **17.** $-5x^4 - 2x^3 + 12x^2$ **18.** $-x^3 + 3x^2 - 9x$ **19.** $2y^3 - y$
20. $7y^5 - 2y^4 - 7y^3$ **21.** $-6b^4 - 20b^2$ **22.** $5a^2 - 8ab + 5ac + 10b^2 - 3bc - c^2$ **23.** $-r^2s + 11r^2t - 33st + rs$
24. $c^3d^3 - 18c^3d^2 + 9c^2d^3 + 3cd^3$ **25.** $21f^3g^3h^3 + f^3g^2h^3 - 16f^2g^3h^3$ **26.** $5x^4y^2 - 6a^2b^2 - 6abxy$

Practice Exercises for Section 8.2, page 200

1. $7a^2 - 3b^2 + c^2$ **2.** $-2y$ **3.** $x - 2y$ **4.** $3x$ **5.** $-3x$ **6.** $-8xz + 2yz$ **7.** $10x$ **8.** $-5x + 5y + 1$
9. $-4a - 12x + 8b$ **10.** $-6y$ **11.** $2xy$ **12.** $a^2b^2 + 6ab$ **13.** $\frac{1}{4}pq$ **14.** $\frac{3}{5}cd - 2st$ **15.** 0 **16.** a^2b^2
17. r^3s^3 **18.** $2.5c^2d$ **19.** $-15abc$ **20.** $10r^2s^2t^3$ **21.** $6u^2p^3$ **22.** $8x^2y^3$ **23.** $12p^2q^4$ **24.** 0
25. $16a^3b^5$ **26.** $7.2m^4p^2$ **27.** 0 **28.** $\frac{9d^3m^4}{2}$ **29.** $3y^2 + 14y + 15$ **30.** $b^2c^3 - 3ab^2c^3$

Review Practice Exercises, page 201

1. $5n + 10$ **2.** $-3x + 30$ **3.** $3u - 15$ **4.** $12p - 21$ **5.** $-2pq + 28p^2$ **6.** $2x - 7y$ **7.** $-14x - 5y$ **8.** $10x + 4$
9. $5a - 8b$ **10.** $-2a$ **11.** $15x + 2y - 6$ **12.** $-3x^3 - 15x^2 + 10x$ **13.** $-3x + 48$ **14.** $2y + 1$ **15.** $6x^3y - 3x^2y$
16. $9a^2b^3 + 9ab^2$ **17.** $-4x^3y^3 + 4x^2y^4$ **18.** $7x^3 + 27x^2$ **19.** $-5x^3 + 28x$ **20.** $28p^2q^2 - 15p^2q$ **21.** $3x - 3y$
22. $10x - 32$ **23.** $-5x - 40y$ **24.** $9a - 5$ **25.** $-44x + 20y + 3$ **26.** $30x - 36y + 10$ **27.** $9x + 30$ **28.** $-3r - 8$
29. $19y - 12$ **30.** $-9x - 2$ **31.** $-2y$ **32.** $5ab + 4a - 3b$ **33.** $-ab - 9ac + 8bc$ **34.** $-3x^2y - 7xy^2 + 10x^2y^2$
35. 0 **36.** $b^2d^5 + 7b^5d^2$ **37.** $-ab$ **38.** $-4xy - 9xyz$ **39.** $-10m^4p^3$ **40.** $-13f^2g^3$ **41.** $2r^6t^5$ **42.** $-15x^2z^3$

Chapter 9

Practice Exercises for Section 9.1, page 208

1. $25 = 5(5) = 5 + 5 + 5 + 5 + 5 = 75 - 50 = 100 - 75$ (There are many other answers.)
2. $\$2.50 = \$1.50 + \$1.00 = \$0.50 + \$0.50 + \$0.50 + \$0.50 + \$0.50 = \$1.25 \times 2 = \$5.00 + 2$
3. $10 = 7 + 2 + 1 = 4 + 5 + 1 = 9 + 0.5 + 0.5$ **4.** $1.5 = 0.5 + 1 = 0.3 + 0.2 + 1 = 0.7 + 0.3 + 0.5$
5. $5 - 3x - 6 = -1 - 3x = -7$ when $x = 2$ **6.** $2x + 4x - 20 = 6x - 20 = -38$ when $x = -3$
7. $4x - 6x + 3x^2 = -2x + 3x^2 = 1$ when $x = 1$ **8.** 5 **9.** $-5y - 3y^2 + 15y + 2y = -3y^2 + 12y = -12 - 24 = -36$ when $y = -2$
10. $4ab^2 - 5ab^2 + 3ab + 2ab = -ab^2 + 5ab = 4 - 10 = -6$ when $a = -1$ and $b = 2$

Practice Exercises for Section 9.2, page 214

1. 10 **2.** 8 **3.** 3 **4.** 3 **5.** 10 **6.** 6 **7.** 12 **8.** 7 **9.** -7 **10.** 0 **11.** 0 **12.** 0

Practice Exercises for Section 9.3, page 216

1. 1 **2.** 1 **3.** 2 **4.** 2 **5.** 1 **6.** 2 **7.** 3 **8.** 3 **9.** 3 **10.** 4

Review Practice Exercises, page 216

1. $50 = 10 + 10 + 10 + 10 + 10 = 20 + 20 + 10 = 5(10) = 25(2)$ (There are many other answers.)
2. $3.5 = 1.5 + 2 = 2 + 1 + 0.5 = 2.5 + 1$ **3.** $10 - 5x - 15 = -5x - 5 = 0$ when $x = -1$ **4.** $12 - 6x + 2x^2 = 8$ when $x = 2$
5. $8x^2 - 3x^2 - 6x = 5x^2 - 6x = 5 \cdot 4 - 6(-2) = 20 + 12 = 32$ when $x = -2$
6. $-3xy^2 + x^2y - 8xy^2 + 10x^2y = -11xy^2 + 11x^2y = 44 + 22 = 66$ when $x = -1$ and $y = 2$ **7.** 3 **8.** 8 **9.** 2 **10.** -16
11. 3 **12.** 2 **13.** 1 **14.** 2 **15.** 3 **16.** 4 **17.** 3 **18.** 2

Chapter 10

Practice Exercises for Section 10.1, page 224

1. 4   **2.** 3   **3.** 27   **4.** 81   **5.** 12   **6.** 6   **7.** 7   **8.** 4   **9.** 14   **10.** 18   **11.** 4   **12.** 10

13. 7   **14.** 7   **15.** 1   **16.** 3   **17.** 1   **18.** 15   **19.** $\frac{1}{4}$ **20.** 18   **21.** –7   **22.** 2   **23.** –3

24. 2   **25.** 20   **26.** 13   **27.** 3   **28.** 16   **29.** 27   **30.** –10

Practice Exercises for Section 10.2, page 227

1. $-\frac{15}{4}$ **2.** 9   **3.** 0.6   **4.** –4   **5.** $\frac{13}{4}$ **6.** $-\frac{6}{5}$ **7.** 20   **8.** 2   **9.** $\frac{1}{2}$ **10.** $\frac{1}{2}$ **11.** 3   **12.** 2

13. 35   **14.** 4   **15.** 9   **16.** 2   **17.** 21   **18.** $\frac{1}{2}$ **19.** $\frac{1}{3}$ **20.** 16   **21.** –4   **22.** 6   **23.** –27

24. –1   **25.** 3   **26.** 18   **27.** 7   **28.** $\frac{1}{6}$ **29.** 16   **30.** –1

Practice Exercises for Section 10.3, page 230

1. 3   **2.** 5   **3.** 5   **4.** 2   **5.** 2   **6.** 3   **7.** 1   **8.** 2   **9.** –5   **10.** 5   **11.** 2   **12.** 4   **13.** 4

14. 2   **15.** 1   **16.** 2   **17.** 3   **18.** $\frac{-1}{5}$ **19.** 2   **20.** –21   **21.** 2   **22.** 7   **23.** 4   **24.** 2   **25.** 8

26. 2   **27.** 3   **28** –10   **29.** –1   **30.** 2

Practice Exercises for Section 10.4, page 232

1. 0.6529   **2.** 1.869   **3.** –0.285   **4.** 0.884   **5.** 1.46   **6.** 0.273   **7.** 6.179   **8.** 0.5   **9.** 0.307

10. 1.347   **11.** 3.6255   **12.** 2.06   **13.** 0.33   **14.** 5   **15.** 6   **16.** –0.3   **17.** –7   **18.** 1.2   **19.** –2.35

20. 4   **21.** 2   **22.** 1.2   **23.** 2.1   **24.** 1   **25.** 1.5   **26.** –2   **27.** 6   **28.** –0.1   **29.** 3   **30.** 5

Practice Exercises for Section 10.5, page 234

1. 6   **2.** 9   **3.** 3   **4.** 8   **5.** 5   **6.** –7   **7.** 18   **8.** –13   **9.** 5   **10.** 9   **11.** –35   **12.** 6

13. 11   **14.** 30   **15.** 6   **16.** 6   **17.** 9   **18.** 2   **19.** 15   **20.** 8   **21.** 18   **22.** 30   **23.** 20

24. –10   **25.** 7   **26.** 18   **27.** 3   **28.** 30   **29.** 2   **30.** –7

Practice Exercises for Section 10.6, page 237

1. $-\frac{9}{2}$ **2.** –13   **3.** $-\frac{3}{2}$ **4.** $\frac{20}{11}$ **5.** 0   **6.** $\frac{4}{13}$ **7.** –3   **8.** –3   **9.** 1   **10.** $\frac{7}{3}$ **11.** 1

12. $\frac{11}{2}$ **13.** 4   **14.** 3   **15.** $\frac{1}{8}$ **16.** 10   **17.** 2.5   **18.** no solution   **19.** $\frac{2}{3}$ **20.** 3   **21.** $\frac{3}{2}$

22. 2.1   **23.** $\frac{1}{6}$ **24.** 8   **25.** –4.4   **26.** 7   **27.** 5   **28.** –2   **29.** 10   **30.** 8

Review Practice Exercises, page 238

1. 22   **2.** 5   **3.** –9   **4.** –4   **5.** 5   **6.** –3   **7.** 2   **8.** 4   **9.** –9   **10.** –4   **11.** 5   **12.** –9

13. 3   **14.** –4   **15.** –10   **16.** 0   **17.** 36   **18.** 56   **19.** 160   **20.** –12   **21.** 21   **22.** 0   **23.** –5

24. 6   **25.** –5   **26.** 45   **27.** $\frac{9}{8}$ **28.** $\frac{1}{2}$ **29.** 0   **30.** 19.5   **31.** –4   **32.** 8   **33.** 9   **34.** 4

35. $\frac{1}{2}$ **36.** $\frac{2}{3}$ **37.** $-\frac{18}{5}$ **38.** $-\frac{7}{6}$ **39.** $-\frac{1}{38}$ **40.** –3   **41.** $\frac{1}{5}$ **42.** 0.3   **43.** 4   **44.** $\frac{6}{5}$

45. $\frac{1}{3}$ **46.** –2   **47.** $\frac{1}{4}$ **48.** $\frac{1}{2}$ **49.** 8   **50.** –5

Chapter 11

Practice Exercises for Section 11.1, page 244

1. $10 - y$ **2.** $r + 3$ **3.** $\dfrac{10}{a}$ **4.** $\dfrac{c}{4}$ **5.** $\dfrac{c}{m}$ **6.** $\dfrac{m}{t}$ **7.** $\dfrac{q}{l}$ **8.** qr **9.** $\dfrac{a}{c}$ **10.** $10 - ex$ **11.** $\dfrac{y - b}{m}$

12. $\dfrac{b - 10}{5}$ **13.** $\dfrac{-10 - n}{5}$ **14.** $\dfrac{7 - c}{5}$ **15.** $\dfrac{15 + T}{3}$ **16.** $\dfrac{7N}{8}$ **17.** $\dfrac{-1}{3BC}$ **18.** $\dfrac{100 - 4A - 3B}{5}$

19. $\dfrac{6 - 3x}{2}$ **20.** $4x - 3$ **21.** $6 - 6y$ **22.** $10 - 3y$ **23.** $\dfrac{y + 6}{2}$ **24.** $\dfrac{12 - 3y}{4}$ **25.** $\dfrac{t - qy}{p}$ **26.** $\dfrac{10 - px}{q}$

27. $\dfrac{3}{2yz}$ **28.** $\dfrac{3x + 2y - 10t}{4}$ **29.** $\dfrac{10t + 4z - 2y}{3}$ **30.** $\dfrac{2p + 3q - w}{4}$

Practice Exercises for Section 11.2, page 246

1. $\dfrac{2A - ha}{h}$ **2.** $\dfrac{P - 2W}{2}$ **3.** $R - N$ **4.** $A - L$ **5.** $N + D$ **6.** $G - N$ **7.** $\dfrac{M}{V}$ **8.** $\dfrac{D}{V}$ **9.** Vt

10. $\dfrac{5}{9}(F - 32)$ **11.** $\dfrac{2A}{b}$ **12.** $\dfrac{3r}{h}$ **13.** $\dfrac{ct - R}{t}$ **14.** $\dfrac{E}{I}$ **15.** $pn + C$ **16.** $\dfrac{PV}{nR}$ **17.** $\dfrac{A}{1 + RT}$

18. $\dfrac{A}{S + 1}$

Practice Exercises for Section 11.3, page 247

1. 5 in. **2.** 3 in. **3.** 3 in. **4.** 2.5 cm. **5.** \$4574.55 **6.** 15 years **7.** \$350 **8.** \$1133 **9.** 240 mi.
10. $0.125 \dfrac{g \cdot}{ml.}$ **11.** 10 cm. **12.** 5 in. **13.** 14 in. **14.** \$2187 **15.** 12 in. **16.** \$68

Review Practice Exercises, page 249

1. $\dfrac{4 - y}{5}$ **2.** $5y - 4$ **3.** $\dfrac{1 + 8x}{3}$ **4.** $9x - 30$ **5.** $\dfrac{2a + b - d}{4}$ **6.** $\dfrac{y + 8}{3}$ **7.** $\dfrac{2 - y}{7}$ **8.** $\dfrac{4y - 1}{3}$

9. $\dfrac{V}{A}$ **10.** $\dfrac{3V}{A}$ **11.** $S - a - c$ **12.** $\dfrac{U}{I}$ **13.** $\dfrac{P - A}{Ak}$ **14.** $\dfrac{ax - 16}{b}$ **15.** $\dfrac{d - s}{t}$ **16.** $\dfrac{f - v}{a}$

17. $\dfrac{I}{PR}$ **18.** $\dfrac{s - a}{s}$ **19.** 3 cm. **20.** 9.2 cm. **21.** 6 cm. **22.** 25 ml. **23.** 4.3 hours or 4 hours and 18 minutes

Chapter 12

Practice Exercises for Section 12.1, page 256

1. Three less than a number is two. **2.** A number subtracted from three is two. **3.** The sum of four and a number is five.
4. The sum of a number and four is five. **5.** The product of five and a number is fifteen. **6.** The product of a number and five is fifteen. **7.** The quotient of a number divided by three is six. **8.** Three divided by a number is six. **9.** The sum of a number and five is seven. **10.** The difference of a number and three (in this order) is negative five. **11.** When six is subtracted from a number, the result is ten. **12.** The sum of nine and a number is negative three. **13.** Twice a number is negative fourteen. **14.** The product of negative thirteen and a number is six. **15.** Twenty is four times a number.
16. One-fourth of a number is three. **17.** Three-fourths of a number is nine. **18.** The opposite of two-fifths of a number is three.

Practice Exercises for Section 12.2, page 258

1. The sum of twice a number and four is six. **2.** Four less than twice a number is six. **3.** The sum of a number divided by two and two is three. **4.** Two divided by a number plus two is three. **5.** Three less than the quotient of a number divided by three is five. **6.** Three subtracted from the quotient of three divided by a number is five. **7.** Four minus the quotient of a number divided by five is three. **8.** Four minus the quotient of five divided by a number is three. **9.** The sum of three times a number and one is ten. **10.** The difference of five times a number and six (in this order) is nine. **11.** Five is equal to the sum of four times a number and nine. **12.** Two is equal to the sum of five times a number plus twelve. **13.** The difference of seven and a number (in this order) is nine. **14.** The sum of six and twice a number is zero.

Practice Exercises for Section 12.3, page 261

1. $x + 7 = 19$; 12   2. $x + 17 = 29$; 12   3. $x + 8 = -12$; -20 4. $x - 13 = -4$; 9   5. $29 - x = 4$; 25

6. $x - 13 = 41$; 54   7. $18 - x = 7$; 11   8. $54 - x = 41$; 13   9. $2x = 5$; 2.5   10. $44x = 132$; 3   11. $\frac{x}{2} = 17$; 34

12. $\frac{32}{x} = 4$; 8   13. $\frac{x}{12} = 21$; 252   14. $3x = -30$; -10 15. $-2x = 0$; 0   16. $x + 12 = 10$; -2 17. $x - 5 = 2$; 7

18. $\frac{2}{3}x = 6$; 9   19. $\frac{x}{7} = -3$; -21 20. $x + 13 = 31$; 18   21. $11x = 187$; 17   22. $\frac{5}{3}x = 20$; 12   23. $\frac{x}{4} = 15$; 60

24. $\frac{50}{x} = 5$; 10

Practice Exercises for Section 12.4, page 264

1. $5x + 7 = 9$; $\frac{2}{5}$ 2. $3x - 21 = 7$; $\frac{28}{3}$ 3. $2x + (-8) = 32$; 20   4. $8x - 1 = 11$; $\frac{3}{2}$ 5. $\frac{x}{6} - 1 = \frac{1}{2}$; 9

6. $5x + 14 = 7x + 8$; 3   7. $2x + 15 = 5x$; 5   8. $\frac{x}{5} - 4 = 2$; 30   9. $\frac{8x}{5} + 3 = 7$; 2.5   10. $\frac{2x}{7} - 2 = 10$; 42

11. $x + 14 = 86$; 72   12. $13x = 91$; 7   13. $\frac{4}{7}x = 44$; 77   14. $x + 6 = 14$; 8   15. $7x + 6 = 62$; 8

16. $\frac{1}{5}x - 9 = 3$; 60   17. $4x + x = 80$; 16   18. $2x - 7 = 11$; 9   19. $3(x + 4) = 15$; 1   20. $20 - 5x = 10$; 2

Practice Exercises for Section 12.5, page 267

1. $2x + 4 = 10$; 3   2. $2x + 3x = 30$; 6   3. $3x = 12 - x$; 3   4. $0.05x + 0.10x = 1.05$; 7   5. $x + 1.48 = 5x$; 0.37

6. $\frac{x}{4} - 5 = 11$; 64   7. $\frac{85 + x}{2} = 90$; 95   8. $4(2x - 3) = 44$; 7   9. $\frac{x}{3} + 4 = 8$; 12   10. $x + 40 = 5x$; 10

11. $4x + 1 = 9$; 2   12. $4(x - 6) = 60$; 21   13. $10x - 2x = 22$; $\frac{22}{8}$ 14. $5 + 4x = 25$; 5   15. $\frac{7 + x}{2} = 10$; 13

16. $3x + 5 = 20$; 5   17. $4 = x + (3x - 8)$; 3   18. $5x + 7 = 47$; 8   19. $15 - (x + 6) = 3$; 6   20. $12 - 5x = 7$; 1

21. $3x + x = 12$; 3

Practice Exercises for Section 12.6, page 269

1. $x - 6 = 2$ 2. $5x = 30$ 3. $x + (3x) = 84$ 4. $9x - x = 56$

Review Practice Exercises, page 270

1. The sum of a number and two is seventeen. Two is added to a number. The result is seventeen.   2. Twice a number plus five equals twenty-nine. The product of two and a number is increased by five. The result is twenty-nine.   3. The sum of five times a number and seven is forty-two. Seven added to the product of five and a number is forty-two.   4. Nine less than a number is thirteen. The difference between a number and nine (in this order) is thirteen.   5. Nine less than the product of three and a number is twenty-one. Three times a number minus nine is twenty-one.   6. Three less than the product of seven and a number is twenty-five. The difference between seven times a number and three (in this order) is twenty-five.   7. The sum of one-third of a number and two is five. Two more than the quotient of a number divided by three is five.   8. One-eighth of a number minus four is two. Four less than the quotient of a number divided by eight is two.   9. The difference between ten and one-fifth of a number (in this order) is six. The quotient of a number divided by five when subtracted from ten is six.
10. Three more than the quotient of six divided by a number is nine. The sum of six divided by a number and three is nine.
11. Five subtracted from three times a number is seven. Five less than three times a number is equal to seven.   12. One-half the sum of three times a number and seven is eight. When three times a number is increased by seven and the sum is divided by two, the result is eight.   13. Six times the difference of a number and five (in this order) is forty-two. The product of six and five less than a number is forty-two.   14. The sum of four times a number and twelve is thirty-two. When four times a number is increased by twelve the result is thirty-two.   15. Six subtracted from the quotient of a number and five is equal to three. Six less than a number that is divided by five is three.   16. Two times seven less than a number is six. The product of two and the difference of a number and seven (in this order) is six.   17. The difference of five times a number and two (in this order) is thirteen. Two less than five times a number is thirteen.   18. Ten is equal to one-half of a number increased by thirteen. Ten is the sum of one-half of a number and thirteen.   19. Fifteen added to the quotient of five times a number divided by four is ten. The sum of five times a number divided by four and fifteen is ten.   20. Three times the sum of a

number divided by four and four is nine. The product of three and the sum of a quarter of a number and four equals nine.
21. Three times the sum of twice a number and nine is equal to nine. When three is multiplied by nine more than twice a number, the result is nine.   **22.** One-fifth of seven less than three times a number equals four. When seven is subtracted from three times a number and the difference is divided by five, the result is four.   **23.** $7 - x = 14; -7$ **24.** $x - 2 = -8; -6$
25. $x - 9 = 51; 60$ **26.** $x - 4 = 28; 32$ **27.** $25 - x = -2; 27$ **28.** $5x + 8 = 33; 5$ **29.** $7x - 3 = 109; 16$
30. $2x - 7 = 43; 25$ **31.** $\frac{x}{16} + 11 = 13; 32$ **32.** $\frac{x}{32} + 20 = 26; 192$

Chapter 13

Practice Exercises for Section 13.1, page 277

1. $\frac{2}{3}$ **2.** $\frac{7}{6}$ **3.** $\frac{3}{1}$ **4.** $\frac{2}{1}$ **5.** $\frac{3}{2}$ **6.** $\frac{3}{4}$ **7.** $\frac{4}{7}$ **8.** $\frac{1}{3}$ **9.** $\frac{2}{5}$ **10.** $\frac{3}{4}$ **11.** 2:7   **12.** 2:3
13. 2:9   **14.** 1:3   **15.** 5:1   **16.** 3:4   **17.** 5   **18.** 24   **19.** 5   **20.** 65   **21.** 2   **22.** 45   **23.** 8
24. 15   **25.** 12   **26.** 31   **27.** 32   **28.** 3   **29.** 11   **30.** 27   **31.** 18 errors   **32.** 60 domestic cars
33. 160 domestic cars   **34.** 360 kilometers   **35.** (a) 40 games   (b) 8 games   **36.** 14 inches   **37.** $5760
38. 9 cups   **39.** 19.5 miles   **40.** 1125

Practice Exercises for Section 13.2, page 284

1. $(0.36)(144) = P; P = 51.84$ **2.** $R(350) = 35; R = 10\%$ **3.** $0.25W = 26; W = 104$ **4.** $0.035(54) = P; P = 1.89$
5. $1(11) = P; P = 11$ **6.** $R(25) = 75; R = 300\%$ **7.** $R(60) = 80; R = 1.33\%$ **8.** $0.7W = 35; W = 50$ **9.** 2.31
10. 400   **11.** 80   **12.** 70   **13.** 12   **14.** 50   **15.** 573.8   **16.** 2.56   **17.** $48   **18.** 17 questions
19. 75%   **20.** 195 tons   **21.** $1170   **22.** $97.30   **23.** 62.5%   **24.** $8.70   **25.** 400 workers   **26.** 25%
27. 20%   **28.** 10%   **29.** 20%   **30.** 25%

Practice Exercises for Section 13.3, page 288

1. length = 12 m., width = 4 m.   **2.** length = 32 m., width = 15 m.   **3.** length = 10 cm., width = 2 cm.
4. length = 7 ft., width = 3 ft.   **5.** length = 24 ft., width = 15 ft.   **6.** length = 10 cm., width = 2 cm.
7. length = 62 ft., width = 51 ft.   **8.** side = 17 cm.   **9.** side = 37 yards   **10.** side = 28 m.
11. length = 6.5 ft., width = 3.5 ft.   **12.** 3 in.   **13.** 4 cm.   **14.** 8 in.   **15.** length = 20 cm., width = 10 cm.
16. 0.9 in.

Practice Exercises for Section 13.4, page 293

1. Deborah: 1; Edward: 9   **2.** Eileen: 28; Irene: 13   **3.** Mark: 6; David: 21   **4.** Pablo: 12; Neil: 4
5. Sarah: 10; Martin: 8   **6.** Juan: 4; Alicia: 11   **7.** 33   **8.** 7   **9.** 14   **10.** Elaine: 35; Betty: 43
11. Tony: 50; Anne: 55   **12.** Ken: 4; Lee: 8

Review Practice Exercises, page 294

1. 12   **2.** 45   **3.** 2   **4.** 6   **5.** 5   **6.** 6.25 pounds   **7.** 15.625   **8.** 600   **9.** 0.7875   **10.** 32
11. $2.75   **12.** 56 oranges   **13.** 40%   **14.** 25%   **15.** 75%   **16.** 20%   **17.** 600 dogs
18. length = 62 ft., width = 51 ft.   **19.** length = 15 m., width = 6 m.   **20.** side = 19 ft.   **21.** Joan: 18; Tom: 12
22. Peter: 12; Tom:18   **23.** 80%   **24.** 40%   **25.** 18 ft.   **26.** 24   **27.** Sigmund: 10; Raymond: 8

Chapter 14

Practice Exercises for Section 14.1, page 302

1. $A = (4, 10)$ **2.** $B = (7, 10)$ **3.** $C = (10, 11)$ **4.** $D = (13, 12)$ **5.** $E = (1, 9)$ **6.** $F = (2, 8)$ **7.** $G = (4, 8)$
8. $H = (6, 9)$ **9.** $I = (9, 7)$ **10.** $J = (11, 7)$ **11.** $K = (1, 6)$ **12.** $L = (3, 6)$ **13.** $M = (5, 5)$ **14.** $N = (6, 6)$
15. $O = (8, 6)$ **16.** $P = (10, 4)$ **17.** $Q = (13, 4)$ **18.** $R = (2, 3)$ **19.** $S = (3, 4)$ **20.** $T = (5, 3)$
21. $U = (7, 2)$ **22.** $V = (8, 2)$ **23.** $W = (11, 3)$ **24.** $X = (1, 1)$ **25.** $Y = (6, 1)$ **26.** $Z = (12, 1)$

Practice Exercises for Section 14.2, page 306

1. $A = (-7, 7)$ **2.** $B = (0, 7)$ **3.** $C = (6, 7)$ **4.** $D = (-6, 6)$ **5.** $E = (-4, 5)$ **6.** $F = (-1, 6)$ **7.** $G = (2, 5)$
8. $H = (5, 4)$ **9.** $I = (-5, 3)$ **10.** $J = (-3, 4)$ **11.** $K = (-2, 3)$ **12.** $L = (0, 3)$ **13.** $M = (3, 2)$
14. $N = (4, 1)$ **15.** $O = (6, 3)$ **16.** $P = (7, 2)$ **17.** $Q = (-7, 0)$ **18.** $R = (-4, 2)$ **19.** $S = (-1, -1)$
20. $T = (1, -1)$ **21.** $U = (5, -2)$ **22.** $V = (-5, -4)$ **23.** $W = (-4, -6)$ **24.** $X = (-3, -3)$ **25.** $Y = (0, -5)$
26. $Z = (3, -6)$

Practice Exercises for Section 14.3, page 310

1. – 10.

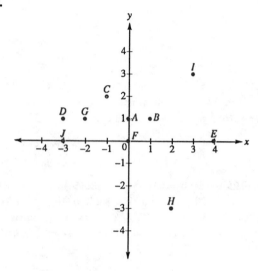

Practice Exercises for Section 14.4, page 312

1.

2.

3.

4.

5.

Practice Exercises for Section 14.5, page 315

1.

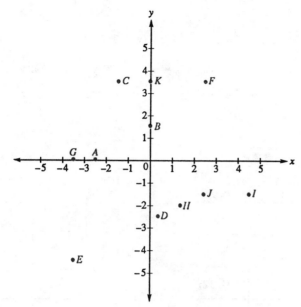

Review Practice Exercises, page 316

1.

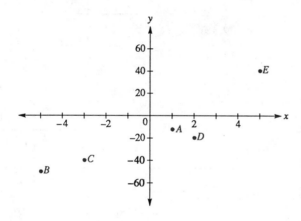

2.

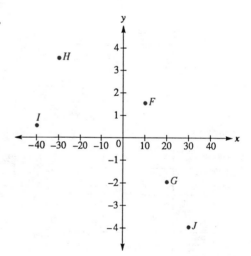

3.

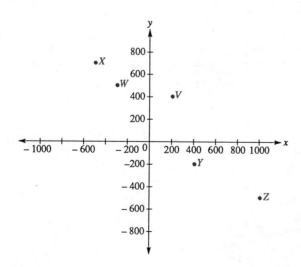

Chapter 15

Practice Exercises for Section 15.1, page 320

1. $x - y = 10$ **2.** $xy = 10$ **3.** $\dfrac{x}{y} = 2$ **4.** $x + 2y = 4$ **5.** $x^2 + y^2 = 25$ **6.** $x + 3y = 10$ **7.** $2x - y = 15$

8. $\dfrac{2x}{3y} = 21$ **9.** $y + x^2 = 16$ **10.** $x - \dfrac{x}{y} = 100$ **11.** $x^2 - y^2 = 12$ **12.** $x = \dfrac{y}{2} + 2$ **13.** $y = x^2$ **14.** $y = x + 2$

15. $y = x^3$ **16.** $y = -2x$ **17.** $y = 2x + 1$ **18.** $y = 2x - 1$ **19.** $y = 2x$ **20.** $y = \sqrt{x}$ **21.** $y = x^2 - 1$

22. $y = x^3$

Practice Exercises for Section 15.2, page 324

1.

x	y	(x, y)
0	1	$(0, 1)$
$-\dfrac{1}{8}$	0	$\left(-\dfrac{1}{8}, 0\right)$
-1	-7	$(-1, -7)$
-2	-15	$(-2, -15)$
2	17	$(2, 17)$
-3	-23	$(-3, -23)$

2.

x	y	(x, y)
0	6	$(0, 6)$
-1	0	$(-1, 0)$
1	12	$(1, 12)$
-2	-6	$(-2, -6)$
$\dfrac{1}{2}$	9	$\left(\dfrac{1}{2}, 9\right)$
2	18	$(2, 18)$

3.

x	y	(x, y)
0	3	$(0, 3)$
$\dfrac{3}{2}$	0	$\left(\dfrac{3}{2}, 0\right)$
-1	5	$(-1, 5)$
1	1	$(1, 1)$
3	-3	$(3, -3)$
2	-1	$(2, -1)$

4.

x	y	(x, y)
0	-2	$(0, -2)$
$\dfrac{1}{2}$	0	$\left(\dfrac{1}{2}, 0\right)$
-1	-6	$(-1, -6)$
$\dfrac{3}{4}$	1	$\left(\dfrac{3}{4}, 1\right)$
2	6	$(2, 6)$
3	10	$(3, 10)$
-2	-10	$(-2, -10)$

5.

x	y	(x, y)
0	-1	$(0, -1)$
$-\dfrac{1}{3}$	0	$\left(-\dfrac{1}{3}, 0\right)$
-2	5	$(-2, 5)$
2	-7	$(2, -7)$
3	-10	$(3, -10)$
-3	8	$(-3, 8)$
-4	11	$(-4, 11)$

6.

x	y	(x, y)
0	2	$(0, 2)$
$-\dfrac{2}{3}$	0	$\left(-\dfrac{2}{3}, 0\right)$
-2	-4	$(-2, -4)$
-1	-1	$(-1, -1)$
2	8	$(2, 8)$
1	5	$(1, 5)$
3	11	$(3, 11)$

7.

x	y	(x, y)
0	−1	(0, −1)
$-\frac{1}{4}$	0	$\left(-\frac{1}{4}, 0\right)$
1	−5	(1, −5)
2	−9	(2, −9)
3	−13	(3, −13)
−2	7	(−2, 7)
−3	11	(−3, 11)

8.

x	y	(x, y)
0	−4	(0, −4)
$\frac{4}{5}$	0	$\left(\frac{4}{5}, 0\right)$
1	1	(1, 1)
2	6	(2, 6)
−1	−9	(−1, −9)
3	11	(3, 11)
−3	−19	(−3, −19)

For Exercises 9–30, the given answers are only some of the solutions.

9. (0, −5), (1, −3), (2.5, 0)   **10.** (0, 3), (1, 2), (3, 0)   **11.** (0, −5), (1, −7), (−2.5, 0)   **12.** (0, 5), (5, 0), (1, 4)

13. (0, −3), (3, 0), (1, −2)   **14.** (0, 2), (3, 0), (−3, 4)   **15.** (0, −2), (3, 0), (−3, −4)   **16.** (0, 2), (5, 0), (−5, 4)

17. (0, −2), (5, 0), (−5, −4)   **18.** (0, −5), (15, 0), (3, −4)   **19.** (0, 1), (2, 5), (3, 7)   **20.** (−1, −5), (0, −2), (1, 1)

21. (−2, −4), (−1, −2), (0, 0)   **22.** (0, 3), (4, 0), (8, −3)   **23.** (0, −2), (5, 0), (10, 2)   **24.** (0, 3), (2, 2), (6, 0)

25. (0, −3), (1, −1), (2, 1)   **26.** (0, 2), (1, 0), (2, −2)   **27.** (0, −3), (1, 0), (2, 3)   **28.** (0, 4), (1, 2), (2, 0)

29. (0, −2), (2, −1), (4, 0)   **30.** (−4, 8), (−2, 5), (0, 2)

Practice Exercises for Section 15.3, page 331

9.

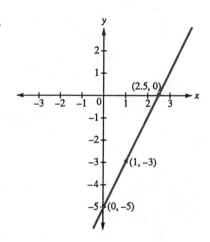

10.

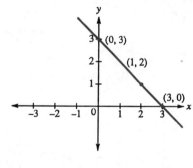

11.

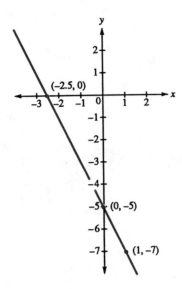

12.

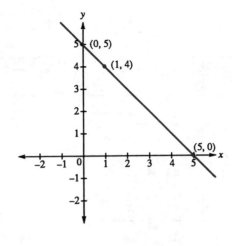

13.

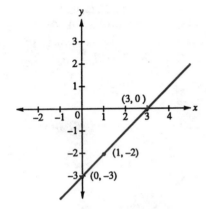

14.

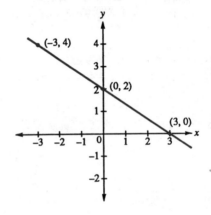

15.

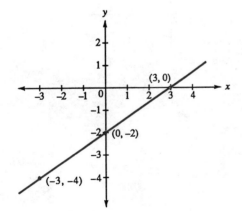

16.

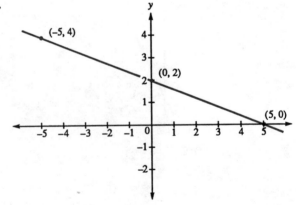

17.

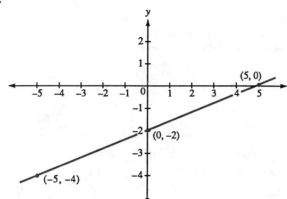

18.

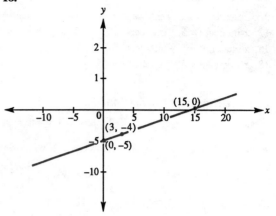

19.

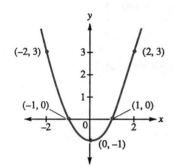

20.

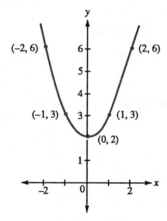

21.

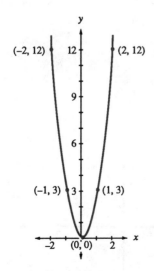

22.

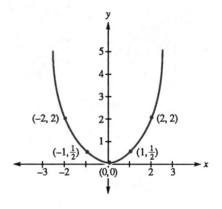

Practice Exercises for Section 15.4, page 334

9. x-intercept = 2.5, y-intercept = –5   **10.** x-intercept = 3, y-intercept = 3   **11.** x-intercept = –2.5, y-intercept = –5

12. x-intercept = 5, y-intercept = 5   **13.** x-intercept = 3, y-intercept = –3   **14.** x-intercept = 3, y-intercept = 2

15. x intercept = 3, y-intercept = –2   **16.** x-intercept = 5, y-intercept = 2   **17.** x-intercept = 5, y-intercept = –2

18. x-intercept = 15, y-intercept = –5

Practice Exercises for Section 15.5, page 337

1. slope = 1, y-intercept = 1   **2.** slope = 3, y-intercept = –2   **3.** slope = $\frac{1}{2}$, y-intercept = 1

4. slope = 1, y-intercept = –3   **5.** slope = 3, y-intercept = 1   **6.** slope = –2, y-intercept = –1

7. slope = $\frac{2}{3}$, y-intercept = –2   **8.** slope = $-\frac{3}{4}$, y-intercept = 1   **9.** slope = $\frac{1}{2}$, y-intercept = $-\frac{3}{2}$

10. slope = $\frac{3}{2}$, y-intercept = $-\frac{5}{2}$ **11.** slope = $-\frac{5}{4}$, y-intercept = $\frac{3}{4}$ **12.** slope = –3, y-intercept = 4

13. slope = 4, y-intercept = –6   **14.** slope = $-\frac{3}{2}$, y-intercept = 3   **15.** slope = –2, y-intercept = 3

16. slope = 3, y-intercept = –1   **17.** slope = $\frac{1}{2}$, y-intercept = –2   **18.** slope = $-\frac{1}{3}$, y-intercept = 2

19. slope = –1, y-intercept = 1   **20.** slope = $\frac{3}{4}$, y-intercept = –3   **21.** slope = $\frac{5}{2}$, y-intercept = –5

22. slope = –4, y-intercept = 2   **23.** slope = $-\frac{4}{5}$, y-intercept = 4   **24.** slope = 1, y-intercept = 3

Practice Exercises for Section 15.6, page 340

1.

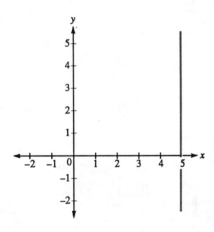

2.

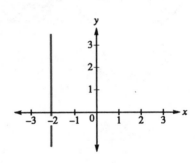

3.

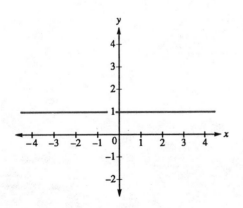

4.

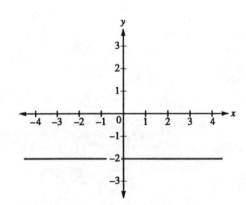

5.

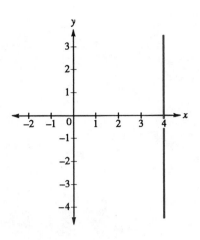

6.

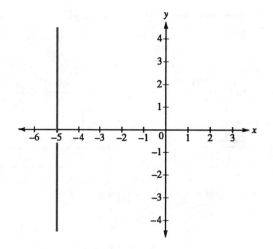

7.

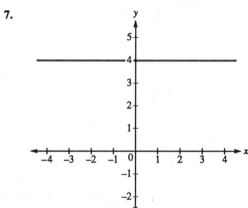

8.

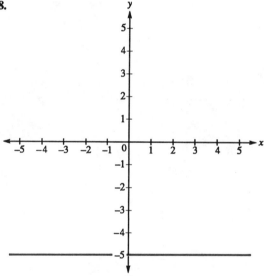

9.

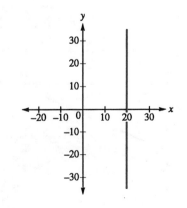

10.

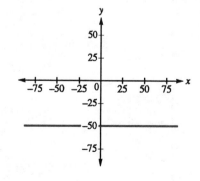

Review Practice Exercises, page 341

1. $2x - 3y = 6$ **2.** $x + 4y = 8$ **3.** $x = -y + 3$ **4.** $\dfrac{y}{4} + \dfrac{x}{2} = 12$ **5.** $y = x - 1$ **6.** $x - y = 1$

7.

x	y	(x, y)
0	0	(0, 0)
1	2	(1, 2)
2	4	(2, 4)
3	6	(3, 6)
5	10	(5, 10)

8.

x	y	(x, y)
0	-2	(0, -2)
1	1	(1, 1)
-1	-5	(-1, -5)
2	4	(2, 4)
3	7	(3, 7)

9. x-intercept = -1; y-intercept = 2; slope = 2 **10.** x-intercept = -4; y-intercept = 12; slope = 3

11. x-intercept = 2; y-intercept = -3; slope = $\dfrac{3}{2}$ **12.** x-intercept = 2; y-intercept = -3; slope = $\dfrac{3}{2}$

13. x-intercept = -2; y-intercept = 1; slope = $\dfrac{1}{2}$ **14.** x-intercept = $\dfrac{1}{3}$; y-intercept = -1; slope = 3

15. x-intercept = 1; y-intercept = $\dfrac{2}{3}$; slope = $-\dfrac{2}{3}$

16.

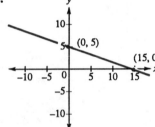

17.

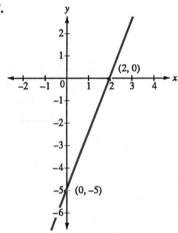

19.

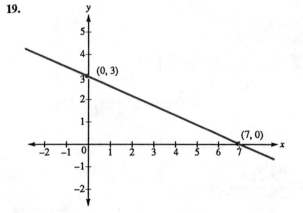

20.

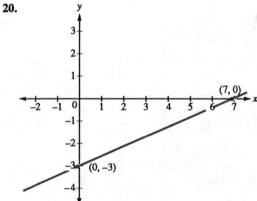

21.

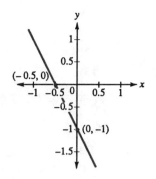

22.

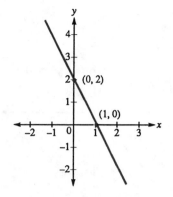

23.

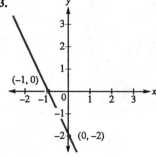

24.

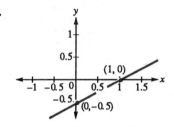

25.

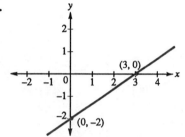

26.

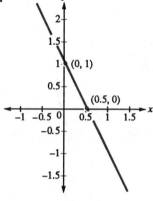

27.

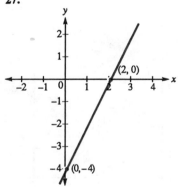

28.

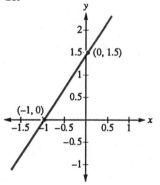

29.

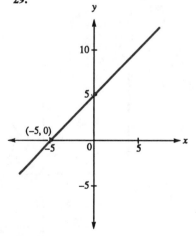

30.

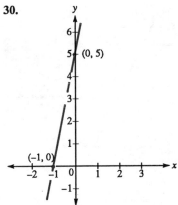

31.

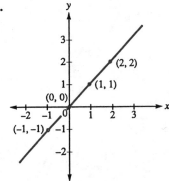

32.

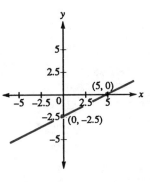

33.

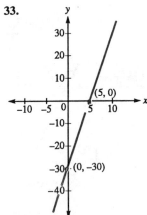

34.

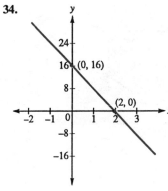

35.

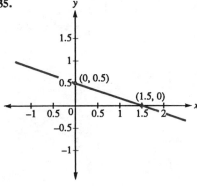

36.

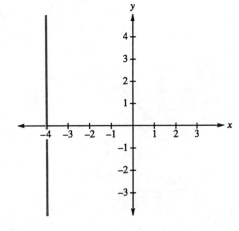

37.

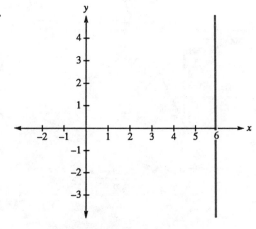

38.

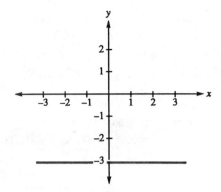

39.

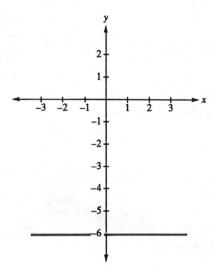

40.

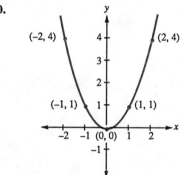

41.

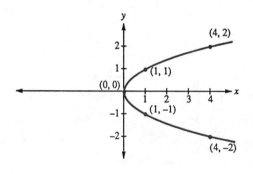

42.

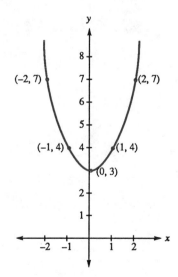

43.

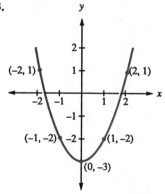

44.

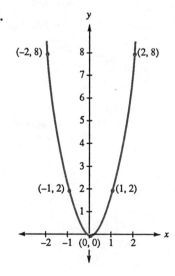

Appendix

Practice Exercises for Section A.1, page 346

1. 1.74   **2.** 20.08   **3.** 12   **4.** 10.76   **5.** 4.42   **6.** 8.76   **7.** 7.08   **8.** 8.64   **9.** 13.63   **10.** 12.95
11. 8.75   **12.** 1.3   **13.** 2.2   **14.** 4.7   **15.** 0.002   **16.** 0.96   **17.** 2.7   **18.** 9.2   **19.** 2.8   **20.** 3.666

Practice Exercises for Section A.2, page 350

1. 30.9014   **2.** 0.8313   **3.** 0.0008   **4.** 4.5   **5.** 130   **6.** 9.45   **7.** 0.36   **8.** 0.0061   **9.** 73.44
10. 2.822   **11.** 44   **12.** 10,000   **13.** 12   **14.** 5   **15.** 0.03   **16.** 30   **17.** 0.05   **18.** 860   **19.** 3.1
20. 21

Practice Exercises for Section A.3, page 351

1. 4000   **2.** 10,000   **3.** 50   **4.** 398,000,000   **5.** 44,000   **6.** 1250   **7.** 35,800   **8.** 135,000
9. 2,460,000   **10.** 3,400,000   **11.** 42.86   **12.** 1.0   **13.** 1.53   **14.** 4.0   **15.** 16   **16.** 6   **17.** 35.89
18. 0.397   **19.** 5937.0508   **20.** 0.05723

Practice Exercises for Section A.4, page 354

1. $6\frac{2}{3}$ **2.** $1\frac{7}{11}$ **3.** $3\frac{1}{4}$ **4.** $2\frac{1}{10}$ **5.** $5\frac{2}{7}$ **6.** $\frac{9}{5}$ **7.** $\frac{11}{4}$ **8.** $\frac{9}{2}$ **9.** $\frac{20}{11}$ **10.** $\frac{20}{3}$ **11.** $\frac{9}{12}$
12. $\frac{15}{33}$ **13.** $\frac{56}{72}$ **14.** $\frac{9}{48}$ **15.** $\frac{55}{65}$ **16.** $\frac{3}{5}$ **17.** $\frac{2}{3}$ **18.** $\frac{3}{5}$ **19.** $\frac{1}{13}$ **20.** $\frac{7}{13}$

Practice Exercises for Section A.5, page 356

1. 0.4   **2.** 0.75   **3.** 1.4   **4.** 0.57   **5.** 1.18   **6.** 0.6   **7.** 3.01   **8.** 0.375   **9.** 2.9   **10.** 0.15
11. $\frac{9}{20}$ **12.** $\frac{3}{2}$ **13.** $\frac{203}{100}$ **14.** $\frac{5}{8}$ **15.** $\frac{3}{500}$ **16.** $\frac{1}{4}$ **17.** $\frac{7}{4}$ **18.** $\frac{1}{8}$ **19.** $\frac{2}{5}$ **20.** $\frac{14}{5}$

Practice Exercises for Section A.6, page 360

1. 1   **2.** $\frac{1}{3}$ **3.** $\frac{3}{2}$ **4.** $5\frac{4}{5}$ **5.** 7   **6.** $\frac{24}{35}$ **7.** $\frac{41}{48}$ **8.** $\frac{49}{36}$ **9.** $10\frac{1}{4}$ **10.** $17\frac{13}{30}$ **11.** $\frac{11}{23}$
12. $\frac{4}{15}$ **13.** $1\frac{1}{2}$ or $\frac{3}{2}$ **14.** $\frac{1}{6}$ **15.** $1\frac{1}{2}$ **16.** $\frac{1}{2}$ **17.** $1\frac{3}{5}$ **18.** $1\frac{7}{8}$ **19.** $\frac{43}{60}$ **20.** $1\frac{2}{5}$ or $\frac{7}{5}$

Practice Exercises for Section A.7, page 363

1. $\frac{1}{2}$ **2.** 1   **3.** $\frac{2}{25}$ **4.** $\frac{4}{3}$ **5.** $\frac{1}{5}$ **6.** 2   **7.** 3   **8.** 5   **9.** 6   **10.** $\frac{1}{2}$ **11.** 8   **12.** $\frac{25}{16}$
13. $\frac{1}{4}$ **14.** $\frac{1}{26}$ **15.** 6   **16.** 12   **17.** $\frac{1}{25}$ **18.** $\frac{3}{2}$ **19.** 3   **20.** 5

Practice Exercises for Section A.8, page 366

1. $\frac{3}{20}$ **2.** $\frac{7}{10}$ **3.** $\frac{3}{2}$ **4.** $\frac{1}{200}$ **5.** $\frac{9}{400}$ **6.** 2   **7.** 0.3   **8.** 0.014   **9.** 0.836   **10.** 0.0005
11. 62.5%   **12.** 7%   **13.** 40%   **14.** 48%   **15.** 0.2%   **16.** 2%   **17.** 170%   **18.** 54.6%   **19.** 4.5%
20. 300%

Index